Die Grundlehren der mathematischen Wissenschaften

in Einzeldarstellungen
mit besonderer Berücksichtigung
der Anwendungsgebiete

Band 183

J. L. Lions · E. Magenes

Non-Homogeneous Boundary Value Problems and Applications

Translated from the French by
P. Kenneth

Volume III

Springer-Verlag Berlin Heidelberg New York 1973

J. L. Lions
University of Paris

E. Magenes
University of Pavia

Title of the French Original Edition:
Problèmes aux limites non homogènes et applications (tome III)
Publisher: S. A. Dunod, Paris 1968

Translator:
P. Kenneth
Paris

AMS Subject Classifications (1970)
Primary 35J20, 35J25, 35J30, 35J35, 35J40, 35K20, 35K35, 35L20,
Secondary 46E35, 46F15, 46F99

ISBN 3-540-05832-X Springer-Verlag Berlin Heidelberg New York
ISBN 0-387-05832-X Springer-Verlag New York Heidelberg Berlin

Library of Congress Catalog Card Number 71-151 407

Preface to the English Translation

The present translation follows the French edition without change, except for some corrections which were suggested to us by the remarks of C. Baiocchi and M. L. Bernardi, to whom we express our sincerest thanks. We have added a complementary bibliography. We also wish to thank P. Kenneth for his excellent work of translation.

Paris/Pavia, March 1972

J. L. Lions E. Magenes

Introduction

1. Our essential objective is the study of the *linear, non-homogeneous* problems:

(1) $Pu = f$ in $\mathcal{O}$, an open set in $\mathbf{R}^N$,

$$(2) \begin{cases} Q_j u = g_j \text{ on } \partial\mathcal{O} \text{ (boundary of } \mathcal{O}\text{)}, \\ \text{or on a subset of the boundary } \partial\mathcal{O} \quad 1 \le j \le \nu, \end{cases}$$

where P is a linear differential operator in $\mathcal{O}$ and where the Q_j's are linear differential operators on $\partial\mathcal{O}$.

In Volumes 1 and 2, we studied, for particular classes of systems $\{P, Q_j\}$, problem (1), (2) in classes of *Sobolev spaces* (in general constructed starting from L^2) of positive integer or (by interpolation) non-integer order; then, by transposition, in classes of Sobolev spaces of *negative order*, until, by passage to the limit on the order, we reached the spaces of distributions of finite order.

In this volume, we study the analogous problems in spaces of *infinitely differentiable* or *analytic functions* or of *Gevrey-type functions* and by duality, in *spaces of distributions*, of *analytic functionals* or of *Gevrey-type ultra-distributions*. In this manner, we obtain a clear vision (at least we hope so) of the various possible formulations of the boundary value problems (1), (2) for the systems $\{P, Q_j\}$ considered here.

2. One difficulty in this direction is connected with the (locally convex) *topologies*, which are "naturally" tied to the spaces of Gevrey-type functions and their duals (a difficulty which did not appear in Volumes 1 and 2, where, for the essential part, all the spaces were Hilbert or Banach spaces). The indispensable *minimum* on this subject is given in Chapter 7 (1). No doubt a number of our results, in particular in Chapters 10 and 11, could be improved (in the sense of strengthening certain topologies) by a tighter topological analysis of the situation; however, the technical difficulties seem incompatible to us with the interest of the eventual complements; for this reason we have reduced the topological considerations to their strict minimum.

3. Once having introduced the Gevrey-type spaces and their duals, as well as their vector-valued analogues, we consider the problems of type (1), (2) in the following order:

1) elliptic problems (Chapter 8),

(1) First chapter of this Volume; we continue the numbering of the chapters from the preceding Volumes.

2) general evolution problems (Chapter 9),

3) parabolic problems (Chapter 10),

4) hyperbolic problems, or problems well-posed in the sense of Petrowski or of Schroedinger (Chapter 11).

We proceed according to the following steps, which are analogous in principle (but with entirely different "technical details") to those of Volumes 1 and 2:

(i) study of the *regularity* of problem (1), (2), i.e.: assuming "regular" data f and g_j (in the sense: C^∞-functions, or Gevrey functions, or analytic functions), we study the *corresponding regularity of* u;

(ii) by *transposition* of the isomorphism established from the results of type (i), we deduce from them (after obtaining *trace theorems*) the solutions of problems (1), (2) in spaces of distributions, of analytic functionals or of Gevrey functionals.

Thus, the "basic tools" are regularity theorems and trace theorems, which are established for each situation considered.

In particular, for elliptic equations, the starting regularity theorem states that the solutions of the boundary value problems with analytic data are analytic (a theorem which, moreover, appears as a particular case of a much more general result on elliptic iterates; see Chapter 8) and the main trace theorem characterizes, as functionals analytic on the boundary, the solutions of the elliptic equations whose right-hand terms are distributions which "do not grow too rapidly at the boundary".

A similar situation for parabolic problems is discussed in Chapter 10, in which we obtain complete characterizations by Gevrey-type functions and functionals.

The other problems (hyperbolic, well-posed in the sense of Petrowski, or of Schroedinger) are studied according to the same principles, but since, here, "optimal" regularity results do not exist, we do not obtain the most general results (certain questions remaining open in this direction).

A very brief Appendix gives some applications to the calculus of variations and to optimal control theory in Gevrey-type spaces.

4. As for the preceding volumes, each Chapter ends with comments and, except for Chapter 7, a list of open problems.

5. The applications of the theory of non-homogeneous boundary value problems given in this volume and in Volumes 1 and 2 are not exhaustive; various other applications relative to numerical analysis are given in Aubin [1], [2], [3], Bossavit [1], Lions [10] and applications to non-linear problems in Lions [9] for example.

The authors warmly thank G. Geymonat for his constructive criticism.

Paris, June 18, 1969.

Contents

Chapter 7

Scalar and Vector Ultra-Distributions.

Chapter 8

Elliptic Boundary Value Problems in Spaces of Distributions and Ultra-Distributions.

Chapter 9

Evolution Equations in Spaces of Distributions and Ultra-Distributions.

Chapter 10

Parabolic Boundary Value Problems in Spaces of Ultra-Distributions.

Chapter 11

Evolution Equations of the Second Order in *t* and of Schroedinger Type.

Contents of Volume I

(Published 1972)

Contents of Volume II

(Published 1972)

Chapter 7

Scalar and Vector Ultra-Distributions

In this chapter we introduce certain spaces of scalar or vector-valued, infinitely differentiable functions and the spaces of ultra-distributions derived from them by duality. These notions are essential to the remaining part of the text. However, in order not to burden the presentation, particularly with techniques from the theory of topological vector spaces which are in a certain sense marginal to the theory of partial differential equations, we state only the definitions and properties of these spaces and refer the reader to the original texts for the proofs.

1. Scalar-Valued Functions of Class M_k

1.1 The Sequences $\{M_k\}$

Our aim is to generalize the notion of distribution of *L. Schwartz* [1] (see also the recapitulation given in Chapter 1, Section 1) on an open set Ω in $\mathbf{R}^n$, by taking as *fundamental space*, instead of $\mathscr{D}(\Omega)$, a "smaller" space of infinitely differentiable functions in Ω whose derivatives are bounded by suitable sequences of positive numbers. Let us therefore introduce, once and for all, the fundamental hypotheses on the sequences to be considered.

Let $\{M_k\}$, $k = 0, 1, 2, \ldots$, be a sequence of positive real numbers such that

(1.1) $$M_k^2 \leq M_{k-1} M_{k+1} \quad \forall k \geq 1$$

(*logarithmic convexity* condition),

(1.2) $$\sum_{k=1}^{\infty} \frac{M_{k-1}}{M_k} < +\infty$$

(*non-quasi-analyticity* condition),

(1.3) $$\begin{cases} \textit{there exists a constant } H \textit{ such that} \\ M_{k+1} \leq H^k M_k \quad \forall k \end{cases}$$

(sufficient condition for *stability with respect to differentiation*),

$$(1.4) \qquad \begin{cases} \text{there exists a constant } c_1 \text{ such that} \\ \binom{k}{j} M_{k-j} M_j \leq c_1 M_k \qquad \forall k \text{ and } \forall j,\ 0 \leq j \leq k \end{cases}$$

(sufficient condition for *stability with respect to multiplication and composition*). □

Essential examples of sequences $\{M_k\}$ are the *Gevrey sequences*:

$$(1.5) \qquad M_k = (k!)^s,\ M_k = k^{ks},\ M_k = \Gamma(sk+1),$$

where s is real and > 1 and Γ is the Euler function.

1.2 The Space $\mathscr{D}_{M_k}(\Omega)$

For all definitions and notations of the theory of topological vector spaces to be used in the sequel, we refer the reader to Bourbaki [1], [2], Grothendieck [4], Garnir-de Wilde-Schmets [1], Horvath [1], Trèves [1].

Let Ω be an arbitrary, non-empty, open set in $\mathbf{R}^n$.

Definition 1.1. $\mathscr{D}_{M_k}(\Omega)$ *denotes the space of infinitely differentiable functions* $x \to \varphi(x)$ *with compact support in* Ω, *such that there exist two positive numbers* c *and* L *(dependent on* φ*) with*

$$(1.6) \qquad \sup_{x \in \Omega} |D^\alpha \varphi(x)| \leq c L^k M_k, \quad |\alpha| = k, \quad k = 0, 1, 2, \ldots \quad \square$$

The following are some properties of $\mathscr{D}_{M_k}(\Omega)$ (see for instance Roumieu [1], [2]):

a) $\mathscr{D}_{M_k}(\Omega)$ does not reduce to $\{0\}$: for every ball S_ε of radius ε contained in Ω, there exists a $\varrho_\varepsilon \in \mathscr{D}_{M_k}(\Omega)$ satisfying the conditions: $\varrho_\varepsilon(x) \geq 0$, support of $\varrho_\varepsilon(x)$ contained in S_ε, $\int_\Omega \varrho_\varepsilon(x)\,dx = 1$;

b) there exists a *partition of unity* by functions of $\mathscr{D}_{M_k}(\Omega)$, for every open covering of Ω;

c) $\mathscr{D}_{M_k}(\Omega)$ is stable with respect to differentiation, that is if $\varphi \in \mathscr{D}_{M_k}(\Omega)$ then $D^\alpha \varphi \in \mathscr{D}_{M_k}(\Omega)$, $\forall \alpha$;

d) $\mathscr{D}_{M_k}(\Omega)$ is an algebra;

e) $\mathscr{D}_{M_k}(\Omega)$ is *dense* in $\mathscr{D}(\Omega)$.

Properties a), b) and e) follow from (1.1) and (1.2), c) from (1.3) and d) from (1.4). □

Remark 1.1. We shall not always use *all* the hypotheses (1.1), ..., (1.4); we shall specify whenever necessary. □

A "natural" topology is introduced in $\mathscr{D}_{M_k}(\Omega)$ as follows.

For $\mathscr{K}$ a compact set contained in Ω and L a positive number, consider the subspace $\mathscr{D}_{M_k}(\Omega; \mathscr{K}, L)$ of $\mathscr{D}_{M_k}(\Omega)$ made up of the functions φ with support in $\mathscr{K}$ and such that (1.6) holds for this fixed L, c still depending on φ. We can easily see that $\mathscr{D}_{M_k}(\Omega; \mathscr{K}, L)$ is a Banach space for the norm

$$\|\varphi\| = \sup_{\substack{|\alpha|=k, k=0,1,\ldots \\ x \in \mathscr{K}}} \left| \frac{D^\alpha \varphi(x)}{L^k M_k} \right|. \tag{1.7}$$

From the algebraic point of view, we evidently have

$$\mathscr{D}_{M_k}(\Omega) = \bigcup_{\mathscr{K}, L} \mathscr{D}_{M_k}(\Omega; \mathscr{K}, L)$$

as $(\mathscr{K}, L)$ varies over the set of couples such that $\mathscr{K}$ is an arbitrary compact set contained in Ω and L an arbitrary positive number; this set is filtering for the relation

$$(\mathscr{K}, L) \leq (\mathscr{K}', L') \text{ if } \mathscr{K} \subset \mathscr{K}',\ L \leq L'.$$

We also have

$$\mathscr{D}_{M_k}(\Omega; \mathscr{K}, L) \subset \mathscr{D}_{M_k}(\Omega; \mathscr{K}', L') \quad \text{if} \quad (\mathscr{K}, L) \leq (\mathscr{K}', L')$$

with continuous injection.

Thus it seems natural to provide the space $\mathscr{D}_{M_k}(\Omega)$ with the *inductive limit topology* of the topologies of the spaces $\mathscr{D}_{M_k}(\Omega; \mathscr{K}, L)$ (that is, the finest locally convex topology for which the injection of $\mathscr{D}_{M_k}(\Omega; \mathscr{K}, L)$ into $\mathscr{D}_{M_k}(\Omega)$ is continuous). This amounts to defining

$$\mathscr{D}_{M_k}(\Omega) = \operatorname{ind} \lim_{\mathscr{K} \to \Omega,\ L \to +\infty} \mathscr{D}_{M_k}(\Omega; \mathscr{K}, L), \tag{1.8}$$

where $\mathscr{K}$ increases monotonically to Ω and L increases monotonically to $+\infty$; and it suffices for $\mathscr{K}$ and L to vary over a sequence $\mathscr{K}_n$ and L_n with the same properties. □

Remark 1.2. Since $\mathscr{K}$ is bounded, we can easily see that the norm (1.7) for $\mathscr{D}_{M_k}(\Omega; \mathscr{K}, L)$ is equivalent to any one of the following norms:

$$\|\varphi\| = \sup_{|\alpha|=k, k=0,1,\ldots} \frac{\|D^\alpha \varphi\|_{L^p(\Omega)}}{L^k M_k}, \quad \text{fixed } p \text{ with } 1 < p < \infty.$$

For $p = 2$, we may also consider the norm

$$\|\varphi\| = \left(\sum_{k=0}^{\infty} \sum_{|\alpha|=k} \frac{\|D^\alpha \varphi\|^2_{L^2(\Omega)}}{(L^k M_k)^2} \right)^{1/2},$$

for which $\mathscr{D}_{M_k}(\Omega; \mathscr{K}, L)$ is a Hilbert space. □

We have

Proposition 1.1. *The injection of* $\mathcal{D}_{M_k}(\Omega; \mathcal{K}, L)$ *into* $\mathcal{D}_{M_k}(\Omega; \mathcal{K}', L')$ *is compact if* $(\mathcal{K}, L) \leq (\mathcal{K}', L')$.

Proof. Let $\{\varphi_n\}$ be a bounded sequence in $\mathcal{D}_{M_k}(\Omega; \mathcal{K}, L)$. By applying the Ascoli-Arzela theorem we see that there exists a subsequence $\{\varphi_{n_i}\}$ of $\{\varphi_n\}$ and a function φ in $\mathcal{D}_{M_k}(\Omega; \mathcal{K}, L)$, and therefore in $\mathcal{D}_{M_k}(\Omega; \mathcal{K}', L')$, such that

$$\lim_{i\to\infty,\ x\in\mathcal{K}} \sup |D^\alpha \varphi_{n_i}(x) - D^\alpha \varphi(x)| = 0, \ \forall \alpha. \tag{1.9}$$

Therefore, it is sufficient to show that

$$\lim_{i\to\infty} \|\varphi_{n_i} - \varphi\|_{\mathcal{D}_{M_k}(\Omega;\mathcal{K}',L')} = 0.$$

Now, thanks to (1.7) and the fact that φ_n and φ have their support in $\mathcal{K}$,

$$\|\varphi_{n_i} - \varphi\|_{\mathcal{D}_{M_k}(\Omega,\mathcal{K}',L')} = \sup_{x\in\mathcal{K}',|\alpha|=k,k=0,1,\dots} \frac{|D^\alpha\varphi_{n_i}(x) - D^\alpha\varphi(x)|}{(L')^k M_k}$$

$$\leq \sup_{0\leq k\leq N}\ \sup_{x\in\mathcal{K}',|\alpha|=k} \frac{1}{(L')^k M_k}(|D^\alpha\varphi_{n_i}(x) - D^\alpha\varphi(x)|) +$$

$$+ \sup_{k>N}\left(\frac{L}{L'}\right)^N \sup_{x\in\mathcal{K}',|\alpha|=k} \frac{1}{L^k M_k}|D^\alpha\varphi_{n_i}(x) - D^\alpha\varphi(x)| \leq$$

$$\leq \sup_{0\leq k\leq N}\ \sup_{x\in\mathcal{K}',|\alpha|=k} \frac{|D^\alpha\varphi_n(x) - D^\alpha\varphi(x)|}{(L')^k M_k} + \left(\frac{L}{L'}\right)^N \|\varphi_{n_i} - \varphi\|_{\mathcal{D}_{M_k}(\Omega;\mathcal{K},L)} \leq$$

$$\leq \sup_{0\leq k\leq N}\ \sup_{x\in\mathcal{K}',|\alpha|=k} \frac{|D^\alpha\varphi_{n_i}(x) - D^\alpha\varphi(x)|}{(L')^k M_k} + \left(\frac{L}{L'}\right)^N c$$

with constant c. For fixed $\varepsilon > 0$, we can therefore take N_ε such that $c(L/L')^{N_\varepsilon} < \varepsilon/2$ and then it follows from (1.9) that there exists an n_ε such that for $n > n_\varepsilon$ we have

$$\|\varphi_{n_i} - \varphi\|_{\mathcal{D}_{M_k}(\Omega;\mathcal{K}',L')} < \varepsilon. \quad \square$$

Thus we see that $\mathcal{D}_{M_k}(\Omega)$ is an inductive limit of a sequence of Banach spaces E_n such that

(1.10) $E_n \subset E_{n+1}$, *with continuous and compact injection.*

Following S. Silva [1], if $E = \operatorname{ind}\lim_{n\to\infty} E_n$ with (1.10), we shall call E *an inductive limit of a regular sequence of Banach spaces.* This type of space has in particular the following topological properties (see Silva [1], Yoshinaga [1], Raikov [1], Matagne [1], Komatsu [5]).

(α) it is *separated*;

(β) every *bounded* set in E is contained in an E_n, for suitable n, and is *bounded in* E_n;

(γ) it is *complete, separable, a Montel space* (and therefore *reflexive*), *the strong dual of a Fréchet and Schwartz space* and therefore also a $(\mathscr{DF})$-space (for the definitions of Schwartz spaces and $(\mathscr{DF})$-spaces see, for example, Horvath [1], Garnir-de Wilde-Schmets [1]).

Therefore $\mathscr{D}_{M_k}(\Omega)$ satisfies conditions α), β) and γ); furthermore it is a *nuclear* space (see Mityagin [1]). ◻

We shall call $\mathscr{D}_{M_k}(\Omega)$ a *space of functions of class* M_k; *when* M_k *is given by one of the sequences* (1.5) *we shall call* $\mathscr{D}_{M_k}(\Omega)$ *the Gevrey space of order* s (it is easy to see that the three sequences (1.5) yield the *same* space) and we shall often denote it by $\mathscr{D}_s(\Omega)$. ◻

Remark 1.3. We have pointed out properties α) and β) of inductive limit spaces of a regular sequence of Banach spaces. We note that, given a space E, inductive limit of an *increasing* sequence of Banach spaces (i.e. $E_n \subset E_{n+1}$, with continuous injection), E also satisfies properties α) and β) in the following two cases:

1) E is a *strict* inductive limit (see Dieudonné-Schwartz [1], Proposition 4);

2) the spaces E_n are *reflexive* (see D. G. Schaeffer [1], Appendix 2).

An immediate consequence of properties α) and β) then is that E *is reflexive if the* E_n*'s are reflexive.*

1.3 The Spaces $\mathscr{D}_{M_k}(\mathscr{K})$ and $\mathscr{E}_{M_k}(\Omega)$

Let Ω be an open set in $\mathbf{R}^n$ and $\mathscr{K}$ a compact set contained in Ω.

Definition 1.2. $\mathscr{D}_{M_k}(\mathscr{K})$ *denotes the space of restrictions to* $\mathscr{K}$ *of the infinitely differentiable functions on* Ω, $x \to \varphi(x)$, *such that there exist two positive numbers* c *and* L (*dependent on* φ) *such that*

$$\sup_{x\in\mathscr{K}} |\mathrm{D}^\alpha \varphi(x)| \leq c L^k M_k, \; |\alpha| = k, \; k = 0, 1, 2, \ldots \tag{1.11}$$

$\mathscr{D}_{M_k}(\mathscr{K})$ is provided with the following topology. For fixed L, consider the subspace $\mathscr{D}_{M_k}(\mathscr{K}; L)$ of $\mathscr{D}_{M_k}(\mathscr{K})$ made up of the elements φ such that (1.11) holds for this fixed L, provided with the norm

$$\|\varphi\| = \sup_{\substack{x\in\mathscr{K}, |\alpha|=k \\ k=0,1,2,\ldots}} \left| \frac{\mathrm{D}^\alpha \varphi(x)}{L^k M_k} \right|$$

(or with the norms of Remark 1.2). It is a Banach space and the *injection of* $\mathcal{D}_{M_k}(\mathcal{K}, L)$ *into* $\mathcal{D}_{M_k}(\mathcal{K}, L')$ *is compact if* $L < L'$ (see Proposition 1.1). It is then natural to provide $\mathcal{D}_{M_k}(\mathcal{K})$ with the inductive limit topology of the spaces $\mathcal{D}_{M_k}(\mathcal{K}, L)$ as L increases monotonically to $+\infty$; thus $\mathcal{D}_{M_k}(\mathcal{K})$ is an inductive limit space of a regular sequence of Banach spaces. ▯

Now let Ω be an arbitrary open set in $\mathbf{R}^n$.

Definition 1.3. $\mathcal{E}_{M_k}(\Omega)$ *denotes the space of infinitely differentiable functions on* Ω, $x \to \varphi(x)$, *such that for every compact set* $\mathcal{K}$ *contained in* Ω, *the restriction of* φ *to* $\mathcal{K}$ *belongs to* $\mathcal{D}_{M_k}(\mathcal{K})$.

If we let $\mathcal{K}$ vary over the compact sets contained in Ω, and if we denote by $r_{\mathcal{K}}$ the restriction of φ to $\mathcal{K}$, then we can provide $\mathcal{E}_{M_k}(\Omega)$ with the *projective limit topology* of the topologies of the spaces $\mathcal{D}_{M_k}(\mathcal{K})$ with respect to the set of mappings $r_{\mathcal{K}}$ (that is, the coarsest locally convex topology which makes the mappings $r_{\mathcal{K}}$ of $\mathcal{E}_{M_k}(\Omega)$ into $\mathcal{D}_{M_k}(\mathcal{K})$ continuous). ▯

We note that no hypothesis on the M_k's has been used for the definition of the spaces $\mathcal{D}_{M_k}(\mathcal{K})$. On the other hand, using (1.3) and (1.4), we see that $\mathcal{D}_{M_k}(\mathcal{K})$ and $\mathcal{E}_{M_k}(\Omega)$ are stable with respect to differentiation and are algebras.

2. Scalar-Valued Ultra-Distributions of Class M_k; Generalizations

2.1 The Space $\mathcal{D}'_{M_k}(\Omega)$

Let Ω be an open set in $\mathbf{R}^n$ and $\{M_k\}$ a sequence of positive numbers satisfying (1.1), (1.2), (1.3) and (1.4).

Definition 2.1. $\mathcal{D}'_{M_k}(\Omega)$ *denotes the dual of* $\mathcal{D}_{M_k}(\Omega)$, *provided with the strong dual topology.*

The elements of $\mathcal{D}'_{M_k}(\Omega)$ (continuous linear forms on $\mathcal{D}_{M_k}(\Omega)$) are called *ultra-distributions on* Ω of class M_k. If M_k is given by any one of the formulas (1.5), then we shall call the elements of $\mathcal{D}'_{M_k}(\Omega)$ *Gevrey ultra-distributions or functionals of order* s and we shall also denote $\mathcal{D}'_{M_k}(\Omega)$ by $\mathcal{D}'_s(\Omega)$. ▯

$\mathcal{D}'_{M_k}(\Omega)$ *is a Frechet and a Schwartz space.* ▯

Applying property e), Section 1.2, of the space $\mathcal{D}_{M_k}(\Omega)$, we may *identify* $\mathcal{D}'(\Omega)$ with a subspace of $\mathcal{D}'_{M_k}(\Omega)$; therefore every distribution on Ω is an ultra-distribution on Ω of class M_k and we have the continuous injection

$$\mathcal{D}'(\Omega) \subset \mathcal{D}'_{M_k}(\Omega). \quad (2.1)$$ ▯

Taking into account property b), Section 1.2, of $\mathscr{D}_{M_k}(\Omega)$, we see that, as for the distributions of L. Schwartz [1], we can show the *localization principle* for ultra-distributions and then define the *support of an ultra-distribution.* □

Differentiation in $\mathscr{D}'_{M_k}(\Omega)$ and multiplication by a function of $\mathscr{E}_{M_k}(\Omega)$ are also defined as for distributions, that is:
for $u \in \mathscr{D}'_{M_k}(\Omega)$, we have

$$\langle D^\alpha u, \varphi\rangle = (-1)^{|\alpha|} \langle u, D^\alpha \varphi\rangle \;\forall \varphi \in \mathscr{D}_{M_k}(\Omega); \tag{2.2}$$

for $u \in \mathscr{D}'_{M_k}(\Omega)$ and $\psi \in \mathscr{E}_{M_k}(\Omega)$, we have

$$\langle \psi u, \varphi\rangle = \langle u, \psi\varphi\rangle \quad \forall \varphi \in \mathscr{D}_{M_k}(\Omega). \quad \square \tag{2.3}$$

The structure of ultra-distributions is specified by the following theorem (see Roumieu [2], Theorem 10).

Theorem 2.1. *Every ultra-distribution $u \in \mathscr{D}'_{M_k}(\Omega)$ can be represented, in a non-unique fashion, by the form*

$$u = \sum_{k=0}^{\infty} \sum_{|\alpha|=k} D^\alpha \mu_\alpha, \tag{2.4}$$

where the μ_α's are measures on Ω such that

$$\sum_{k=0}^{\infty} \sum_{|\alpha|=k} M_k L^k \int_{\mathscr{K}} |d\mu_\alpha| < +\infty \tag{2.5}$$

for every $L > 0$ and every compact set $\mathscr{K}$ contained in Ω.

Conversely, if the μ_α's are measures on Ω satisfying (2.5), *then u given by* (2.4) (*that is*

$$\langle u, \varphi\rangle = \sum_{k=0} \sum_{|\alpha|=k} (-1)^{|\alpha|} \int_{\Omega} D^\alpha \varphi \, d\mu_\alpha)$$

defines an element of $\mathscr{D}'_{M_k}(\Omega)$. □

Remark 2.1. The same type of structure results, where the μ_α's are $L^{p'}(\Omega)$-functions, can be obtained by using the (equivalent) definition of $\mathscr{D}_{M_k}(\Omega; \mathscr{K}, L)$ given in Remark 1.2, with $1/p + 1/p' = 1$.

2.2 Non-Symmetric Spaces of Class M_k

For the definitions of the spaces $\mathscr{D}_{M_k}(\Omega)$ and $\mathscr{E}_{M_k}(\Omega)$ we have considered the variables $x_1, \ldots, x_n$ in *symmetric* fashion; but, in view of certain applications, it is useful to treat them in *non-symmetric* fashion; and this can be realized (see for example Roumieu [2]) by imposing, instead of (1.6), bounds of the type

$$\sup_{x \in \Omega} |D^\alpha \varphi(x)| \leq c L^{|\alpha|} M_\alpha, \;\forall \alpha, \tag{2.6}$$

where the sequence $\{M_\alpha\}$ depends on the multi-index of differentiation α.

We shall restrict ourselves to a discussion of a particular case which arises in connection with partial differential *evolution* equations. □

Again let Ω be an arbitrary open set in $\mathbf{R}^n$ and let $]t_0, t_1[$ be an arbitrary open interval, finite or infinite, in $\mathbf{R}^1$; consider the cylinder $Q = \Omega \times]t_0, t_1[$ in $\mathbf{R}^{n+1}$ and two sequences of positive numbers $\{M_k\}$ and $\{N_h\}$ satisfying conditions (1.1)—(1.4).

We then can introduce the spaces $\mathscr{D}_{N_h,M_k}(Q)$, $\mathscr{D}_{N_h,M_k}(\mathscr{K})$ ($\mathscr{K}$ a compact set in Q) and $\mathscr{E}_{N_h,M_k}(Q)$ in an obvious way; for example, if we denote by (x, t) the point in $\mathbf{R}^{n+1}$ $(x = (x_1, \ldots, x_n) \in \mathbf{R}^n, t \in \mathbf{R}^1)$: $\mathscr{D}_{N_h,M_k}(Q)$ *is the space of infinitely differentiable functions* $(x, t) \to \varphi(x, t)$ *with compact support in* Q, *such that there exist two positive numbers* c *and* L *(depending on* φ*) with*

$$(2.7) \qquad \sup_{(x,t)\in Q} |D_x^\alpha D_t^h \varphi(x, t)| \leq c L^{k+h} M_k N_h, \quad |\alpha| = k, \; k, h = 0, 1, \ldots$$

$\mathscr{D}_{N_h,M_k}(Q)$ is provided with a topology of inductive limit of Banach spaces in a completely analogous way to what was done for $\mathscr{D}_{M_k}(\Omega)$. □

In this manner, we arrive at the notion of *ultra-distribution on* Q of class $\{N_h, M_k\}$, as a continuous linear form on $\mathscr{D}_{N_h,M_k}(Q)$; more precisely, we define

$$\mathscr{D}'_{N_h,M_k}(Q) = \text{strong dual of } \mathscr{D}_{N_h,M_k}(Q). \quad \square$$

In the following chapters we shall make use of the particular case for which $M_k = (k!)^r$, $N_h = (h!)^s$, with real $r, s > 1$; for this case, we shall sometimes denote the spaces $\mathscr{D}_{N_h,M_k}(Q)$ and $\mathscr{D}'_{N_h,M_k}(Q)$ by $\mathscr{D}_{s,r}(Q)$ and $\mathscr{D}'_{s,r}(Q)$ and call them *Gevrey spaces of order* (s, r).

2.3 Scalar Ultra-Distributions of Beurling-Type

Let us also call attention to another generalization of the notion of distribution which is close to the one given in Section 2.1. We still consider an open set Ω in $\mathbf{R}^n$ and a sequence $\{M_k\}$ satisfying (1.1)—(1.4), and, instead of $\mathscr{D}_{M_k}(\Omega)$, we use as our *fundamental space*, $\mathscr{B}_{M_k}(\Omega)$ defined by

Definition 2.2. $\mathscr{B}_{M_k}(\Omega)$ *is the space of infinitely differentiable functions* $x \to \varphi(x)$ *with compact support in* Ω, *such that for every* $L > 0$ *there exists a positive number* c *(depending on* L *and* φ*) such that*

$$(2.8) \qquad \sup_{x\in\Omega} |D^\alpha \varphi(x)| \leq c L^k M_k, \quad |\alpha| = k, \; k = 0, 1, 2, \ldots$$

$\mathscr{B}_{M_k}(\Omega)$ is provided with the following topology:

$$(2.9) \qquad \mathscr{B}_{M_k}(\Omega) = \operatorname*{ind\,lim}_{\mathscr{K}\to\Omega} \big(\operatorname*{project\,lim}_{L\to 0} \mathscr{D}_{M_k}(\Omega; \mathscr{K}, L)\big),$$

where $\mathscr{K}$ increases monotonically to Ω and L decreases monotonically to zero (compare with definition (1.8) of $\mathscr{D}_{M_k}(\Omega)$).

By applying Proposition 1.1., we see that $\mathscr{B}_{M_k}(\Omega)$ is a *strict inductive limit of Fréchet and Schwartz spaces,* whose topological properties we therefore know (see for example Yoshinaga [1]).

Evidently $\mathscr{B}_{M_k}(\Omega) \subset \mathscr{D}_{M_k}(\Omega)$.

By applying Lemma 1 on page 66 of Roumieu [1], we can show that there exists another sequence $\{M_k^*\}$ satisfying conditions (1.1) and (1.2) such that

$$\mathscr{D}_{M_k^*}(\Omega) \subset \mathscr{B}_{M_k}(\Omega); \tag{2.10}$$

from which we have the fact that properties a) and b) of Section 1.2 hold for $\mathscr{B}_{M_k}(\Omega)$ as well (see Remark 1.1).

Furthermore, applying Proposition 9, page 54 of Roumieu [1] ($\mathscr{D}_{M_k^*}(\Omega)$ is dense in $\mathscr{D}_{M_k}(\Omega)$) we see that $\mathscr{B}_{M_k}(\Omega)$ is *dense* in $\mathscr{D}_{M_k}(\Omega)$.

Finally we see that $\mathscr{B}_{M_k}(\Omega)$ is a stable algebra with respect to differentiation. □

The space of ultra-distributions of Beurling-type relative to the sequence $\{M_k\}$ is defined by

$$\mathscr{B}'_{M_k}(\Omega) = \text{strong dual of } \mathscr{B}_{M_k}(\Omega). \tag{2.11}$$

From the properties of $\mathscr{B}_{M_k}(\Omega)$, we deduce that $\mathscr{D}'_{M_k}(\Omega)$ may be identified with a subspace of $\mathscr{B}'_{M_k}(\Omega)$; for a fixed sequence $\{M_k\}$, we therefore have

$$\mathscr{D}'(\Omega) \subset \mathscr{D}'_{M_k}(\Omega) \subset \mathscr{B}'_{M_k}(\Omega). \quad \square \tag{2.12}$$

For the development of the theory of Beurling-type ultra-distributions, see Beurling [1] and Björck [1] (see also the Comments to this chapter). The notions given here are sufficient for the rest of this text. □

Remark 2.2. Clearly we can also consider *non-symmetric* Beurling spaces with respect to $x_1, \ldots, x_n$; in particular we can consider the spaces $\mathscr{B}_{N_h,M_k}(\Omega)$ and $\mathscr{B}'_{N_h,M_k}(\Omega)$ (compare with Section 2.2).

3. Spaces of Analytic Functions and of Analytic Functionals

3.1 The Spaces $\mathscr{H}(\mathscr{K})$ and $\mathscr{H}'(\mathscr{K})$

Let $\mathscr{K}$ be a compact set in $\mathbf{R}^n$. In the sequel, we shall use the space $\mathscr{H}(\mathscr{K})$ of analytic functions on $\mathscr{K}$. One of the usual ways to define $\mathscr{H}(\mathscr{K})$ is the following.

We imbed $\mathbf{R}^n$ in the n-dimensional complex space $\mathbf{C}^n$. For $\mathcal{O}$ an open set in $\mathbf{C}^n$, we denote by $\mathscr{H}(\mathcal{O})$ the space of holomorphic functions on $\mathcal{O}$, provided with the topology of uniform convergence on the compact subsets of $\mathcal{O}$. Then we define $\mathscr{H}(\mathscr{K})$ by

$$\mathscr{H}(\mathscr{K}) = \operatorname*{ind\,lim}_{\mathcal{O}} \mathscr{H}(\mathcal{O}) \tag{3.1}$$

as $\mathcal{O}$ varies over the set of (complex) neighborhoods of $\mathscr{K}$. $\mathscr{H}(\mathscr{K})$ is complete, separable, a Montel space (and therefore reflexive), the dual of a Fréchet-Schwartz space, nuclear (see for example Grothendieck [1] [3]).

The space of analytic functionals on $\mathscr{K}$ is by definition the space

$$\mathscr{H}'(\mathscr{K}) = \text{strong dual of } \mathscr{H}(\mathscr{K}) \tag{3.2}$$

(see the Comments for references).

3.2 The Spaces $\mathscr{H}(\Gamma)$ and $\mathscr{H}'(\Gamma)$

For the sequel we shall require the particular case where $\mathscr{K}$ is an $(n-1)$-dimensional, real analytic variety Γ, the boundary of a bounded open set Ω in $\mathbf{R}^n$ (Ω is considered as a variety with boundary, the boundary being Γ).

For this case it is of interest to introduce $\mathscr{H}(\Gamma)$ without "leaving" the variety Γ, by using the Laplace-Beltrami operator Δ_Γ on Γ (for the definition of Δ_Γ, see for example de Rham [1]).

For fixed $L > 0$, we define $\mathscr{H}_L(\Gamma)$ as the space of infinitely differentiable functions φ on Γ, such that there exists a positive number c (depending on φ) such that

$$\sup_{x\in\Gamma} |\Delta_\Gamma^k \varphi(x)| \leq c L^k (2k)! \quad k = 0, 1, \ldots; \tag{3.3}$$

$\mathscr{H}_L(\Gamma)$ is a Banach space for the norm

$$\|\varphi\| = \sup_{x\in\Gamma, k=0,1,\ldots} \left| \frac{\Delta_\Gamma^k \varphi(x)}{L^k(2k)!} \right| . \tag{3.4}$$

It will follow from Chapter 8 (theorem on "elliptic iterates") that

$$\mathscr{H}(\Gamma) = \operatorname*{ind\,lim}_{L\to+\infty} \mathscr{H}_L(\Gamma) . \tag{3.5}$$

From which we deduce (see Lions-Magenes [1], Proposition 1.3) a theorem on the structure of analytic functionals on Γ:

Theorem 3.1. *Every element u of $\mathscr{H}'(\Gamma)$ may be represented, non-uniquely by the form*

$$\langle u, \varphi \rangle = \sum_{k=0}^{\infty} \int_{\Gamma} \Delta_{\Gamma}^{k} \varphi \, \mathrm{d}\mu_k, \tag{3.6}$$

where the μ_k's are measures on Γ such that

$$\sum_{k=0}^{\infty} L^k (2k)! \, |\mu_k| < +\infty \text{ for every } L > 0. \tag{3.7}$$

Conversely, if the measures μ_k are given with (3.7), *then u, defined by* (3.6), *is an element of $\mathscr{H}'(\Gamma)$.* ☐

Remark 3.1. The norm (3.4) may be replaced by the norm

$$\|\varphi\| = \sup_{k=0,1,\ldots} \frac{\|\Delta_{\Gamma}^{k} \varphi\|_{L^p(\Gamma)}}{L^k (2k!)}, \quad 1 < p < +\infty.$$

Another structure theorem for analytic functionals on Γ can be deduced from this, by replacing the measures μ_k in Theorem 3.1 with functions $g_k \in L^{p'}(\Gamma)$ $(1/p + 1/p' = 1)$. Similarly, $\mathscr{H}_L(\Gamma)$ becomes a Hilbert space if we provide it with the norm

$$\|\varphi\| = \left(\sum_{k=0}^{\infty} \frac{1}{L^{2k}((2k)!)^2} \|\Delta_{\Gamma}^{k} \varphi\|_{L^2(\Gamma)}^2 \right)^{1/2}.$$

4. Vector-Valued Functions of Class M_k

4.1 The Space $\mathscr{D}_{M_k}(\mathscr{I}; F)$

For the applications to the boundary value problems studied in the following chapters, the above definitions must be extended to the case of functions and ultra-distributions with values in a topological vector space. We shall restrict ourselves to the consideration of functions and ultra-distributions of a single real variable, in an open interval of $\boldsymbol{R}^1$; this is in fact the case of interest for the evolution equations we have in mind.

Thus let F be a *locally convex, separated, topological vector space.*

Let $\mathscr{I} =]t_0, t_1[$ be an open interval, *finite or infinite, in* $\boldsymbol{R}^1$.

$\mathscr{D}(\mathscr{I}; F)$ shall denote the space of infinitely differentiable functions, $t \to \varphi(t)$, on $\mathscr{I}$, with values in F, and with compact support, provided with the topology of L. Schwartz [2] (this is the inductive limit topology of the spaces $\mathscr{D}(\mathscr{I}, \mathscr{K}; F)$ as $\mathscr{K}$ varies over the set of compact intervals contained in $\mathscr{I}$, where $\mathscr{D}(\mathscr{I}, \mathscr{K}; F)$ is the subspace of $\mathscr{D}(\mathscr{I}; F)$ of functions with support contained in $\mathscr{K}$, provided with the topology of uniform convergence on $\mathscr{K}$ of φ and of each of its derivatives). ☐

Again, let $\{M_k\}$ be a sequence of positive numbers satisfying (1.1)—(1.4).

Definition 4.1. *$\mathscr{D}_{M_k}(\mathscr{I}; F)$ denotes the space of functions $t \to \varphi(t)$ defined on $\mathscr{I}$ and with values in F, such that*

$$\varphi \in \mathscr{D}(\mathscr{I}; F) \tag{4.1}$$

and such that there exists a number $L > 0$ and a bounded set $\mathscr{B}$ in F (both depending on φ) such that

$$\frac{\varphi^{(k)}(t)}{L^k M_k} \in \mathscr{B},\ \forall t \in \mathscr{I},\ k = 0, 1, 2, \ldots \tag{4.2}$$

If $F = \mathbf{C}$ (scalar case) we recover (algebraically) the space $\mathscr{D}_{M_k}(\mathscr{I})$ defined in Section 1.2.

The topology on $\mathscr{D}_{M_k}(\mathscr{I}; F)$ is defined as follows. Let $\mathscr{K}$ be a compact interval contained in $\mathscr{I}$ and let $L > 0$ be fixed; we consider the subspace $\mathscr{D}_{M_k}(\mathscr{I}, \mathscr{K}, L; F)$ of $\mathscr{D}_{M_k}(\mathscr{I}; F)$ made up of the elements φ with support in $\mathscr{K}$ and such that (4.2) holds for this fixed L (the bounded set $\mathscr{B}$ depending on φ); and we provide $\mathscr{D}_{M_k}(\mathscr{I}, \mathscr{K}, L; F)$ with the topology defined by the fundamental system of neighborhoods of the origin given by

$$\mathscr{V} = \left\{ \varphi \,\middle|\, \frac{\varphi^{(k)}(t)}{L^k M_k} \in \mathscr{V}_F,\ \forall t,\ k = 0, 1, 2, \ldots \right\},$$

where $\mathscr{V}_F$ describes a fundamental system of neighborhoods of the origin in F.

As for the scalar case (see Section 1.2), we provide $\mathscr{D}_{M_k}(\mathscr{I}; F)$ with the inductive limit topology of the topologies of the spaces $\mathscr{D}_{M_k}(\mathscr{I}, \mathscr{K}, L; F)$ as $\mathscr{K}$ increases monotonically to $\mathscr{I}$ and L increases monotonically to $+\infty$; therefore

$$\mathscr{D}_{M_k}(\mathscr{I}; F) = \operatorname*{ind\ lim}_{\mathscr{K} \to \mathscr{I}, L \to +\infty} \mathscr{D}_{M_k}(\mathscr{I}, \mathscr{K}, L; F). \tag{4.3}$$

Then, if $F = \mathbf{C}$, we have $\mathscr{D}_{M_k}(\mathscr{I}, \mathbf{C}) = \mathscr{D}_{M_k}(\mathscr{I})$ algebraically and topologically.

4.2 The Spaces $\mathscr{D}_{M_k}(\mathscr{K}; F)$ and $\mathscr{E}_{M_k}(\mathscr{I}; F)$

Let $\mathscr{K}$ be a compact interval in $\mathbf{R}^1$.

Definition 4.2. *$\mathscr{D}_{M_k}(\mathscr{K}; F)$ denotes the space of infinitely differentiable functions, $t \to \varphi(t)$, defined on $\mathscr{K}$, with values in F, such that there exists* *a positive number* L *and a bounded set* $\mathscr{B}$ *in* F *(both depending on* φ*) such that*

$$\frac{\varphi^{(k)}(t)}{L^k M_k} \in \mathscr{B}, \ \forall t \in \mathscr{K}, \ k = 0, 1, 2, \ldots \tag{4.4}$$

$\mathscr{D}_{M_k}(\mathscr{K}, F)$ is provided with the following topology. For every fixed $L > 0$, we consider the subspace $\mathscr{D}_{M_k}(\mathscr{K}, L; F)$ of $\mathscr{D}_{M_k}(\mathscr{K}; F)$ made up of the elements φ for which (4.4) holds for this fixed L, provided with the topology defined by the fundamental system of neighborhoods of the origin given by

$$\mathscr{V} = \left\{ \varphi \,\middle|\, \frac{\varphi^k(t)}{L^k M_k} \in \mathscr{V}_F, \ \forall t \in \mathscr{K}, \ k = 0, 1. 2 \ldots \right\},$$

where $\mathscr{V}_F$ describes a fundamental system of neighborhoods of the origin in F.

We then provide $\mathscr{D}_{M_k}(\mathscr{K}; F)$ with the inductive limit topology of the spaces $\mathscr{D}_{M_k}(\mathscr{K}, L; F)$ as L increases monotonically to $+\infty$. ◻

Now let $\mathscr{I}$ be an open, finite or infinite, interval in $\mathbf{R}^1$.

Definition 4.3. $\mathscr{E}_{M_k}(\mathscr{I}; F)$ *denotes the space of infinitely differentiable functions* $t \to \varphi(t)$, *on* $\mathscr{I}$, *with values in* F, *such that for every compact interval* $\mathscr{K}$ *contained in* $\mathscr{I}$ *the restriction of* φ *to* $\mathscr{K}$ *belongs to* $\mathscr{D}_{M_k}(\mathscr{K}; F)$.

By letting $\mathscr{K}$ vary over the set of compact intervals contained in $\mathscr{I}$, we provide $\mathscr{E}_{M_k}(\mathscr{I}; F)$ with the projective limit topology of the topologies of the spaces $\mathscr{D}_{M_k}(\mathscr{K}; F)$ with respect to the set of restrictions $r_{\mathscr{K}}$ of φ to $\mathscr{K}$.

4.3 The Spaces $\mathscr{D}_{\pm,M_k}(\mathscr{I}; F)$

We still let $\mathscr{I} =]t_0, t_1[$ be an open, finite or infinite, interval in $\mathbf{R}^1$.

Definition 4.4. $\mathscr{D}_{+,M_k}(\mathscr{I}; F)$ *denotes the space of infinitely differentiable functions* $t \to \varphi(t)$, *defined on* $\mathscr{I}$, *with values in* F, *and with support bounded on the left in* $\mathscr{I}$ *(that is, zero in a neighborhood of* t_0, *depending on* φ*) and such that*

$$\left\{\begin{array}{l} \text{for each } b \text{ with } t_0 < b < t_1, \text{ there exists a} \\ \text{number } L > 0 \text{ and a bounded set } \mathscr{B} \text{ in } F \text{ (both} \\ \text{depending on } \varphi \text{ and } b) \text{ such that} \\ \dfrac{\varphi^{(k)}(t)}{L^k M_k} \in \mathscr{B}, \ \forall t \in]t_0, b], \ k = 0, 1,2 \ \ldots \end{array}\right. \tag{4.5}$$

The topology of $\mathscr{D}_{+M_k}(\mathscr{I}; F)$ is defined in the following way.

For each closed sub-interval $[a, b]$ of $\mathcal{I}$ ($t_0 < a < b < t_1$) and for each fixed $L > 0$, we consider the space

$$(4.6)\qquad \begin{cases} \mathcal{D}_{a,M_k}([a\ b]\ L; F) = \text{(closed) subspace of } \mathcal{D}_{M_k}([a, b], L; F) \\ \text{made up of the functions } \varphi \text{ such that } \varphi^{(k)}(a) = 0, \\ k = 0\ 1, 2, \ldots \end{cases}$$

Next we define the space

$$(4.7)\qquad \mathcal{D}_{a,M_k}([a\ b]; F) = \operatorname*{ind\,lim}_{L\to+\infty} \mathcal{D}_{a,M_k}([a, b], L; F),$$

where L increases monotonically to $+\infty$, and the space

$$(4.8)\qquad \mathcal{D}_{a,M_k}([a, t_1[; F) = \operatorname*{project\,lim}_{b\to t_1} \mathcal{D}_{a,M_k}([a, b]; F),$$

where b increases monotonically to t_1 and where the projective limit is taken with respect to the mappings $r_{b',b'}{}'$, the restrictions of the φ's defined in $[a, b'']$ to the interval $[a, b']$ for $b' < b''$.

Finally, if we identify the functions φ of $\mathcal{D}_{a,M_k}([a, t_1[;F)$ with their extensions by zero in $]t_0, a[$, we can provide $\mathcal{D}_{+,M_k}(\mathcal{I}; F)$ with the inductive limit topology of the topologies of the spaces $\mathcal{D}_{a,M_k}([a, t_1[; F)$ as a decreases monotonically to t_0; that is

$$(4.9)\qquad \mathcal{D}_{+,M_k}(\mathcal{I}; F) = \operatorname*{indlim}_{a\to t_0} \mathcal{D}_{a,M_k}([a, t_1[; F).$$

In short, we have defined $\mathcal{D}_{+,M_k}(\mathcal{I}; F)$ by

(4.10)

$$\mathcal{D}_{+,M_k}(\mathcal{I}; F) = \operatorname*{ind\,lim}_{a\to t_0}\Big(\operatorname*{project\,lim}_{b\to t_1}\big(\operatorname*{ind\,lim}_{L\to+\infty} \mathcal{D}_{a,M_k}([a, b], L; F)\big)\Big). \qquad \square$$

We define $\mathcal{D}_{-,M_k}(\mathcal{I}; F)$ in a completely analogous way (we interchange the roles of t_0 and t_1; the functions of $\mathcal{D}_{-,M_k}(\mathcal{I}; F)$ have their supports bounded on the right in $\mathcal{I}$, that is they vanish in a neighborhood of t_1). $\square$

Remark 4.1. The spaces $\mathcal{D}_{M_k}(\mathcal{I}; F)$, $\mathcal{E}_{M_k}(\mathcal{I}; F)$ and $\mathcal{D}_{\pm,M_k}(\mathcal{I}; F)$ are stable with respect to differentiation. By using the known properties of the space of continuous functions on $\mathcal{I}$ with values in F (see Bourbaki [3], Chapter III, § 1, No. 1) we also see that the spaces under consideration are separated. $\square$

Remark 4.2. In connection with the spaces $\mathcal{D}_{\pm,M_k}(\mathcal{I}; F)$ and $\mathcal{E}_{M_k}(\mathcal{I}, F)$, we note that in Lions-Magenes [3], [4] we have defined the topologies in a slightly different manner. The definitions given here are easier to use, and in any case the topologies are *equivalent*[1].

Remark 4.3. We shall also use the space $\mathscr{D}_+(\mathscr{I}; F)$ of infinitely differentiable functions on $\mathscr{I}$, with values in F, *and with support bounded on the left in* $\mathscr{I}$, provided with the topology of L. Schwartz [3]; in the same notation as for the definition of $\mathscr{D}_{+,M_k}(\mathscr{I}; F)$, this means that

$$\mathscr{D}_+(\mathscr{I}; F) = \operatorname*{ind\,lim}_{a \to t_0} \big(\operatorname*{project\,lim}_{b \to t_1} \mathscr{D}_a([a,b]; F)\big),\ t_0 < a < b < t_1,$$

where $\mathscr{D}_a([a, b]; F)$ is the space of infinitely differentiable functions φ on $[a, b]$, such that $\varphi^{(k)}(a) = 0$, $k = 0, 1, 2, \ldots$, provided with the topology of uniform convergence on $[a, b]$ for φ and each of its derivatives.

Similarly, we shall consider the space $\mathscr{D}_-(\mathscr{I}; F)$ of functions *with support bounded on the right in* $\mathscr{I}$. ☐

4.4 Remarks on the Topological Properties of the Spaces $\mathscr{D}_{M_k}(\mathscr{I}; F)$, $\mathscr{E}_{M_k}(\mathscr{I}; F)$, $\mathscr{D}_{\pm,M_k}(\mathscr{I}; F)$

Several questions may come up in connection with the spaces $\mathscr{D}_{M_k}(\mathscr{I}; F)$, $\mathscr{E}_{M_k}(\mathscr{I}; F)$ and $\mathscr{D}_{\pm,M_k}(\mathscr{I}; F)$: what are the topological properties of these spaces relative to the properties of the space F? For example, are these spaces complete if F is complete? Furthermore: what is the relation between these spaces and the spaces defined by using topological tensor products (for example $\mathscr{D}_{M_k}(\mathscr{I}) \otimes F$, provided with one of the topologies of Grothendieck [1])?

These are questions which we shall avoid in so far as it is possible. As we go along, we shall give the topological properties required for the applications, for the concrete spaces which will be used.

Here we just call attention to a problem which will come up in Chapters 10 and 11: let $F = \operatorname*{ind\,lim}_{n \to \infty} F_n$, where the F_n's are separated, topological vector spaces; is the space $\mathscr{D}_{M_k}(\mathscr{I}; F), \ldots$ the inductive limit of the spaces $\mathscr{D}_{M_k}(\mathscr{I}; F_n), \ldots$?

---

[1] Let us verify this for $\mathscr{D}_{+,M_k}(\mathscr{I}; F)$, for example. In Lions-Magenes [3], [4], $\mathscr{I} = \mathbf{R}$ and $\mathscr{D}_{+,M_k}(\mathbf{R}; F)$ is denoted by $\mathscr{D}_{+,M_k}(F)$. From the definition given in Lions-Magenes [3], [4] we immediately deduce the fact that the topology of $\mathscr{D}_{+,M_k}(F)$ is finer than the topology of $\mathscr{D}_{+,M_k}(\mathbf{R}; F)$. We must therefore show that it is also coarser, i.e. that the identity is continuous from $\mathscr{D}_{+,M_k}(\mathbf{R}; F)$ into $\mathscr{D}_{+,M_k}(F)$, and for this it is sufficient that the identity be continuous from $\mathscr{D}_{a,M_k}(\mathbf{R}, F)$ (space obtained by extending the functions of $\mathscr{D}_{a,M_k}([a, +\infty[; F)$ by zero for $t < a$) into $\mathscr{E}(F)$ (space of infinitely differentiable functions on $\mathbf{R}$ with values in F, according to the notation of Lions-Magenes [3], [4]), which is obvious from the definitions.

In Chapters 10 and 11, we shall make use of the following two results due to G. Geymonat [1], [2]:

$$\text{(4.11)} \quad \begin{cases} \text{if } F \text{ and } F_n \text{ are } (\mathscr{DF})\text{-spaces in the sense} \\ \text{of Grothendieck [2], and if they are complete, then} \\ \mathscr{D}_{a,M_k}([a,b];F) = \operatorname*{ind\,lim}_{n\to+\infty, L\to+\infty} \mathscr{D}_{a,M_k}([a,b],L;F_n) \end{cases}$$

(see Geymonat [2], Corollary to Theorem 4.2).

$$\text{(4.12)} \quad \begin{cases} \text{if } F \text{ and } F_n \text{ are complete nuclear spaces, then} \\ \mathscr{D}_{a,M_k}([a,b],L;F) = \operatorname*{ind\,lim}_{n\to+\infty} \mathscr{D}_{a,M_k}([a,b],L;F_n) \end{cases}$$

(see Geymonat [2], Corollary to Theorem 4.1).

We note that (4.11) applies with $F = \mathscr{H}(\Gamma)$ and $F = \mathscr{D}_{M_k}(\Omega)$, and (4.12) with $F = \mathscr{D}(\Omega)$, because of the properties of $\mathscr{H}(\Gamma)$ (see Section 3) and of $\mathscr{D}(\Omega)$ (see Grothendieck [1], Schwartz [1]). □

Remark 4.4. From (4.11) and (4.12) we can also deduce conditions to obtain Suslin-type spaces (see Geymonat [2]); for example $\mathscr{D}_{a,M_k}([a,b];\mathscr{H}(\Gamma))$ is a Suslin-type space. □

5. Vector-Valued Ultra-Distributions of Class M_k; Generalizations

5.1 Recapitulation on Vector-Valued Distributions

From now on we shall assume that

(5.1) *F is a reflexive, separated, locally convex vector space.*

We still let $\mathscr{I}$ be an open interval, finite or infinite, of $\mathbf{R}^1$. We denote by $\mathscr{D}'(\mathscr{I}, F)$ (resp. $\mathscr{D}'_+(\mathscr{I}; F)$, resp. $\mathscr{D}'_-(\mathscr{I}; F)$) the *strong dual* of $\mathscr{D}(\mathscr{I}; F')$ (resp. $\mathscr{D}_-(\mathscr{I}; F')$, resp. $\mathscr{D}_+(\mathscr{I}; F')$) (where F' = strong dual of F) and we call it *space of distributions on $\mathscr{I}$ with values in F.* □

Remark 5.1. These definitions are different and more restrictive than the usual definitions of L. Schwartz [3]: according to L. Schwartz, the space of distributions on $\mathscr{I}$ with values in F is the space $\mathscr{L}(\mathscr{D}(\mathscr{I}); F)$ of continuous linear mappings of $\mathscr{D}(\mathscr{I})$ into F.

Similarly, one introduces the spaces $\mathscr{L}(\mathscr{D}_-(\mathscr{I}); F)$ and $\mathscr{L}(\mathscr{D}_+(\mathscr{I}); F)$ which correspond to $\mathscr{D}'_+(\mathscr{I}; F)$ and $\mathscr{D}'_-(\mathscr{I}; F)$ respectively.

However, the above definition will be more convenient for the applications we have in mind. In any case, we have

$$\text{(5.2)} \qquad \mathscr{D}'(\mathscr{I}; F) \subset \mathscr{L}(\mathscr{D}(\mathscr{I}); F)$$

(resp. $\mathscr{D}'_+(\mathscr{I}; F) \subset \mathscr{L}(\mathscr{D}_-(\mathscr{I}); F)$, resp. $\mathscr{D}'_-(\mathscr{I}; F) \subset \mathscr{L}(\mathscr{D}_+(\mathscr{I}); F)$) and under certain conditions on F (see Schwartz [3], Lions-Magenes [2], § 4.2) the first and second terms of (5.2) coincide, for example if F is a reflexive Banach space. ☐

Differentiation in $\mathscr{D}'(\mathscr{I}; F)$ and multiplication by a (scalar) function ψ of $\mathscr{E}(\mathscr{I})$ are defined in the usual manner by the formulas:

$$\left\langle \frac{\mathrm{d}u}{\mathrm{d}t}, \varphi \right\rangle = -\left\langle u, \frac{\mathrm{d}\varphi}{\mathrm{d}t} \right\rangle, u \in \mathscr{D}'(\mathscr{I}; F), \varphi \in \mathscr{D}(\mathscr{I}; F'), \tag{5.3}$$

$$\langle \psi u, \varphi \rangle = \langle u, \varphi\psi \rangle, u \in \mathscr{D}'(\mathscr{I}; F), \psi \in \mathscr{E}(\mathscr{I}), \varphi \in \mathscr{D}(\mathscr{I}; F'), \tag{5.4}$$

where the brackets denote the duality between $\mathscr{D}'(\mathscr{I}; F)$ and $\mathscr{D}(\mathscr{I}; F')$. ☐

The following property holds:

(5.5) $\begin{cases} \mathscr{D}'_+(\mathscr{I}; F) \ (\textit{resp. } \mathscr{D}'_-(\mathscr{I}; F)) \textit{ can be identified with} \\ \textit{the subspace of distributions } u \in \mathscr{D}'(\mathscr{I}; F) \textit{ with} \\ \textit{support bounded on the left (resp. on the right).} \end{cases}$

Indeed the space $\mathscr{D}(\mathscr{I}; F')$ is dense in $\mathscr{D}_-(\mathscr{I}; F')$ and consequently $\mathscr{D}'_+(\mathscr{I}; F)$ may be identified with a subspace of $\mathscr{D}'(\mathscr{I}; F)$. Next, if $u \in \mathscr{D}'_+(\mathscr{I}; F)$, then u vanishes for $t < t_u$, for a suitable t_u; for otherwise, there would exist a sequence of intervals $[a_n, b_n]$ contained in $\mathscr{I} =]t_0, t_1[$ ($\mathscr{I}$ finite or infinite), with $b_n \to t_0$, and a sequence of functions $\varphi_n \in \mathscr{D}(\mathscr{I}; F')$ with

$$\text{support of } \varphi_n \subset [a_n, b_n] \tag{5.6}$$

such that $\langle u, \varphi_n \rangle \neq 0$. Then, by replacing φ_n with $K_n\varphi_n$, $K_n \in \mathbf{C}$, we could always assume that

$$\langle u, \varphi_n \rangle = 1. \tag{5.7}$$

But (5.6) implies $\varphi_n \to 0$ in $\mathscr{D}_-(\mathscr{I}; F')$, which contradicts (5.7).

Finally, let $u \in \mathscr{D}'(\mathscr{I}; F)$ and have its support bounded on the left; let $\theta \in \mathscr{E}(\mathscr{I}; \mathbf{C})$, with $\theta(t) = 1$ in a neighborhood of the support of u and $\theta(t) = 0$ in a neighborhood of t_0. Then $\theta u = u$ and, if $\varphi \in \mathscr{D}(\mathscr{I}; F')$, we have

$$\langle u, \varphi \rangle = \langle u, \theta\varphi \rangle. \tag{5.8}$$

If $\varphi \to 0$ (in $\mathscr{D}(\mathscr{I}; F')$) for the topology of $\mathscr{D}_-(\mathscr{I}; F')$, then $\theta\varphi \to 0$ in $\mathscr{D}(\mathscr{I}; F')$ and therefore $\langle u, \theta\varphi \rangle \to 0$; therefore, thanks to (5.8), u defines a continuous linear form on $\mathscr{D}(\mathscr{I}; F')$ provided with the topology induced by $\mathscr{D}_-(\mathscr{I}; F')$, so that $u \in \mathscr{D}'_+(\mathscr{I}; F)$, which completes the proof of (5.5). ☐

Remark 5.2. The analogue to (5.5) for $\mathscr{L}(\mathscr{D}_-(\mathscr{I}); F)$ may be *incorrect*; for example if $\mathscr{I} = \mathbf{R}$, $F = \mathscr{D}'(\mathbf{R})$, the mapping $u \in \mathscr{L}(\mathscr{D}_-(\mathbf{R}); \mathscr{D}'(\mathbf{R}))$ defined by

$$u = \delta(x - t) \ \left(\text{i.e. } u(\varphi)\ (x) = \varphi(x),\ x \in \mathbf{R}\right), \tag{5.9}$$

does not have its support bounded on the left in t, since the support of u in $\mathbf{R}^2 = \mathbf{R}_x \times \mathbf{R}_t$ is the line $x = t$. This example also shows that in (5.2) the inclusion may be *strict*, since u given by (5.9), cannot, according to (5.5), also belong to $\mathscr{D}'_+(\mathbf{R}; \mathscr{D}'(\mathbf{R}))$, as its support is not bounded on the left.

5.2 The Space $\mathscr{D}'_{M_k}(\mathscr{I}; F)$

Let $\mathscr{I}$ and F be defined as in Section 5.1 and let $\{M_k\}$ be a sequence of positive numbers satisfying (1.1), (1.2), (1.3) and (1.4).

Definition 5.1 $\mathscr{D}'_{M_k}(\mathscr{I}; F)$ *denotes the dual space of* $\mathscr{D}_{M_k}(\mathscr{I}; F')$ *provided with the strong dual topology.*

The elements of $\mathscr{D}'_{M_k}(\mathscr{I}; F)$ are called *ultra-distributions of class* M_k *on* $\mathscr{I}$, *with values in* F. ▯

Remark 5.3. The analogue to Remark 5.1 holds: by analogy with the definition of vector-valued distributions of Schwartz [3], we could call $\mathscr{L}(\mathscr{D}_{M_k}(\mathscr{I}); F)$ the space of vector-valued ultra-distributions of class M_k, but definition 5.1, although more restrictive, is more convenient for the sequel.

In fact, we have

$$\mathscr{D}'_{M_k}(\mathscr{I}; F) \subset \mathscr{L}\left(\mathscr{D}_{M_k}(\mathscr{I}); F\right) \tag{5.10}$$

and under certain hypotheses on F the equality holds in (5.10) (for example if F is a Frechet space; G. Geymonat [2], Prop. 5.1). ▯

$\mathscr{D}'(\mathscr{I}; F)$ (defined as in 5.1) can be identified to a subspace of $\mathscr{D}'_{M_k}(\mathscr{I}; F)$ by using the fact that

$$\mathscr{D}_{M_k}(\mathscr{I}; F') \text{ is dense in } \mathscr{D}(\mathscr{I}; F'). \tag{5.11}$$

Indeed, let $\varphi \in \mathscr{D}(\mathscr{I}; F')$; let ϱ_n be a regularizing sequence of scalar functions, *of class* M_k ($\varrho_n \in \mathscr{D}_{M_k}(\mathbf{R})$, $\varrho_n \geq 0$, $\int_{-\infty}^{+\infty} \varrho_n(t)\, dt = 1$, ϱ_n has support in $[\alpha_n, \beta_n]$, $\alpha_n, \beta_n \to 0$),; such a sequence exists (see Section 1.2, property a)). But, according to (5.1), F' is quasi-complete, see for example Bourbaki [2], page 88.

Then we can *regularize* φ *by* ϱ_n: $\varphi * \varrho_n = \varphi_n$ is defined by

$$\varphi_n(t) = \int_{\mathscr{I}} \varrho_n(t - \sigma)\, \varphi(\sigma)\, d\sigma$$

(which is well-defined for sufficiently large n); we have: $\varphi_n \in \mathscr{D}_{M_k}(\mathscr{I}; F')$, the φ_n's have support in a fixed compact set;
$\varphi_n \to \varphi$ in F', uniformly in t, as well as each derivative; whence (5.11).

Thus we have

$$\mathscr{D}'(\mathscr{I}; F) \subset \mathscr{D}'_{M_k}(\mathscr{I}; F). \quad \square \tag{5.12}$$

Differentiation in $\mathscr{D}'_{M_k}(\mathscr{I}; F)$ and multiplication by a scalar function ψ of $\mathscr{E}_{M_k}(\mathscr{I})$ are defined, as for ordinary distributions, by formulas analogous to (5.3) and (5.4). $\square$

The *structure* of ultra-distributions of $\mathscr{D}'_{M_k}(\mathscr{I}; F)$ can be specified, at least under the additional hypothesis:

$$\mathscr{D}^0(\mathscr{I}; F') \textit{ is barrelled}, \tag{5.13}$$

where $\mathscr{D}^0(\mathscr{I}; F')$ is the space of continuous functions with compact support in $\mathscr{I}$, with values in F', provided with the topology of L. Schwartz [2] (that is, the inductive limit topology of the spaces $\mathscr{D}^0(\mathscr{I}, \mathscr{K}; F')$, as $\mathscr{K}$ varies over the set of compact intervals contained in $\mathscr{I}$, where $\mathscr{D}^0(\mathscr{I}; \mathscr{K}, F')$ is the subspace of $\mathscr{D}^0(\mathscr{I}; F')$ made up of the functions with support contained in $\mathscr{K}$ and provided with the topology of uniform convergence on $\mathscr{K}$).

Then we have (see Lions-Magenes [3], Theorem 7.1, Chapter 1):

Theorem 5.1. *Under hypotheses* (5.1) *and* (5.13) *every* $u \in \mathscr{D}'_{M_k}(\mathscr{I}; F)$ *may be represented, non-uniquely, by the form*

$$u = \sum_{k=0}^{\infty} \frac{\mathrm{d}^k}{\mathrm{d}t^k} \mu_k, \tag{5.14}$$

where

$$\mu_k \in \big(\mathscr{D}^0(\mathscr{I}; F')\big)' \text{ (measures on } \mathscr{I} \text{ with values in } F) \tag{5.15}$$

and

$$\left\{\begin{array}{l}\textit{for every continuous (scalar) function } \theta \textit{ with} \\ \textit{compact support in } \mathscr{I}, \textit{ and every } L > 0, \textit{ the} \\ \textit{series } \sum_{k=0}^{\infty} L^k M_k \theta \mu_k \textit{ converges in} \big(\mathscr{D}^0(\mathscr{I}; F')\big)'\end{array}\right. \tag{5.16}$$

and where (5.14) *means that*

$$\langle u, \varphi \rangle = \sum_{k=0}^{\infty} (-1)^k \langle \mu_k, \varphi^{(k)} \rangle, \ \forall \varphi \in \mathscr{D}_{M_k}(\mathscr{I}; F') \tag{5.17}$$

(the brackets in the summation denoting the duality between $\mathscr{D}^0(\mathscr{I}, F')$ and $(\mathscr{D}^0(\mathscr{I}; F'))'$ and $\langle u, \varphi \rangle$ denoting the duality between $\mathscr{D}'_{M_k}(\mathscr{I}; F)$ and $\mathscr{D}_{M_k}(\mathscr{I}; F')$). *Conversely, every* u *in the form* (5.14), (5.17) *defines an element of* $\mathscr{D}'_{M_k}(\mathscr{I}; F)$. $\square$

Remark 5.4. Hypothesis (5.13) is satisfied if F' is a Frechet space (see Bourbaki [3], Chapter III, § 1, no. 1) and therefore, according to (5.1), if F is the dual of a Frechet space (this is the case for $F = \mathscr{D}_{M_k}(\Omega)$ and $F = \mathscr{H}(\Gamma)$). Other cases have been pointed out to us by G. Geymonat: (5.13) is satisfied if F' is a barrelled, nuclear $(\mathscr{D}\mathscr{F})$-space or if F' is a strict inductive limit of a sequence of nuclear Frechet spaces (thus (5.13) holds if $F = \mathscr{D}'_{M_k}(\Omega)$ or $F = \mathscr{H}'(\Gamma)$ or $F = \mathscr{D}'(\Omega)$). □

5.3 The Space $\mathscr{D}'_{\pm,M_k}(\mathscr{I}; F)$

The hypotheses on $\mathscr{I}$, F and $\{M_k\}$ are the same as in Section 5.2.

Definition 5.2. $\mathscr{D}'_{+,M_k}(\mathscr{I}; F)$ (resp. $\mathscr{D}'_{-,M_k}(\mathscr{I}; F)$) *denotes the strong dual of* $\mathscr{D}_{-,M_k}(\mathscr{I}; F')$ (resp. $\mathscr{D}_{+,M_k}(\mathscr{I}; F')$). □

Remark 5.5. The analogue to Remark 5.3 holds. □

In the same way as for (5.5), it can be shown (see Lions-Magenes [3], Theorem 9.1, Chapter I) that

$$(5.18)\qquad \begin{cases} \mathscr{D}'_{+,M_k}(\mathscr{I}; F) (\textit{resp. } \mathscr{D}'_{-,M_k}(\mathscr{I}; F)) \textit{ can be identified} \\ \textit{with the subspace of } \mathscr{D}'_{M_k}(\mathscr{I}; F) \textit{ of} \\ \textit{ultra-distributions of class } M_k \textit{ with support} \\ \textit{bounded on the left (resp. on the right).} \end{cases}$$

From which we deduce (see Lions-Magenes [3], Theorem 9.2, Chapter I) the following theorem on the structure of the elements of $\mathscr{D}'_{+,M_k}(\mathscr{I}; F)$ (and analogously of $\mathscr{D}'_{-,M_k}(\mathscr{I}; F)$):

Theorem 5.2. *Under hypotheses* (5.1) *and* (5.13), *if u is given in* $\mathscr{D}'_{+,M_k}(\mathscr{I}; F)$ *with $t_* =$ the left endpoint of the support of u, then, for all $\tilde{t} < t_*$, there exists a (non-unique) decomposition of u in the form*

$$(5.19)\qquad u = \sum_{k=0}^{\infty} \frac{\mathrm{d}^k}{\mathrm{d}t^k} \mu_k,$$

where

$$(5.20)\qquad \mu_k \in (\mathscr{D}^0(\mathscr{I}; F'))',$$

$$(5.21)\qquad \mu_k = 0 \text{ for } t < \tilde{t},$$

$$(5.22)\qquad \begin{cases} \textit{for every continuous (scalar) function } \chi, \textit{ with} \\ \textit{compact support in } \mathscr{I}, \textit{ and every } L > 0, \textit{ the} \\ \textit{series } \sum_{k=0}^{\infty} L^k M_k \chi \mu_k \textit{ converges in } (\mathscr{D}^0(\mathscr{I}; F'))', \end{cases}$$

and (5.19) *is taken in a sense analogous to* (5.17).

Conversely, if u is given in $\mathscr{D}'_{M_k}(\mathscr{I}; F)$ *and if for every* $\tilde{t} < t_*$ *there exists a decomposition of u in the form* (5.19), *with* (5.20), (5.21) *and* (5.22), *then u belongs to* $\mathscr{D}'_{+,M_k}(\mathscr{I}; F)$ *and vanishes for* $t < t_*$. □

Remark 5.6. If $M_k = (k!)^s$, $s > 1$, we shall write $\mathscr{D}_s(\mathscr{I}; F)$, $\mathscr{D}'_s(\mathscr{I}; F)$, ... instead of $\mathscr{D}_{M_k}(\mathscr{I}; F)$, $\mathscr{D}'_{M_k}(\mathscr{I}; F)$, ... and speak of *Gevrey ultra-distributions of order s, with values in F.*

5.4 Vector-Valued Ultra-Distributions of Beurling-Type

The generalization of scalar ultra-distributions given in Section 2.3 can be extended to the vector case.

We still have $\mathscr{I} =]t_0, t_1[$, a finite or infinite, open interval in $\mathbf{R}^1$, F a space satisfying (5.1), $\{M_k\}$ a sequence of numbers satisfying (1.1), (1.2), (1.3) and (1.4). As fundamental space, we take $\mathscr{B}_{M_k}(\mathscr{I}; F)$, defined, in analogy to (2.9), by

(5.23) $$\mathscr{B}_{M_k}(\mathscr{I}; F) = \operatorname*{ind\,lim}_{\mathscr{K}\to\mathscr{I}} \left(\operatorname*{project\,lim}_{L\to 0} \mathscr{D}_{M_k}(\mathscr{I}, \mathscr{K}, L; F)\right),$$

where $\mathscr{K}$ (compact interval contained in $\mathscr{I}$) increases monotonically to $\mathscr{I}$ and L decreases monotonically to zero (compare with (4.3)).

Then the space of *Beurling-type ultra-distributions on* $\mathscr{I}$, *with values in* F, *relative to the sequence* M_k is by definition

$$\mathscr{B}'_{M_k}(\mathscr{I}; F) = \text{strong dual of } \mathscr{B}_{M_k}(\mathscr{I}; F') \quad \square$$

Analogously, we define the space

(5.24)

$$\mathscr{B}_{+,M_k}(\mathscr{I}; F) = \operatorname*{ind\,lim}_{a\to t_0} \left(\operatorname*{project\,lim}_{b\to t_1} \left(\operatorname*{project\,lim}_{L\to 0} \mathscr{D}_{a,M_k}([a, b], L; F)\right)\right),$$

where $t_0 < a < b < t_1$ and $L > 0$ (compare with (4.10)), and then the space

(5.25) $$\mathscr{B}'_{-,M_k}(\mathscr{I}; F) = \text{strong dual of } \mathscr{B}_{+,M_k}(\mathscr{I}; F').$$

In the same way, we introduce the spaces $\mathscr{B}_{-,M_k}(\mathscr{I}; F)$ and $\mathscr{B}'_{+,M_k}(\mathscr{I}; F)$. □

We restrict our discussion to pointing out that

(5.26) $$\mathscr{B}_{M_k}(\mathscr{I}; F) \textit{ is dense in } \mathscr{D}_{M_k}(\mathscr{I}; F)$$

(apply the remarks of Sections 2.3 and 5.2, see (5.11)), from which, after suitable identifications, it follows that

(5.27) $$\mathscr{D}'(\mathscr{I}; F) \subset \mathscr{D}'_{M_k}(\mathscr{I}; F) \subset \mathscr{B}'_{M_k}(\mathscr{I}; F). \quad \square$$

Remark 5.7. If $M_k = (k!)^s$, $s > 1$, we shall write $\mathscr{B}_s(\mathscr{I}; F)$, $\mathscr{B}'_s(\mathscr{I}; F)$,... instead of $\mathscr{B}_{M_k}(\mathscr{I}; F)$, $\mathscr{B}'_{M_k}(\mathscr{I}, F)$, ...

5.5 The Particular Case: F = Banach Space

Let us only consider the case of $\mathscr{D}_{M_k}(\mathscr{I}; F)$ and $\mathscr{D}'_{M_k}(\mathscr{I}; F)$ (analogous considerations are also valid for the other spaces).

Thus let $\mathscr{I} =]t_0, t_1[$ be an interval in $\mathbf{R}^1$, $\{M_k\}$ a sequence of numbers satisfying (1.1), (1.2), (1.3) and (1.4), and F a Banach space. Then, definition 4.1 of the space $\mathscr{D}_{M_k}(\mathscr{I}; F)$ is equivalent to: *$\mathscr{D}_{M_k}(\mathscr{I}; F)$ is the space of functions $\varphi \in \mathscr{D}(\mathscr{I}; F)$ such that there exist two positive numbers c and L (both depending on φ) such that*

$$\sup_{t \in \mathscr{I}} \|\varphi^{(k)}(t)\|_F \leq c L^k M_k, \; k = 0, 1, 2, \ldots \tag{5.28}$$

The space $\mathscr{D}_{M_k}(\mathscr{I}, \mathscr{K}, L; F)$, which intervenes in (4.3) for the definition of the topology of $\mathscr{D}_{M_k}(\mathscr{I}; F)$, is now a Banach space with norm

$$\sup_{t \in \mathscr{K}, k=0,1,\ldots} \left\| \frac{\varphi^{(k)}(t)}{L^k M_k} \right\|_F . \tag{5.29}$$

Therefore $\mathscr{D}_{M_k}(\mathscr{I}; F)$ is an *inductive limit of Banach spaces.* □

Another equivalent definition of the space $\mathscr{D}_{M_k}(\mathscr{I}; F)$ can be given by using the space $L^p(\mathscr{I}; F)$ (of classes) of pth-power $(1 < p < +\infty)$ integrable functions on $\mathscr{I}$, with values in F, provided with the norm

$$\|\varphi\|_{L^p(\mathscr{I}; F)} = \left(\int_{t_0}^{t_1} \|\varphi(t)\|_F^p \, dt \right)^{1/p} . \tag{5.30}$$

Condition (5.28) then becomes

$$\|\varphi^{(k)}\|_{L^p(\mathscr{I}; F)} \leq c L^k M_k, \quad \forall k.$$

Using this presentation of the space $\mathscr{D}_{M_k}(\mathscr{I}; F)$ we obtain (see Lions-Magenes [3], Theorem 2.1, Chapter I) a new theorem on the structure of ultra-distributions of $\mathscr{D}'_{M_k}(\mathscr{I}; F)$, which is analogous to Theorem 5.1, but where the measures μ_k are replaced by locally p'^{th}-power integrable $\left(\frac{1}{p} + \frac{1}{p'} = 1\right)$ functions g_k, with values in F.

6. Comments

The notion of distribution on an open set Ω in $\mathbf{R}^n$ introduced by L. Schwartz [1] has been generalized in several ways. The main idea of these generalizations has been to choose a suitable "fundamental space" of functions, different from the space $\mathscr{D}(\Omega)$ of Schwartz, and to consider the continuous linear forms on this space. In this manner, one has obtained "objects" which have been called *generalized functions* or *distributions, ultra-distributions, ...*: see Gelfand-Shilov [1] and the bibliography of this work.

It was Gevrey [1] who introduced the function spaces $\mathscr{E}_{M_k}(\Omega)$ for the case $M_k = (k!)^s$, with $s > 1$ (for $s = 2$, functions of this type had already been considered for the study of solutions of the heat equation by Holmgren [1] and E. E. Levi [1]). The generalization to sequences $\{M_k\}$ satisfying (1.1), (1.2), (1.3), (1.4) takes its inspiration from the theory of *non-quasi-analytic* functions and from the Denjoy-Carleman theorem (see the book of Mandelbrojt [1] and its bibliography); under hypothesis (1.1), condition (1.2) is necessary and sufficient for the space $\mathscr{D}_{M_k}(\Omega)$ not to be reduced to $\{0\}$. For classes of non-quasi-analytic functions, we also call attention to the work of Friberg [1], Friedman [1], Rudin [1], Boman [1], Talenti [1], Carleson [1], Dzanasija [1], [2], Mityagin [2], Leray-Waelbroeck [1] ... For interpolation between Gevrey spaces, see Goulaouic [1], [2].

For the theory of $\mathscr{D}_{M_k}(\Omega)$-spaces, we have followed and used the work of Roumieu [1], [2], to which we refer the reader for developments of the theory; see also Gelfand-Shilov [1] (see the S-type spaces in Volume 2 and in particular the space S_0, which coincides with $\mathscr{D}_{M_k}(\mathbf{R})$ for $M_k = k^{\beta k}$, $\beta > 1$, and the Appendix to Volume 2).

The spaces of ultra-distributions $\mathscr{B}'_{M_k}(\Omega)$ were introduced by Beurling in [1] (see Björck [1], Larsson [1] for the development of the theory and applications), where the definition is given in a different, way (instead of the estimates of type (2.8) on the derivatives, a condition on the Fourier transform of φ is used). For the spaces $\mathscr{B}_{M_k}(\Omega)$ with $M_k = (k!)^s$ and their applications to the Cauchy problem, see Hörmander [1].

Analytic functionals were introduced by Fantappié (see [1], [2] and also Pellegrino [1]). The bibliography on analytic functionals and their applications is very extensive: in addition to the studies noted in the text and above, we indicate in particular the work of Leray [1], [2], Silva [1], [6], da Silva Dias [1], Malgrange [3], Martineau [3], Mantovani-Spagnolo [1], Köthe [1], Tillman [3] ...

Analytic functionals and distributions are also closely connected with the problem of *boundary values* of holomorphic functions: see Köthe [2], Tillman [1], [2], Zerner [1], Martineau [2], Ehrenpreis [1], Beltrami-Wohlers [1]. In fact, it is through his development of the studies on this problem that M. Satô [1] arrived at his theory of *hyperfunctions*, which generalize both distributions and analytic functionals (roughly speaking, according to the presentation of Martineau [1], a hyperfunction on $\mathbf{R}^n$ is a locally finite series of analytic functionals with compact support, which "stick" together). On the theory of hyperfunctions, see also Bengel [1], [2], [3], Harvey [1], Harvey-Komatsu [1], Komatsu [2], Martineau [1], [2], Boutet de Monvel-Krée [1], Schapira [2], [3], [5], Kantor [1]. Along similar lines, we draw attention to the theory of tempered ultra-distributions of Silva [3], [4] (see also Hasumi [1], Yoshinaga [2]) of which an axiomatic generalization has recently been announced by Sousa Menderes [1] (see also Silva [4]).

Finally, let us note the ultra-distributions introduced by Trèves [2] (see also Steinberg-Trèves [1]) in the study of the Cauchy problem and the axiomatic formalization of ultra-distributions given by Schapira [1].

The theory of vector-valued distributions was founded by L. Schwartz in [2], [3]; an axiomatic theory appears in Silva [5]. We also note the work of Yoshinaga [3], [4]. For the spaces of infinitely differentiable, vector-valued functions, also see de Wilde [1], Garnir-de Wilde-Schmets [1].

For the spaces of vector-valued functions of class M_k (Section 4), see Geymonat [1], [2], Lions-Magenes [3], [4], [5].

Finally, the vector-valued ultra-distributions of class M_k and of Beurling-type were introduced by Lions-Magenes [3], [4], [5].

Chapter 8

Elliptic Boundary Value Problems in Spaces of Distributions and Ultra-Distributions

This chapter requires only the knowledge of the *scalar-valued* distributions and ultra-distributions of Chapter 7. From the point of view of boundary value problems, we assume the essential parts of Chapter 2 (elliptic problems) to be known.

1. Regularity of Solutions of Elliptic Boundary Value Problems in Spaces of Analytic Functions and of Class M_k; Statement of the Problems and Results

1.1 Recapitulation on Elliptic Boundary Value Problems

In Chapter 2 we have studied boundary value problems for linear elliptic equations in the Sobolev spaces $H^s(\Omega)$; as an immediate consequence of the fact that $\bigcap_{s=0}^{\infty} H^s(\Omega) = \mathscr{D}(\overline{\Omega})$ (see Corollary 9.2, Chapter 1), the study of the same problems, in the space $\mathscr{D}(\overline{\Omega})$ of infinitely differentiable functions on $\overline{\Omega}$, follows. More precisely, from Theorem 5.2, Chapter 2 and the first Remark of Section 8.2, Chapter 2, we deduce the following result:

Theorem 1.1. *Let:*

(I) *Ω be a bounded open set in $\mathbf{R}^n$ with boundary Γ, an $(n-1)$-dimensional, infinitely differentiable variety, Ω being locally on one side of Γ;*

(II) *A be a differential operator given by*

$$Au = \sum_{|p|,|q| \leq m} (-1)^{|p|} \mathrm{D}^p(a_{pq}(x)\, \mathrm{D}^q u), \tag{1.1}$$

where $a_{pq} \in \mathscr{D}(\overline{\Omega})$, A being properly elliptic in $\overline{\Omega}$ (in the sense of Definition 1.2 of Chapter 2);

(III) $\{B_j\}_{j=0}^{m-1}$ *be a system of boundary operators given by*

$$B_j u = \sum_{|h| \leq m_j} b_{jh}(x)\, \mathrm{D}^h u, \; j = 0, 1, \ldots, m-1, \tag{1.2}$$

where $b_{jh} \in \mathscr{D}(\Gamma)$, $0 \leq m_j \leq 2m-1$, *the system* $\{B_j\}_{j=0}^{m-1}$ *covering* A *on* Γ *(in the sense of Definition* 1.5 *of Chapter* 2).
Then the boundary value problem

$$\begin{cases} Au = f \text{ in } \Omega, \\ B_j u = g_j \text{ on } \Gamma, \; j = 0, \ldots, m-1, \end{cases} \tag{1.3}$$

with f *given in* $\mathscr{D}(\overline{\Omega})$ *and the* g_j*'s given in* $\mathscr{D}(\Gamma)$, *admits a solution* u *belonging to* $\mathscr{D}(\overline{\Omega})$ *and determined up to addition of a function* w *of the space* N *defined by*: $N = \{w \mid w \in \mathscr{D}(\overline{\Omega}),\; B_j w = 0,\; j = 0, \ldots, m-1,\; Aw = 0\}$, *if and only if*

$$\int_\Omega f\bar{v}\, \mathrm{d}x + \sum_{j=0}^{m-1} \int_\Gamma g_j \bar{\varphi}_j\, \mathrm{d}\sigma = 0 \tag{1.4}$$

for every element $\Phi = \{v; \varphi_0, \ldots, \varphi_{m-1}\}$ *of the space* $= \{\Phi \mid v \in \mathscr{D}(\overline{\Omega}),\; \varphi_j \in \mathscr{D}(\Gamma),\; j = 0, \ldots, m-1,\; \mathscr{P}^*\Phi = 0\}$, $\mathscr{P}^*$ *being defined by* (5.2), *Chapter* 2.

Now we ask whether, if the "data" f, g and Γ are analytic or more generally of class M_k, (the solution or) the solutions of problem (1.3), furnished by Theorem 1.1, are also analytic or of class M_k in $\overline{\Omega}$.

This is a new *regularity* question, which, as we shall see, can be answered affirmatively.

1.2 Statement of the M_k-Regularity Results

We shall work with spaces of functions of class $\{M_k\}$. For the applications we have in mind, we always have

$$M_k = (k!)^\beta, \; \beta \geq 1 \tag{1.5}$$

($\beta = 1$ is the "analytic case", $\beta > 1$ is the "case Gevrey of order β").

However we do not necessarily introduce the sequence $\{M_k\}$ in the form (1.5), in order to show the generality of the question. *We assume* $\{M_k\}$ *to have the following properties* (see also Remark 2.6):

$$M_k > 0 \quad \forall k \geq 0, \tag{1.6}$$

$$M_k^2 \leq M_{k-1} M_{k+1} \quad \forall k \geq 1, \tag{1.7}$$

$$(1.8)\qquad \begin{cases} \text{there exists a constant } c_1 \text{ such that} \\ \binom{k}{t} M_{k-t} M_t \leq c_1 M_k,\ 0 \leq t \leq k, \quad \forall k \geq 0, \end{cases}$$

$$(1.9)\qquad M_k \leq M_{k+1} \quad \forall k \geq 0,$$

$$(1.10)\qquad \begin{cases} \text{there exists a constant } d \text{ such that} \\ M_{t+s} \leq d^{t+s} M_t M_s \quad \forall t, s \geq 0, \end{cases}$$

$$(1.11)\qquad \begin{cases} \text{there exists a positive } d_m \text{ (depending on } m\text{) such that} \\ (M_{2ms})^{2m-t} (M_{2ms+2m})^t \leq d_m^{2m-t} (M_{2ms+t})^{2m} \ \forall s > 0,\ 0 \leq t < 2m. \end{cases}$$

Let us verify that sequence (1.5) *has these properties.*

It can immediately be seen that (1.6), (1.7), (1.8) and (1.9) hold; condition (1.10) is satisfied with $d = 2^\beta$, since $(t+s)! \leq 2^{t+s} t!\, s!$; finally (1.11) is satisfied with $d_m = (2m)^{2m\beta}$, since

$$(1.12)\qquad \left[\frac{((2ms+2m)!)^\beta}{((2ms+t)!)^\beta}\right]^t = [(2ms+t+1)\cdots(2ms+2m)]^{\beta t} \leq$$

$$\leq (2sm+2m)^{\beta t(2m-t)} \leq$$

$$\leq 2m^{2m\beta(2m-t)} (2ms+1)^{\beta t(2m-t)} \leq$$

$$\leq 2m^{2m\beta(2m-t)} [(2ms+1)\cdots(2ms+t)]^{\beta(2m-t)} \leq$$

$$\leq 2m^{2m\beta(2m-t)} \left[\frac{((2ms+t)!)^\beta}{((2ms)!)^\beta}\right]^{2m-t},$$

which is equivalent to (1.11). □

Note that hypotheses (1.2), (1.3) of Chapter 7 on the sequence M_k will only be imposed from Section 4 of this chapter on.

Also note that from (1.7) we deduce

$$(1.13)\qquad \frac{M_{t+s}}{M_t} \leq \frac{M_{k+s}}{M_k} \text{ for } t \leq k \text{ and } \forall s > 0,$$

whence

$$(1.14)\qquad \frac{M_{t+s}}{M_t M_s} \leq \frac{M_{k+i}}{M_k M_i} \text{ for } t \leq k,\ s \leq i.$$

Furthermore from (1.8) we deduce

$$(1.15)\qquad k M_{k-1} \leq \frac{c_1}{M_1} M_k \quad \forall k \geq 1$$

and therefore also

$$M_{k-t} \leq \left(\frac{c_1}{M_1}\right)^t \frac{(k-t)!}{k!} M_k \text{ for } t \leq k. \quad \square \tag{1.16}$$

For $\mathscr{K}$ a compact set in $\mathbf{R}^n$ and $\{M_k\}$ a sequence of numbers satisfying (1.6), ..., (1.11), we shall use the notation $\mathscr{D}_{M_k}(\mathscr{K})$ (see Definition 1.2 of Chapter 7).

If $M_k = k!$, then we recover the space $\mathscr{H}(\mathscr{K})$ of analytic functions on $\mathscr{K}$ (see Section 3.1, Chapter 7).

As we have already recalled in Section 1.2, Chapter 7, the (multiplicative) product of two functions of $\mathscr{D}_{M_k}(\mathscr{K})$ still belongs to $\mathscr{D}_{M_k}(\mathscr{K})$, thanks to (1.8), and the composition $f \circ g$ of two functions still belongs to $\mathscr{D}_{M_k}(\mathscr{K})$, again thanks to (1.8). $\square$

A bounded open set Ω in $\mathbf{R}^n$ will be called of class $\{M_k\}$ if its boundary Γ is an $(n-1)$-dimensional, infinitely differentiable variety, whose "local maps" are given by functions of class $\{M_k\}$, Ω being locally on one side of Γ.

In the next Section we shall prove the following theorem.

Theorem 1.2. *Let $M_k = (k!)^\beta$, with real $\beta \geq 1$ (or more generally $\{M_k\}$ satisfying (1.6), ..., (1.11)).*

Assume the hypotheses of Theorem 1.1 to be satisfied and that $a_{pq} \in \mathscr{D}_{M_k}(\bar{\Omega})$, $b_{jh} \in \mathscr{D}_{M_k}(\Gamma)$ and finally that Ω is of class $\{M_k\}$.

Then, if $u \in \mathscr{D}(\bar{\Omega})$ and if there exist two constants c_0 and L_0 (depending on u) such that

$$\| A^i u \|_{L^2(\Omega)} \leq c_0 L_0^i M_{2mi} \qquad \forall i \geq 0, \tag{1.17}$$

$$\sum_{j=0}^{m-1} \| B_j(A^i u) \|_{H^{2m+2mk-m_j-1/2}(\Gamma)} \leq c_0 L_0^{k+i+1} M_{(k+i+1)2m} \quad \forall i, k \geq 0, \tag{1.18}$$

(where the differential operator A^i is the i-th iterate of A and $A^0 u = u$), the function u belongs to $\mathscr{D}_{M_k}(\bar{\Omega})$, and more precisely $u \in \mathscr{D}_{M_k}(\bar{\Omega}; \chi(L_0))$[(1)]*, where $\chi(L_0)$ is a function of L_0 independent of u and c_0, and $\chi(L_0) \geq L_0$.* $\square$

This theorem, which we shall call the *theorem on "elliptic iterates"*, and which is useful for several points, as we shall see in the following chapters (see also the Comments), contains as a corollary the answer to the question posed at the end of Section 1.1; indeed the following corollary can be deduced from it.

[(1)] For the definition of $\mathscr{D}_{M_k}(\mathscr{K}; L)$ see Chapter 7, Section 1.3.

Corollary 1.1. *Let* $M_k = (k!)^\beta$, *real* $\beta \geq 1$ *(or more generally* $\{M_k\}$ *satisfying* (1.6), ..., (1.11)); *under the hypotheses of Theorem* 1.1, *if furthermore* Ω *is of class* M_k *and* $a_{pq} \in \mathcal{D}_{M_k}(\bar{\Omega})$, $b_{jh} \in \mathcal{D}_{M_k}(\Gamma)$, *then every solution* u *of problem* (1.3) *belongs to* $\mathcal{D}_{M_k}(\bar{\Omega})$ *if* $f \in \mathcal{D}_{M_k}(\bar{\Omega})$ *and* $g_j \in \mathcal{D}_{M_k}(\Gamma)$; *and more precisely* $u \in \mathcal{D}_{M_k}(\bar{\Omega}; \chi(L_0))$ *if* $f \in \mathcal{D}_{M_k}(\bar{\Omega}; L_0)$, $g_j \in \mathcal{D}_{M_k}(\Gamma; L_0)$, *where* $\chi(L_0)$ *is a function of* L_0, *with* $\chi(L_0) \geq L_0$.

1.3 Reduction of the Problem to the Case of the Half-Ball

Via "local maps" and by using the property, already noted, that the composition $f \circ g$ of Gevrey type functions (of class M_k with (1.6), ..., (1.11)) is still a function of the same type, the proof of Theorem 1.2 can be reduced to proving Theorems 1.3 and 2.4 below.

We denote by $x = (x', y) = (x_1, \dots, x_{n-1}, y)$ the point in $\mathbf{R}^n$ and by Ω_ϱ ($\varrho > 0$) the half-ball $\{(x', y) \mid x_1^2 + \dots + x_{n-1}^2 + y^2 \mid < \varrho^2, y > 0\}$ and furthermore by Γ_ϱ the subset of its boundary such that $y = 0$; $\mathrm{D}^\alpha u$, $\mathrm{D}_x^\alpha u$, $\mathrm{D}_y^\alpha u$ denote respectively an arbitrary derivative with respect to all the variables, the variables $x_1, \dots, x_{n-1}$, the variable y. We have

Theorem 1.3. *Let* ϱ_0 *be fixed with* $0 < \varrho_0 < 1$ *and* $M_k = (k!)^\beta$, *with real* $\beta \geq 1$ *(or more generally* $\{M_k\}$ *satisfying* (1.6), ..., (1.11)); *let*

$$\mathcal{A}u = \sum_{|p| \leq 2m} a_p(x)\, \mathrm{D}^p u \tag{1.19}$$

be a properly elliptic operator in $\bar{\Omega}_{\varrho_0}$ *with* $a_p \in \mathcal{D}_{M_k}(\bar{\Omega}_{\varrho_0})$; *let*

$$\mathcal{B}_j u = \sum_{|h| \leq m_j} b_{jh}(x') \mathrm{D}^h u, \; j = 0, \dots, m-1, \tag{1.20}$$

be m *boundary operators with* $b_{jh} \in \mathcal{D}_{M_k}(\Gamma_{\varrho_0})$, $0 \leq m_j \leq 2m - 1$, *the system* $\{\mathcal{B}_j\}_{j=0}^{m-1}$ *covering* $\mathcal{A}$ *on* Γ_{ϱ_0}. *If* $u \in \mathcal{D}(\bar{\Omega}_{\varrho_0})$ *and if there exist two constants* c_0 *and* L_0 *(depending on* u*) such that*

$$\| \mathcal{A}^i u \|_{L^2(\Omega_{\varrho_0})} \leq c_0 L_0^i M_{2mi} \qquad \forall i \geq 0, \tag{1.21}$$

$$\sum_{|\alpha| = 2km} \sum_{j=0}^{m-1} \| \mathrm{D}_{x'}^\alpha (\mathcal{B}_j (\mathcal{A}^i u)) \|_{H^{2m - m_j - 1/2}(\Gamma_{\varrho_0})} \leq \tag{1.22}$$

$$\leq c_0 L_0^{k+i+1} M_{(k+i+1)2m} \qquad \forall i, k \geq 0,$$

then there exists $\varrho' < \varrho_0$ *(depending on* L_0, $\mathcal{A}$, $\mathcal{B}_j$*) such that* $u \in \mathcal{D}_{M_k}(\bar{\Omega}_{\varrho'})$; *and if* c_0 *remains bounded,* u *belongs to a bounded set of* $\mathcal{D}_{M_k}(\bar{\Omega}_{\varrho'})$.

Remark 1.1. Noting the properties of functions of $\mathcal{D}_{M_k}(\bar{\Omega}_{\varrho_0})$ and the ellipticity of $\mathcal{A}$, we may assume that the coefficient of the term $\mathrm{D}_y^{2m} u$ is 1 in (1.19), that is

$$a_{(0,0,\dots,0,2m)} = 1. \tag{1.23}$$

We may also assume, without loss in generality, that the coefficients b_{jh} ar the restrictions to Γ_{ϱ_0} of functions, still denoted by b_{jh}, defined on $\bar{\Omega}_{\varrho_0}$ and belonging to $\mathscr{D}_{M_k}(\bar{\Omega}_{\varrho_0})$; and we may also assume that

$$\sup_{x\in\bar{\Omega}_{\varrho_0}} \sum_{|\alpha|=k} |\mathrm{D}^\alpha a_p(x)| \leq cL^k M_k \quad \forall k \geq 0, \tag{1.24}$$

$$\sup_{x\in\bar{\Omega}_{\varrho_0}} \sum_{|\alpha|=k} |\mathrm{D}^\alpha b_{jh}(x)| \leq cL^{k-2m} M_{k-2m} \quad \forall k \geq 0, \tag{1.25}$$

where c and L are two suitable positive numbers with $L \geq 1$ and where we set $M_{-1} = M_{-2} = \cdots = M_{-2m} = M_0$.

2. The Theorem on "Elliptic Iterates": Proof

2.1 Some Lemmas

We introduce the following notation: for $0 < \varrho < \varrho_0$, k *and* $s = 0, 1, 2, \ldots,$

$$\|u\|_{s,k,\varrho} = \sum_{|q|=s} \sum_{|p|=k} \|\mathrm{D}^q(\mathrm{D}^p_{x'} u)\|_{L^2(\Omega_\varrho)},$$

$$|||\varphi\mathscr{B}u|||_{k,\varrho} = \sum_{j=0}^{m-1} \sum_{|p|=k} \|\varphi \mathrm{D}^p_{x'}(\mathscr{B}_j u)\|_{H^{2m-mj-1/2}(\Gamma_\varrho)},$$

where φ is a fixed function belonging to $\mathscr{D}(\Gamma_{\varrho_0})$.

Let us also agree to set

$$\|u\|_{s,-s,\varrho} = \|u\|_{0,0,\varrho} \; (= \|u\|_{L^2(\Omega_\varrho)}),$$

so that

$$\|u\|_{s-t,k+t,\varrho} \leq \|u\|_{s,k,\varrho} \tag{2.1}$$

for s, t, k such that the two terms of (2.1) are defined.

The following lemma adds some precision to Theorem 16.3, Chapter 1:

Lemma 2.1. *Let integer $s > 0$ be fixed and t be an integer with $0 \leq t < s$ and $0 < \varrho \leq \varrho_0$; there exists a constant c_s (depending only on s and ϱ_0) such that for every $\varepsilon > 0$ and every $u \in H^s(\Omega_\varrho)$ we have*

$$\|u\|_{t,0,\varrho} \leq \varepsilon \|u\|_{s,0,\varrho} + c_s \varepsilon^{-t/(s-t)} \|u\|_{0,0,\varrho}. \tag{2.2}$$

Proof. 1) First we note that it is sufficient to show (2.2) for $\varrho = \varrho_0$, since by homothetic mappings it can easily be deduced for $0 < \varrho < \varrho_0$ as well.

2) Now let us prove (2.2) for $\varrho = \varrho_0$. Applying the *extension methods* introduced in Chapter 1 (see in particular Section 8.1) we are led to the

case of the entire space $\mathbf{R}^n$; $u \in H^s(\Omega_{\varrho_0})$ is extended to a function $v \in H^s(\mathbf{R}^n_+)$ (where $\mathbf{R}^n_+$ is the half-space of x's with $x_n > 0$); next we extend the v's to functions $w \in H^s(\mathbf{R}^n)$. In the course of these extension operations, the semi-norms of type

$$(2.3)\qquad \left(\int_{\Omega_{\varrho_0}} |D^q u|^2 \, dx\right)^{1/2}, \left(\int_{\mathbf{R}^n_+} |D^q v|^2 \, dx\right)^{1/2}, \left(\int_{\mathbf{R}^n} |D^q w|^2 \, dx\right)^{1/2}, \quad |q| \leq s$$

remain equivalent, the constants of equivalence depending only on s and ϱ_0 (it suffices to look at the construction of the extensions as given in Chapter 1, in particular formula (8.4)).

But then if $w \in H^s(\mathbf{R}^n)$ we have, by Fourier transformation ($\hat{w}$ denoting the Fourier transform of w and $\xi = (\xi_1, \ldots, \xi_n)$ the dual variable of $x = (x_1, \ldots, x_n)$),

$$(2.4)\qquad \int_{\mathbf{R}^n} |D^q w(x)|^2 \, dx = \int_{\mathbf{R}^n} |\xi^{2q} \hat{w}(\xi)|^2 \, d\xi$$

and the result then follows from the inequality: for every $\varepsilon > 0$ we have

$$(2.5)\qquad |\xi|^t \leq \varepsilon \, |\xi|^s + \gamma_s \varepsilon^{-t/(t-s)} \qquad \forall \xi \in \mathbf{R}^n \text{ and } t < s, \text{ with constant } \gamma_s,$$

which is a consequence of the inequality

$$|\xi|^t = \varepsilon^\theta \, |\xi|^t \, \varepsilon^{-\theta} \leq \frac{\varepsilon^{\theta p}}{p} |\xi|^{tp} + \frac{1}{p'} \varepsilon^{-\theta p'}, \; tp = s, \; \theta p = 1, \frac{1}{p} + \frac{1}{p'} = 1. \quad \square$$

(2.6)

By an application of Lemma 2.1 we obtain

Lemma 2.2. *Let t be an integer with $0 < t < 2m$ and let $\varrho \leq \varrho_0$; there exists a constant c_m (depending only on m and ϱ_0) such that for every $\varepsilon > 0$ and every function $u \in \mathcal{D}(\bar{\Omega}_{\varrho_0})$ we have*

$$(2.7)\qquad \|u\|_{t,k,\varrho} \leq \varepsilon \, \|u\|_{2m,k,\varrho} + c_m \varepsilon^{-t/(2m-t)} \, \|u\|_{0,k,\varrho} \qquad \forall k \geq 0,$$

$$(2.8)\qquad \|u\|_{k,t,\varrho} \leq \varepsilon \, \|u\|_{k,2m,\varrho} + c_m \varepsilon^{-t/(2m-t)} \, \|u\|_{k,0,\varrho} \qquad \forall k \geq 0,$$

$$(2.9)\qquad \|D_y^t u\|_{0,k,\varrho} \leq \varepsilon \, \|D_y^{2m} u\|_{0,k,\varrho} + c_m \varepsilon^{-t/(2m-t)} \, \|u\|_{0,k,\varrho} \qquad \forall k \geq 0. \quad \square$$

Remark 2.1. Let s be an integer ≥ 0; applying (2.7) to $D^p u$ with $|p| = 2sm$ and summing up, we obtain

$$(2.10)\qquad \|u\|_{2sm+t,k,\varrho} \leq \varepsilon \, \|u\|_{2(s+1)m,k,\varrho} + c_m \varepsilon^{-t/(2m-t)} \, \|u\|_{2sm,k,\varrho}$$

$$\forall s, k \geq 0, \; 0 \leq t < 2m.$$

Analogous inequalities can be deduced from (2.8) and (2.9). $\square$

We shall also use

Lemma 2.3. *For every* $\varepsilon > 0$ *there exists* $c_m(\varepsilon)$, *depending on* ε *(and on* m *and* ϱ_0*) such that for every* $u \in \mathscr{D}(\bar{\Omega}_{\varrho_0})$ *and* $0 < \varrho \leq \varrho_0$ *we have*

$$(2.11) \qquad \sum_{t=1}^{2m} \| \mathrm{D}_y^{2m-t} u \|_{0,t,\varrho} \leq \varepsilon \| \mathrm{D}_y^{2m} u \|_{0,0,\varrho} + c_m(\varepsilon) \| u \|_{0,2m,\varrho}.$$

Proof. Following the same methods as for the proof of Lemma 2.1 (parts 1 and 2), we are led to the case of the entire space $\mathbf{R}^n$ and then, by Fourier transformation (this time denoting the dual variable of $(x_1, \ldots, x_{n-1}, y)$ by $(\xi_1, \ldots, \xi_{n-1}, \eta)$), it all comes down to showing that, for $\varepsilon > 0$ and $0 < t < 2m$, we have

$$\eta^{4m-2t} |\xi|^{2t} \leq \varepsilon |\eta|^{4m} + \gamma_m(\varepsilon) |\xi|^{4m}, \quad \forall \eta \in \mathbf{R}^1 \text{ and } \xi = (\xi_1 \cdots \xi_{n-1}) \in \mathbf{R}^{n-1},$$

with $\gamma_m(\varepsilon)$ depending on m and ε, which is well-known.

Remark 2.2. As for Remark 2.1, it follows from (2.11) that

$$(2.12) \quad \sum_{t=1}^{2m} \| \mathrm{D}_y^{2m+2ms-t} u \|_{0,k,\varrho} \leq \varepsilon \| \mathrm{D}_y^{2m+2ms} u \|_{0,k,\varrho} + c_m(\varepsilon) \| \mathrm{D}^{2ms} u \|_{0,2m+k,\varrho}$$
$$\forall s, k > 0. \quad \square$$

We also note

Lemma 2.4. *Let* $\varphi \in \mathscr{D}(\bar{\Omega}_{\varrho_0})$ *be fixed, then for every integer* $k \geq 0$:

$$(2.13) \quad \| \varphi u \|_{H^k(\Omega_{\varrho_0})} \leq c_k' \left(\max_{\bar{\Omega}_{\varrho_0}} \max_{|p| \leq k} | \mathrm{D}^p \varphi(x) | \right) \| u \|_{H^k(\Omega_{\varrho_0})}, \quad \forall u \in H^k(\Omega_{p_0}),$$

$$(2.14) \quad \| \varphi u \|_{H^{k+1/2}(\Gamma_{\varrho_0})} \leq c_k'' \left(\max_{\Gamma_{\varrho_0}} \max_{|p| \leq k+1} | \mathrm{D}_{x'}^p \varphi(x) | \right) \| u \|_{H^{k+1/2}(\Gamma_{\varrho_0})},$$
$$\forall u \in H^{k+1/2}(\Gamma_{\varrho_0})$$

with c_k' *and* c_k'' *constants depending only on* k.

Proof. The proof of (2.13) is immediate, since k is an integer; (2.14) can be obtained, for example, by interpolation of the mapping $u \to \varphi u$ between $H^k(\Gamma_{\varrho_0})$ and $H^{k+1}(\Gamma_{\varrho_0})$. $\square$

Remark 2.3. (2.14) can be tightened by not including the derivatives of φ up to the order $k + 1$ in the second member; but (2.14) is sufficient for the sequel. For the study of the multipliers φ in H^s-spaces see, for example, Hörmander [1], Peetre [2].

2.2 The Preliminary Estimate

Let ϱ and δ be given positive numbers such that $0 < \varrho < \varrho + \delta < \varrho_0$; let $\chi(t)$ be a fixed infinitely differentiable function on $\mathbf{R}^1$ such that $\chi(t) = 1$ for $t \leq 0$ and $\chi(t) = 0$ for $t \geq 1$.

Set

$$\varphi_{\varrho,\delta}(x) = \chi\left(\frac{|x| - \varrho}{\delta}\right), \quad x \in \bar{\Omega}_{\varrho_0}. \tag{2.15}$$

Thus we have a function $\varphi_{\varrho,\delta} \in \mathscr{D}(\bar{\Omega}_{\varrho_0})$, with support contained in $\bar{\Omega}_{\varrho+\delta}$ and with $\varphi_{\varrho,\delta}(x) = 1$ in $\bar{\Omega}_{\varrho}$. Furthermore, there exists a $\gamma_{|p|}$, which depends only on $|p|$, such that

$$|\mathrm{D}^p \varphi_{\varrho,\delta}(x)| \leq \gamma_{|p|} \delta^{-|p|} \quad \forall x \in \bar{\Omega}_{\varrho_0}. \tag{2.16}$$

Theorem 2.1. *Under the hypotheses of Theorem* 1.3, *there exist two positive constants* ϱ_1 *and* C_1 *such that, if* $0 < \varrho < \varrho + \delta < \varrho_1$ *and if* $u \in \mathscr{D}(\bar{\Omega}_{\varrho_0})$, *we have*

$$\|u\|_{2m,0,\varrho} \leq C_1 \Big\{ \|\varphi_{\varrho,\delta} \mathscr{A} u\|_{0,0,\varrho+\delta} + |||\varphi_{\varrho,\delta} \mathscr{B} u|||_{0,\varrho+\delta} + \\ + \sum_{l=0}^{2m-1} \frac{1}{\delta^{2m-l}} \|u\|_{l,0,\varrho+\delta} \Big\} \tag{2.17}$$

Proof. Thanks to the hypotheses on $\mathscr{A}$ and $\mathscr{B}_j$ we may apply the a priori estimates (4.39) of Chapter 2 (see Theorem 4.3 and Section 8.3 of Chapter 2) to the function $v = \varphi_{\varrho,\delta} u$; thus there exists a $\varrho_1 < \varrho_0$ such that, if $0 < \varrho < \varrho + \delta < \varrho_1$, we have

$$\|\varphi_{\varrho,\delta} u\|_{2m,0,\varrho+\delta} \leq \|\varphi_{\varrho,\delta} u\|_{H^{2m}(\Omega_{\varrho+\delta})} \leq C\{\|\mathscr{A}(\varphi_{\varrho,\delta} u)\|_{0,0,\varrho+\delta} + \\ + |||\mathscr{B}(\varphi_{\varrho,\delta} u)|||_{0,\varrho+\delta} + \|\varphi_{\varrho,\delta} u\|_{H^{2m-1}(\Omega_{\varrho+\delta})}\} \tag{2.18}$$

where the constant C depends only on $\mathscr{A}$ and the $\mathscr{B}_j$'s (and by looking at the proof of Theorem 4.3 of Chapter 2, we easily see that ϱ_1 and C depend on the values of $a_p(0)$ for $|p| = 2m$ and $b_{jh}(0)$ for $|h| = m_j$, on the moduli of continuity of $a_p(x)$ for $|p| = 2m$ and $b_{jh}(x)$ for $|h| = m_j$ at the origin, and on the quantities $\max_{x \in \bar{\Omega}_{\varrho_0}} |a_p(x)|$ and $\max_{y \in \Gamma_{\varrho_0}} |b_{jh}(y)|$ for $|p| < 2m$ and $|h| < m_j$),

But we have

$$\mathscr{A}(\varphi_{\varrho,\delta} u) = \varphi_{\varrho,\delta} \mathscr{A} u + [\mathscr{A}(\varphi_{\varrho,\delta}\, u) - \varphi_{\varrho,\delta} \mathscr{A} u] = \varphi_{\varrho,\delta} \mathscr{A} u + \\ + \sum_{1 \leq |q| \leq 2m} (\mathrm{D}^q \varphi_{\varrho,\delta})\, \mathscr{A}_q u$$

where the $\mathscr{A}_q$'s are differential operators constructed in an obvious way from $\mathscr{A}$ and of order $\leq 2m - |q|$; and therefore, thanks to (2.16), we have

$$\|\mathscr{A}(\varphi_{\varrho,\delta}u)\|_{0,0,\varrho+\delta} \leq C'\{\|\varphi_{\varrho,\delta}\mathscr{A}u\|_{0,0,\varrho+\delta} + \sum_{|p|\leq 2m-1} \delta^{-(2m-|p|)} \|D^p u\|_{0,0,\varrho+\delta}\}, \tag{2.19}$$

where C' depends only on m and on $\max_{x\in\bar{\Omega}_{\varrho_0}} |a_p(x)|$ for $|p| \leq 2m$. In analogous fashion, we have

$$\|\varphi_{\varrho,\delta}u\|_{H^{2m-1}(\Omega_{\varrho+\delta})} \leq \sum_{|q|\leq 2m-1} \|D^q(\varphi_{\varrho,\delta}u)\|_{0,0,\varrho+\delta} \leq \leq C'' \sum_{|p|\leq 2m-1} \frac{1}{\delta^{2m-|p|-1}} \|D^p u\|_{0,0,\varrho+\delta}, \tag{2.20}$$

with C'' depending only on m.

Finally, we have

$$\mathscr{B}_j(\varphi_{\varrho,\delta}u) = \varphi_{\varrho,\delta}\mathscr{B}_j u + [\mathscr{B}_j(\varphi_{\varrho,\delta}u) - \varphi_{\varrho,\delta}\mathscr{B}_j u] = \varphi_{\varrho,\delta}\mathscr{B}_j u + \sum_{1\leq|q|\leq m_j} (D^q\varphi_{\varrho,\delta})\, \mathscr{B}_{j,q}u,$$

where the $\mathscr{B}_{j,q}$'s are differential operators of order $\leq m_j - |q|$, constructed in an obvious way from the $\mathscr{B}_j$'s, with infinitely differentiable coefficients in $\bar{\Omega}_{\varrho_0}$ (see Remark 1.1).

From the above, also applying the (trace) Theorem 7.5 of Chapter 1, we deduce

$$\begin{aligned}\|\mathscr{B}_j(\varphi_{\varrho,\delta}u)\|_{H^{2m-m_j-1/2}(\Gamma_{\varrho+\delta})} &\leq \|\varphi_{\varrho,\delta}\mathscr{B}_j u\|_{H^{2m-m_j-1/2}(\Gamma_{\varrho+\delta})} + \\ &+ \sum_{1\leq|q|\leq m_j} \|(D^q\varphi_{\varrho,\delta})\, \mathscr{B}_{j,q}u\|_{H^{2m-m_j-1/2}(\Gamma_{\varrho+\delta})} \leq \\ &\leq \|\varphi_{\varrho,\delta}\mathscr{B}_j u\|_{H^{2m-m_j-1/2}(\Gamma_{\varrho+\delta})} + \\ &+ C''' \sum_{1\leq|q|\leq m_j} \|D^q(\varphi_{\varrho,\delta})\, \mathscr{B}_{j,q}u\|_{H^{2m-m_j}(\Omega_{\varrho+\delta})} \leq \\ &\leq \|\varphi_{\varrho,\delta}\mathscr{B}_j u\|_{H^{2m-m_j-1/2}(\Gamma_{\varrho+\delta})} + \\ &+ C''' \sum_{1\leq|q|\leq m_j} \sum_{|\alpha|\leq 2m-m_j} \|D^\alpha(D^q\varphi_{\varrho,\delta}\mathscr{B}_{j,q}u)\|_{0,0,\varrho+\delta} \leq \\ &\leq \|\varphi_{\varrho,\delta}\mathscr{B}_j u\|_{H^{2m-m_j-1/2}(\Gamma_{\varrho+\delta})} + C^{\mathrm{IV}} \sum_{|p|\leq 2m-1} \frac{1}{\delta^{2m-|p|}} \|D^p u\|_{0,0,\varrho+\delta}\Big\}\end{aligned} \tag{2.21}$$

with C^{IV} depending on m and on $\max_{x\in\bar{\Omega}_{\varrho_0}} |D^h b_{jh}(x)|$ for $|h| \leq 2m - m_j$.

(2.17) then follows immediately from (2.18), (2.19), (2.20) and (2.21).

2.3 Bounds for the Tangential Derivatives

Let us first show

Lemma 2.5. *Let a be a given function with $a \in \mathscr{D}_{M_k}(\bar{\Omega}_{\varrho_0})$ and*

$$\sup_{x \in \bar{\Omega}_{\varrho_0}} \sum_{|\alpha|=r} |\mathrm{D}^\alpha a(x)| \le c L^r M_r \qquad \forall r \ge 0. \tag{2.22}$$

Then, for every $r > 0$, $s \ge 0$, $0 < \varrho < \varrho + \delta < \varrho_0$ and $u \in \mathscr{D}(\bar{\Omega}_{\varrho_0})$, we have

$$\sum_{|\beta|=s} \sum_{|\alpha|=r} \| \mathrm{D}^\beta [\varphi_{\varrho,\delta}(a\, \mathrm{D}_{x'}^\alpha u - \mathrm{D}_{x'}^\alpha (au))] \|_{0,0,\varrho+\delta} \le \tag{2.23}$$

$$\le C_s^* \sum_{l=0}^{s} \frac{1}{\delta^l} \sum_{t=0}^{s-l} L^{r+t} M_{r+t} \sum_{i=0}^{r-1} \frac{1}{L^i M_i} \| u \|_{s-l-t,i,\varrho+\delta},$$

with the constant C_s^* depending only on s and c.

Proof. We recall that (Leibniz formula)

$$\mathrm{D}^\gamma(uv) = \sum_{\eta \le \gamma} \binom{\gamma_1}{\eta_1} \cdots \binom{\gamma_n}{\eta_n} \mathrm{D}^\eta u\, \mathrm{D}^{\gamma-\eta} v$$

where $\gamma = (\gamma_1, \ldots, \gamma_n)$, $\eta = (\eta_1, \ldots, \eta_n)$ and $\eta \le \gamma$ means $\eta_1 \le \gamma_1, \ldots, \eta_n \le \gamma_n$.

We also recall that

$$\binom{\gamma_1}{\eta_1} \cdots \binom{\gamma_n}{\eta_n} \le \binom{\gamma_1 + \cdots + \gamma_n}{\eta_1 + \cdots + \eta_n}.$$

Then we obtain

$$\sum_{|\beta|=s} \sum_{|\alpha|=r} \| \mathrm{D}^\beta [\varphi_{\varrho,\delta}(a\, \mathrm{D}_{x'}^\alpha u - \mathrm{D}_{x'}^\alpha (au))] \|_{0,0,\varrho+\delta} \le$$

$$\le \sum_{l=0}^{s} \binom{s}{l} \sum_{q=1}^{r} \binom{r}{q} \sum_{t=0}^{s-l} \binom{s-l}{t} \sum_{|\gamma|=l} \sum_{|\nu|=q} \sum_{|\lambda|=r-q} \sum_{|\eta|=t} \sum_{|\mu|=s-l-t} \| |\mathrm{D}^\gamma \varphi_{\varrho,\delta}| \times$$

$$\times |\mathrm{D}^\eta \mathrm{D}_{x'}^\nu a|\, |\mathrm{D}^\mu \mathrm{D}_{x'}^\lambda u| \|_{0,0,\varrho+\delta} \le$$

$$\le \sum_{l=0}^{s} \binom{s}{l} \sum_{q=1}^{r} \binom{r}{q} \sum_{t=0}^{s-l} \binom{s-l}{t} \sum_{|\gamma|=l} \sum_{|\nu|=q} \sum_{|\lambda|=r-q} \sum_{|\eta|=t} \sum_{|\mu|=s-l-t} \max_{x \in \bar{\Omega}_{\varrho_0}} |\mathrm{D}^\gamma \varphi_{\varrho,\delta}| \times$$

$$\times \max_{x \in \bar{\Omega}_{\varrho_0}} |\mathrm{D}^\eta \mathrm{D}_{x'}^\nu a|\, \| \mathrm{D}^\mu \mathrm{D}_{x'}^\lambda u \|_{0,0,\varrho+\delta} \le$$

$$\le C_s' \sum_{l=0}^{s} \binom{s}{l} \sum_{q=1}^{r} \binom{r}{q} \sum_{t=0}^{s-l} \binom{s-l}{t} \frac{1}{\delta^l} c L^{q+t} M_{q+t} \| u \|_{s-l-t,r-q,\varrho+\delta}$$

$$\le C_s' c \sum_{l=0}^{s} \binom{s}{l} \sum_{t=0}^{s-l} \binom{s-l}{t} \frac{1}{\delta^l} \sum_{i=0}^{r-1} \binom{r}{r-i} L^{r-i+t} M_{r-i+t} \| u \|_{s-l-t,i,\varrho+\delta}$$

$$\leq C_s'' \sum_{l=0}^{s} \sum_{t=0}^{s-l} \frac{1}{\delta^l} \sum_{i=0}^{r-1} \binom{r+t}{r+t-i} L^{r+t} M_{r+t} \frac{L^{r-i+t} M_{r-i+t}}{L^{r+t} M_{r+t}} \|u\|_{s-l-t,i,\varrho+\delta}$$

$$\leq \big(\text{thanks to (1.8)}\big) \leq C_s^* \sum_{l=0}^{s} \sum_{t=0}^{s-l} \frac{1}{\delta^l} L^{r+t} M_{r+t} \sum_{i=0}^{r-1} \frac{1}{L^i M_i} \|u\|_{s-l-t,i,\varrho+\delta}. \quad \square$$

Remark 2.4. Assume that the function a, instead of (2.22), satisfies the estimate

$$\sup_{x\in\overline{\Omega}_{\varrho_0}} \sum_{|\alpha|=r} |\mathrm{D}^\alpha a(x)| \leq c L^{r-2m} M_{r-2m} \quad \forall r \geq 0 \tag{2.24}$$

where we set $M_{-1} = \cdots = M_{-2m} = M_0$; and also assume that $s \leq 2m$. Then, with the same proof, we obtain

$$\sum_{|\beta|=s} \sum_{|\alpha|=r} \|\mathrm{D}^\beta[\varphi_{\varrho,\delta}(a\, \mathrm{D}_{x'}^\alpha u - \mathrm{D}_{x'}^\alpha(au))]\| \leq C_s' \sum_{l=0}^{s} \binom{s}{l} \sum_{q=1}^{r} \binom{r}{q} \times$$

$$\times \sum_{t=0}^{s-l} \binom{s-l}{t} \frac{1}{\delta^l} c L^{q+t-2m} M_{q+t-2m} \|u\|_{s-l-t,r-q,\varrho+\delta} \leq C_s'' \sum_{l=0}^{s} \sum_{t=0}^{s-l} \frac{1}{\delta^l} \times$$

$$\times \sum_{i=0}^{r-1} \binom{r}{r-i} L^{r-i+t-2m} M_{r-i+t-2m} \|u\|_{s-l-t,i,\varrho+\delta} \leq$$

$$\leq \big(\text{considering that } t \leq s \leq 2m \text{ and that we have (1.9)}\big) \leq$$

$$\leq C_s''' \sum_{l=0}^{s} \sum_{t=0}^{s-l} \frac{1}{\delta^l} \sum_{i=0}^{r-1} \binom{r}{r-i} L^{r-i} \times M_{r-i} \|u\|_{s-l-t,i,\varrho+\delta} \leq$$

$$\leq \big(\text{thanks to (1.8)}\big) \leq C_s \sum_{l=0}^{s} \sum_{t=0}^{s-l} \frac{1}{\delta^l} M_r L^r \sum_{i=0}^{r-1} \frac{1}{L^i M_i} \|u\|_{s-l-t,i,\varrho+\delta}$$

and therefore finally

$$\sum_{|\beta|=s} \sum_{|\alpha|=r} \|\mathrm{D}^\beta[\varphi_{\varrho,\delta}(a\, \mathrm{D}_{x'}^\alpha u - \mathrm{D}_{x'}^\alpha(au))]\|_{0,0,\varrho+\delta} \tag{2.25}$$

$$\leq \tilde{C}_s \sum_{l=0}^{s} \frac{1}{\delta^l} \sum_{t=0}^{s-l} L^r M_r \sum_{i=0}^{r-1} \frac{1}{L^i M_i} \|u\|_{s-l-t,i,\varrho+\delta}$$

with the constant $\tilde{C}_s$ depending only on s, c and L. $\square$

Remark 2.5. If $\beta = 0$, then (2.23) is also satisfied setting $\varphi_{\varrho,\delta} = 1$. $\square$

Lemma 2.6. *Under the hypotheses of Theorem* 1.3, *if* ϱ_1 *and* C_1 *are the constants of Theorem* 2.1 *and* $0 < \varrho < \varrho + \delta < \varrho_1$, *for every* $\varepsilon > 0$ *there* *exists a $\gamma(\varepsilon)$ such that for every integer $k > 0$ and every $u \in \mathscr{D}(\bar{\Omega}_{\varrho_0})$ we have*

$$(2.26)\quad \|u\|_{2m,2km,\varrho} \leq C_1^*\Big\{\|\mathscr{A}u\|_{0,2km,\varrho+\delta} + \frac{1}{\delta^{2m}}|||\mathscr{B}u|||_{2km,\varrho+\delta} +$$

$$+ \frac{1}{\varepsilon^{2m}\delta^{2m}}\|u\|_{0,2km,\varrho+\delta} + \varepsilon\|u\|_{2m,2km,\varrho+\delta} + M_{2km}\sum_{s=0}^{k-1}\frac{(\gamma(\varepsilon))^{k-s}}{M_{2sm}}\|u\|_{2m,2sm,\varrho+\delta} +$$

$$+ \frac{M_{2km}}{\delta^{2m}}\sum_{s=-1}^{k-1}\frac{(\gamma(\varepsilon))^{k-s-1}}{M_{2(s+1)m}}\|u\|_{2m,2sm,\varrho+\delta}\Big\},$$

with the constant C_1 independent of ε and of u.

Proof. Let us apply (2.17) to $D_{x'}^\alpha u$, with $\alpha = 2km$:

$$(2.27)\quad \|u\|_{2m,2km,\varrho} = \sum_{|\alpha|=2km}\|D_{x'}^\alpha u\|_{2m,0,\varrho} \leq C_1\{\|\mathscr{A}u\|_{0,2km,\varrho+\delta} +$$

$$+ \sum_{|\alpha|=2mk}\|\mathscr{A}(D_{x'}^\alpha u) - D_{x'}^\alpha(\mathscr{A}u)\|_{0,0,\varrho+\delta} + |||\varphi_{\varrho,\delta}\mathscr{B}u|||_{2km,\varrho+\delta} +$$

$$+ \sum_{j=0}^{m-1}\sum_{|\alpha|=2km}\|\varphi_{\varrho,\delta}[\mathscr{B}_j(D_{x'}^\alpha u) - D_{x'}^\alpha(\mathscr{B}_j u)]\|_{H^{2m-m_j-1/2}(\Gamma_{\varrho+\delta})} +$$

$$+ \sum_{l=0}^{2m-1}\frac{1}{\delta^{2m-l}}\|u\|_{l,2km,\varrho+\delta}\}.$$

Then thanks to Lemma 2.5 and Remark 2.5 (with $s = 0$ and $r = 2km$) we obtain

$$\sum_{|\alpha|=2km}\|\mathscr{A}(D_{x'}^\alpha u) - D_{x'}^\alpha(\mathscr{A}u)\|_{0,0,\varrho+\delta} =$$

$$= \sum_{|p|\leq 2m}\sum_{|\alpha|=2km}\|a_p D_{x'}^\alpha D^p u - D_{x'}^\alpha(a_p D^p u)\|_{0,0,\varrho+\delta}$$

$$\leq C_0^* L^{2km} M_{2km}\sum_{|p|\leq 2m}\sum_{i=0}^{2km-1}\frac{1}{L^i M_i}\|D^p u\|_{0,i,\varrho+\delta}$$

$$\leq \sum_{t=0}^{m}\sum_{i=0}^{2km-1}\frac{M_{2km}}{M_i}D_1^{2km-i}\|u\|_{t,i,\varrho+\delta},$$

with the constant $D_1 \geq 1$ and depending on C_0^*, c and L. Now we apply (2.7) with $\varepsilon = 1$ in order to estimate the terms $\|u\|_{t,i,\varrho+\delta}$; we obtain

$$(2.28)\quad \sum_{|\alpha|=2km}\|\mathscr{A}(D_{x'}^\alpha u) - D_{x'}^\alpha(\mathscr{A}u)\|_{0,0,\varrho+\delta} \leq \sum_{i=0}^{2km-1}\frac{M_{2km}}{M_i}D_2^{2km-i}\times$$

$$\times\{\|u\|_{2m,i,\varrho+\delta} + \|u\|_{0,i,\varrho+\delta}\},$$

with $D_2 \geq 1$ and depending on C_0^*, c and L.

Then let $i = 2sm + t$ with $0 \leq s \leq k-1$, $0 \leq t < 2m$; applying (2.8) (and Remark 2.1), we have for every fixed $\varepsilon' > 0$:

$$(2.29)\quad \|u\|_{2m,2sm+t,\varrho+\delta} \leq \varepsilon' \|u\|_{2m,2(s+1)m,\varrho+\delta} + c_m(\varepsilon')^{-t/(2m-t)} \|u\|_{2m,2sm,\varrho+\delta},$$

$$(2.30)\quad \|u\|_{0,2sm+t,\varrho+\delta} \leq \varepsilon' \|u\|_{0,2(s+1)m,\varrho+\delta} + c_m(\varepsilon')^{-t/(2m-t)} \|u\|_{0,2sm,\varrho+\delta},$$

where we note that ε' may be chosen arbitrarily and in particular different for each s and t. Thus, having fixed $\varepsilon > 0$ arbitrarily, $0 < \varepsilon < 1$, we choose, for fixed s and t,

$$(2.31)\qquad \varepsilon' = \varepsilon \frac{M_{2sm+t}}{M_{2(s+1)m}\, \mathrm{D}_2^{2m-t}}.$$

Applying (1.11), it follows that

$$(\varepsilon')^{-t/(2m-t)} \leq \frac{1}{\varepsilon^{2m}} \left(\frac{M_{2(s+1)m}}{M_{2sm+t}}\right)^{t/(2m-t)} \mathrm{D}_2^t \leq \frac{1}{\varepsilon^{2m}}\, \mathrm{d}_m \frac{M_{2sm+t}}{M_{2sm}} \mathrm{D}_2^t$$

and therefore

$$(2.32)\qquad \frac{M_{2km}}{M_{2sm+t}} \mathrm{D}_2^{2km-2sm-t} (\varepsilon')^{-t/(2m-t)} \leq \frac{M_{2km}}{M_{2sm}} (\gamma_1(\varepsilon))^{k-s}$$

with $\gamma_1(\varepsilon) = \mathrm{D}_3 \varepsilon^{2m}$ and D_3 a suitable constant.

We also see that

$$(2.33)\qquad \frac{M_{2km}}{M_{2sm+t}} \mathrm{D}_2^{2km-2sm-t} \varepsilon' = \varepsilon \frac{M_{2km}}{M_{2(s+1)m}} (\mathrm{D}_2^{2m})^{k-s-1}.$$

Then, using (2.29), (2.32) and (2.33), we obtain

$$(2.34)\qquad \sum_{i=0}^{2km-1} \frac{M_{2km}}{M_i} \mathrm{D}_2^{2km-i} \|u\|_{2m,i,\varrho+\delta} \leq 2m\varepsilon \|u\|_{2m,2km,\varrho+\delta} +$$
$$+ 2m\varepsilon \sum_{s=0}^{k-2} \frac{M_{2km}}{M_{2(s+1)m}} (\mathrm{D}_2^{2m})^{k-s-1} \|u\|_{m,2(s+1)m,\varrho+\delta} +$$
$$+ 2mc_m \sum_{s=0}^{k-1} \frac{M_{2km}}{M_{2sm}} (\gamma_1(\varepsilon))^{k-s} \|u\|_{2m,2sm,\varrho+\delta} \leq 2m\varepsilon \|u\|_{2m,2km,\varrho+\delta} +$$
$$+ M_{2km} \sum_{s=0}^{k-1} \frac{(\gamma(\varepsilon))^{k-s}}{M_{2sm}} \|u\|_{2m,2sm,\varrho+\delta},$$

where $\gamma(\varepsilon) = 2m\varepsilon \mathrm{D}_2^{2m} + \gamma_1(\varepsilon)\, 2mc_m$.

Applying (2.30), (2.32), (2.33), we also have

$$(2.35)\qquad \sum_{i=0}^{2km-1} \frac{M_{2km}}{M_i} D_2^{2km-i} \| u \|_{0,i,\varrho+\delta} \le 2m\varepsilon \sum_{s=0}^{k-1} \frac{M_{2km}}{M_{2(s+1)m}} (D_2^{2m})^{k-s-1} \times$$

$$\times \| u \|_{0,2(s+1)m,\varrho+\delta} + 2mc_m \sum_{s=0}^{k-1} \frac{M_{2km}}{M_{2sm}} \big(\gamma_1(\varepsilon)\big)^{k-s} \| u \|_{0,2sm,\varrho+\delta}$$

$$\le \text{(thanks to } (2.1)\text{)} \le 2m\varepsilon \sum_{s=0}^{k-1} \frac{M_{2km}}{M_{2(s+1)m}} (D_2^{2m})^{k-s-1} \| u \|_{2m,2sm,\varrho+\delta} +$$

$$+ 2mc_m \sum_{s=0}^{k-1} \frac{M_{2km}}{M_{2sm}} \big(\gamma_1(\varepsilon)\big)^{k-s} \| u \|_{2m,2(s-1)m,\varrho+\delta} \le$$

$$\le M_{2km} \sum_{s=-1}^{k-1} \frac{\big(\gamma(\varepsilon)\big)^{k-s-1}}{M_{2(s+1)m}} \| u \|_{2m,2sm,\varrho+\delta},$$

still with $\gamma(\varepsilon) = 2m\varepsilon D_2^{2m} + 2mc_m \gamma_1(\varepsilon)$.

Thus finally it follows from (2.28), (2.34), (2.35) that

$$(2.36)\qquad \sum_{|\alpha|=2km} \| \mathscr{A}(D_{x'}^{\alpha} u) - D_{x'}^{\alpha}(\mathscr{A} u) \|_{0,0,\varrho+\delta} \le 2m\varepsilon \| u \|_{2m,km,\varrho+\delta} +$$

$$+ M_{2km} \sum_{s=0}^{k-1} \frac{\big(\gamma(\varepsilon)\big)^{k-s}}{M_{2sm}} \| u \|_{2m,2sm,\varrho+\delta} + M_{2km} \sum_{s=-1}^{k-1} \frac{\gamma(\varepsilon)^{k-s-1}}{M_{2(s+1)m}} \| u \|_{2m,2sm,\varrho+\delta}$$

with $\gamma(\varepsilon)$ a function of ε, c, c_1, L which goes to $+\infty$ as $\varepsilon \to 0$.

Let us now study the term $|||\varphi_{\varrho,\delta} \mathscr{B} u|||_{2km,\varrho+\delta}$ in (2.27). Applying Lemma 2.4 and formula (2.16) we immediately obtain

$$(2.37)\qquad |||\varphi_{\varrho,\delta} \mathscr{B} u|||_{2km,\varrho+\delta} \le \frac{D_4}{\delta^{2m}} |||\mathscr{B} u|||_{2km,\varrho+\delta}$$

with the costant D_4 depending only on γ_i with $i = 0, \ldots, 2m$ (see (2.16)).

Next we have

$$\sum_{j=0}^{m-1} \sum_{|\alpha|=2km} \| \varphi_{\varrho,\delta} [\mathscr{B}_j (D_{x'}^{\alpha} u) - D_{x'}^{\alpha} \mathscr{B}_j u] \|_{H^{2m-m_j-1/2}(\Gamma_{\varrho+\delta})}$$

$\le$ (thanks to Theorems 7.5, Chapter 1)

$$\le \sum_{j=0}^{m-1} \sum_{|\alpha|=2km} \| \varphi_{\varrho,\delta} [\mathscr{B}_j (D_{x'}^{\alpha} u) - D_{x'}^{\alpha} \mathscr{B}_j u] \|_{H^{2m-m_j}(\Omega_{\varrho+\delta})}$$

$$\le \sum_{j=0}^{m-1} \sum_{s=0}^{2m-m_j} \sum_{|\beta|=s} \sum_{|\alpha|=2km} \sum_{|h|\le m_j} \| D^{\beta} \big(\varphi_{\varrho,\delta} \big(b_{jh}\, D_{x'}^{\alpha} D^h u - D_{x'}^{\alpha} (b_{jh}\, D^h u)\big)\big) \|_{0,0,\varrho+\delta}$$

$\le$ (using Remark 2.4)

$$\leq \sum_{j=0}^{m-1} \sum_{s=0}^{2m-m_j} \sum_{|h|\leq m_j} \tilde{C} \sum_{l=0}^{s} \frac{1}{\delta^l} \sum_{t=0}^{s-l} L^{2km} M_{km} \sum_{i=0}^{2km-1} \frac{1}{L^i M_i} \| D^h u \|_{s-l-t,i,\varrho+\delta}$$

$$\leq \tilde{C} \sum_{j=0}^{m-1} \sum_{s=0}^{2m-m_j} \sum_{v=0}^{m_j} \sum_{l=0}^{s} \sum_{t=0}^{s-l} \frac{1}{\delta^l} L^{2km} M_{2km} \sum_{i=0}^{2km-1} \frac{1}{L^i M_i} \| u \|_{s-l-t+v,i,\varrho+\delta}$$

$\leq$ (using (2.7) with $\varepsilon = 1$)

$$\leq \tilde{\tilde{C}} \sum_{l=0}^{2m} \frac{1}{\delta^l} L^{2km} M_{km} \sum_{i=0}^{2km-1} \frac{1}{L^i M_i} \{\| u \|_{2m-l,i,\varrho+\delta} + \| u \|_{0,i,\varrho+\delta}\}.$$

Ans therefore we finally have

$$(2.38) \quad \sum_{j=0}^{m-1} \sum_{|\alpha|=2km} \| \varphi_{\varrho,\delta} [\mathscr{B}_j (D^\alpha_{x'} u) - D^\alpha_{x'} \mathscr{B}_j u] \|_{H^{2m-m_j-1/2}(\Gamma_{\varrho+\delta})} \leq$$

$$\leq \sum_{l=0}^{m} \sum_{i=0}^{2km-1} \frac{M_{2km}}{M_i} D_5^{2km-i} \frac{1}{\delta^l} \{\| u \|_{2m-l,i,\varrho+\delta} + \| u \|_{0,i,\varrho+\delta}\}$$

with the constant D_5 depending on m, c, L and c_1.

Now we note that, thanks to (2.7), for every $\varepsilon' > 0$ we have

$$\| u \|_{2m-l,i,\varrho+\delta} \leq \varepsilon' \| u \|_{2m,i,\varrho+\delta} + c_m (\varepsilon')^{-(2m-l)/l} \| u \|_{0,i,\varrho+\delta}$$

and therefore, taking $\varepsilon' = \delta^l$, we have

$$(2.39) \quad \frac{1}{\delta^l} \| u \|_{2m-l,i,\varrho+\delta} \leq \| u \|_{2m,i,\varrho+\delta} + c_m \frac{1}{\delta^{2m}} \| u \|_{0,i,\varrho+\delta}.$$

Let us use (2.39) in (2.38) and note that $\frac{1}{\delta^l} \leq \frac{1}{\delta^{2m}}$ for $0 \leq l \leq 2m$; we obtain

$$(2.40) \quad \sum_{j=0}^{m-1} \sum_{|\alpha|=2km} \| \varphi_{\varrho,\delta} [\mathscr{B}_j (D^\alpha_{x'} u) - D^\alpha_{x'} \mathscr{B}_j u] \|_{H^{2m-m_j-1/2}(\Gamma_{\varrho+\delta})} \leq$$

$$\leq \sum_{i=0}^{2km-1} \frac{M_{2km}}{M_i} D_6^{2km-i} \{\| u \|_{2m,i,\varrho+\delta} + \frac{1}{\delta^{2m}} \| u \|_{0,i,\varrho+\delta}\},$$

with the constant D_6 depending on m, c, c_1, L.

We note that the second member of (2.40) differs from the second member of (2.28) only by the presence of the factor $1/\delta^{2m}$. Thus, following the same procedure as was used to estimate (2.28), we easily see that we obtain the following estimate (which is the analogue to (2.36)):

$$(2.41)\quad \sum_{j=0}^{m-1} \sum_{|\alpha|=2km} \| \varphi_{\varrho,\delta}[\mathscr{B}_j(\mathrm{D}_{x'}^\alpha u) - \mathrm{D}_{x'}^\alpha(\mathscr{B}_j u)] \|_{H^{2m-m_j-1/2}(\Gamma_{\varrho+\delta})}$$

$$\leq 2m\varepsilon \| u \|_{2m,2km,\varrho+\delta} + M_{2km} \sum_{s=0}^{k-1} \frac{(\gamma(\varepsilon))^{k-s}}{M_{2sm}} \| u \|_{2m,2sm,\varrho+\delta} +$$

$$+ \frac{M_{2km}}{\delta^{2m}} \sum_{s=-1}^{k-1} \frac{(\gamma(\varepsilon))^{k-s-1}}{M_{2(s+1)m}} \| u \|_{2m,2sm,\varrho+\delta}.$$

As far as the last term in (2.27) is concerned, we have, applying (2.7), that for every $\varepsilon' > 0$:

$$\| u \|_{l,2km,\varrho+\delta} \leq \varepsilon' \| u \|_{2m,2km,\varrho+\delta} + c_m(\varepsilon')^{-l/(2m-l)} \| u \|_{0,2km,\varrho+\delta}$$

and therefore, if we take $\varepsilon' = \varepsilon\, \delta^{2m-l}$, we have

$$\frac{1}{\delta^{2m-l}} \| u \|_{l,2km,\varrho+\delta} \leq \varepsilon \| u \|_{2m,2km,\varrho+\delta} + c_m \frac{1}{\delta^{2m}} \varepsilon^{-l/(2m-l)} \| u \|_{0,2km,\varrho+\delta},$$

from which we deduce

$$(2.42)\quad \sum_{l=0}^{2m-1} \frac{1}{\delta^{2m-l}} \| u \|_{l,2km,\varrho+\delta} \leq 2m\varepsilon \| u \|_{2m,2km,\varrho+\delta} + \frac{2mc_m}{\delta^{2m}\varepsilon^{2m}} \| u \|_{0,2km,\varrho+\delta}.$$

The Lemma then follows from (2.27), (2.36), (2.37), (2.41) and (2.42). □

We now introduce the following definitions: for every real $\lambda > 0$ and real R such that $0 < R < \varrho_1$,

$$(2.43)\quad \sigma^k(u, \lambda, R) = \frac{1}{M_{2km}\lambda^{k+1}} \sup_{R/2 \leq \varrho < R} (R - \varrho)^{(k+1)2m} \| u \|_{2m,2km,\varrho}, \qquad k \geq -1;$$

$$(2.44)\quad \sigma_0^k(u, \lambda, R) = \frac{1}{M_{2(k-1)m}\lambda^k} \sup_{R/2 \leq \varrho < R} (R - \varrho)^{2km} \| u \|_{0,2km,\varrho}, \quad k \geq 0;$$

$$(2.45)\quad \psi^k(u, \lambda, R) = \frac{1}{M_{2(k-1)m}\lambda^k} \sup_{R/2 \leq p < R} (R - \varrho)^{2km} ||| \mathscr{B}u |||_{2km,\varrho}, \quad k \geq 0.$$

Note that we have

$$(2.46)\qquad \sigma_0^k(u, \lambda, R) \leq \sigma^{k-1}(u, \lambda, R) \quad \forall k \geq 0.$$

Lemma 2.7. *Under the hypotheses of Theorem* 1.3, *there exists a* λ_1, *independent of* u, *such that for every integer* $k \geq 0$, *every* $\lambda \geq \lambda_1$, *every*

$R < \frac{1}{2}\varrho_1$ *and every* $u \in \mathscr{D}(\bar{\Omega}_{\varrho_0})$, *we have*

$$(2.47)\quad \sigma^k(u, \lambda, R) \leq \frac{1}{4}\frac{M_{2(k-1)m}}{M_{2km}}\sigma^{k-1}(\mathscr{A}u, \lambda, R) + \frac{1}{4M_{2m}}\psi^k(u, \lambda, R) + \\ + \frac{1}{4}\sum_{s=-1}^{k-1}\sigma^s(u, \lambda, R).$$

Proof. First, assume that $k \geq 1$. Multiply (2.26) with $\frac{(R-\varrho)^{2(k+1)m}}{M_{2km}\lambda^{k+1}}$ and take the upper bound for $R/2 \leq \varrho < R$ choosing $\delta = (R-\varrho)/(k+1)$. For the first member we obtain $\sigma^k(u, \lambda, R)$. The second member is equal to $C_1^*(I_1 + I_2 + I_3 + I_4 + I_5 + I_6)$. with

$$I_1 = \frac{1}{M_{2km}\lambda^{k+1}}\sup\,(R-\varrho)^{2(k+1)m}\,\|\mathscr{A}u\|_{0,2km,\varrho+\delta},$$

$$I_2 = \frac{1}{M_{2km}\lambda^{k+1}}\sup\frac{(R-\varrho)^{2(k+1)m}}{\delta^{2m}}\,|||\mathscr{B}u|||_{2km,\varrho+\delta},$$

$$I_3 = \frac{1}{M_{2km}\lambda^{k+1}}\sup\frac{(R-\varrho)^{2(k+1)m}}{\varepsilon^{2m}\delta^{2m}}\,\|u\|_{0,2km,\varrho+\delta},$$

$$I_4 = \frac{\varepsilon}{M_{2km}\lambda^{k+1}}\sup\,(R-\varrho)^{2(k+1)m}\,\|u\|_{2m,2km,\varrho+\delta},$$

$$I_5 = \frac{1}{\lambda^{k+1}}\sum_{s=0}^{k-1}\frac{(\gamma(\varepsilon))^{k-s}}{M_{2sm}}\sup\,(R-\varrho)^{2(k+1)m}\,\|u\|_{2m,2sm,\varrho+\delta},$$

$$I_6 = \frac{1}{\lambda^{k+1}}\sum_{s=-1}^{k-1}\frac{\gamma(\varepsilon)^{k-s-1}}{M_{2(s+1)m}}\sup\frac{(R-\varrho)^{2(k+1)m}}{\delta^{2m}}\,\|u\|_{2m,2sm,\varrho+\delta},$$

where the sup is taken for $R/2 \leq \varrho < R$ and $\delta = (R-\varrho)/(k+1)$.

Then we have

$$(2.48)\ I_1 = \frac{M_{2(k-1)m}}{\lambda M_{2km}}\sup\frac{(R-\varrho)^{2(k+1)m}}{(R-\varrho-\delta)^{2km}}\frac{(R-\varrho-\delta)^{2km}}{M_{2(k-1)m}\lambda^k}\|\mathscr{A}u\|_{0,2km,\varrho+\delta}$$

$$\left(\text{by using the fact that } \frac{(R-\varrho)^{2(k+1)m}}{(R-\varrho-\delta)^{2km}} = (R-\varrho)^{2m}\left(1+\frac{1}{k}\right)^{2km}\right)$$

$$\leq G_1\frac{M_{2(k-1)m}}{\lambda M_{2km}}\,\sigma_0^k(\mathscr{A}u, \lambda, R),$$

with the constant G_1 depending on ϱ_1 and m.

Next we have

$$I_2 = \frac{M_{2(k-1)m}}{\lambda M_{2km}} \sup \frac{(R-\varrho)^{2(k+1)m}}{\delta^{2m}(R-\varrho-\delta)^{2km}} \frac{(R-\varrho-\delta)^{2km}}{M_{2(k-1)m}\lambda^k} |||\mathscr{B}u|||_{2km,\varrho}$$

$$\leq \frac{M_{2(k-1)m}}{\lambda M_{2km}} (k+1)^{2m}\left(1+\frac{1}{k}\right)^{2km} \psi^k(u, \lambda, R).$$

But, thanks to (1.16), we have

$$\frac{M_{2(k-1)m}}{M_{2km}} (k+1)^{2m} \leq \left(\frac{c_1}{M_1}\right)^{2m} \frac{(2km-2m)!}{(2km)!} (k+1)^{2m} \leq$$

$$\leq \left(\frac{c_1}{M_1}\right)^{2m} \left(\frac{k+1}{(k-1)\,2m+1}\right)^{2m} \leq G_2$$

with G_2 depending only on c_1, M_1 and m, and therefore

$$I_2 \leq \frac{G_3}{\lambda} \psi^k(u, \lambda, R) \tag{2.49}$$

with G_3 depending on c_1, M_1 and m.

Similarly, we have

$$I_3 = \frac{M_{2(k-1)m}}{\lambda M_{2km}\varepsilon^{2m}} \sup \frac{(R-\varrho)^{2(k+1)m}}{\delta^{2m}(R-\varrho-\delta)^{2km}} \frac{(R-\varrho-\delta)^{2km}}{M_{2(k-1)m}\lambda^k} \|u\|_{0,2km,\varrho+\delta} \tag{2.50}$$

$$\leq \frac{G_3}{\lambda\varepsilon^{2m}} \sigma_0^k(u, \lambda, R).$$

Also

$$I_4 \leq \varepsilon G_4 \sigma^k(u, \lambda, R), \tag{2.51}$$

with G_4 depending only on m.

Next:

$$I_5 = \frac{\gamma(\varepsilon)}{\lambda} \sum_{s=0}^{k-1} \left(\frac{\gamma(\varepsilon)}{\lambda}\right)^{k-s-1} \sup \frac{(R-\varrho)^{2(k+1)m}}{(R-\varrho-\delta)^{2(s+1)m}} \times \tag{2.52}$$

$$\times \frac{(R-\varrho-\delta)^{2(s+1)m}}{M_{2sm}\lambda^{s+1}} \|u\|_{2m,2sm,\varrho+\delta} \leq G_5 \frac{\gamma(\varepsilon)}{\lambda} \sum_{s=0}^{k-1} \left(\frac{\gamma(\varepsilon)}{\lambda}\right)^{k-s-1} \sigma^s(u,\lambda, R),$$

with G_5 depending only on ϱ_1 and m.

And finally we have

$$I_6 = \frac{1}{\lambda} \sum_{s=-1}^{k-1} \left(\frac{\gamma(\varepsilon)}{\lambda}\right)^{k-s-1} \sup \frac{M_{2sm}(R-\varrho)^{2(k+1)m}}{\delta^{2m} M_{2(s+1)m}(R-\varrho-\delta)^{2(s+1)m}} \times$$

$$\times \frac{(R-\varrho-\delta)^{2(s+1)m}}{M_{2sm}\lambda^{s+1}} \|u\|_{2m,2sm,\varrho+\delta} =$$

$$= \frac{1}{\lambda} \sum_{s=-1}^{k-1} \left(\frac{\gamma(\varepsilon)}{\lambda}\right)^{k-s-1} \frac{M_{2sm}}{M_{2(s-1)m}} \times$$

$$\times \sup \frac{(R-\varrho)^{2(k+1)m} (k+1)^{2m}}{(R-\varrho)^{2m} (R-\varrho)^{2(s+1)m} \left(1 - \frac{1}{k+1}\right)^{2(s+1)m}} \times$$

$$\times \frac{(R-\varrho-\delta)^{2(s+1)m}}{M_{2sm}\lambda^{s+1}} \|u\|_{2m,2sm,\varrho+\delta} \leq \frac{1}{\lambda} \sum_{s=-1}^{k-1} \left(\frac{\gamma(\varepsilon)}{\lambda}\right)^{k-s-1} \frac{M_{2sm}}{M_{2(s+1)m}} \times$$

$$\times \sup (R-\varrho)^{2m(k-s-1)} (k+1)^{2m} \left(1 + \frac{1}{k}\right)^{2km} \times$$

$$\times \frac{(R-\varrho-\delta)^{2(s+1)m}}{M_{2sm}\lambda^{s+1}} \|u\|_{2m,2sm,\varrho+\delta}$$

$$\leq \frac{1}{\lambda} \sum_{s=-1}^{k-1} \left(\frac{\gamma(\varepsilon)}{\lambda}\right)^{k-s-1} \frac{M_{2sm}}{M_{2(s+1)m}} \left(\frac{R}{2}\right)^{2m(k-s-1)} \times$$

$$\times (k+1)^{2m} \left(1 + \frac{1}{k}\right)^{2km} \sigma^s(u, \lambda, R).$$

But the expression

$$\frac{M_{2sm}}{M_{2(s+1)m}} \left(\frac{R}{2}\right)^{2m(k-s-1)} (k+1)^{2m}$$

is bounded by a constant G_2 as $k = 1, 2, \ldots$, and $-1 \leq s \leq k-1$. Indeed it suffices to note that $R < 1$ and that:

a) for $s = -1, 0, 1$,

$$\frac{M_{2sm}}{M_{2(s+1)m}} \left(\frac{1}{2}\right)^{2m(k-s-1)} (k+1)^{2m}$$

is bounded as k varies from 1 to $+\infty$;

b) if $s \geq 2$ (and therefore $k \geq 3$), then, thanks to (1.16), we have

$$\frac{M_{2sm}}{M_{2(s+1)m}} \left(\frac{1}{2}\right)^{2m(k-s-1)} \leq \left(\frac{c_1}{M_1}\right)^{2m} \frac{(2sm)!}{(2sm+2m)!} \left(\frac{1}{2}\right)^{2m(k-s-1)};$$

but for the same s's and k's the function of s:

$$\chi(s) = \frac{(2sm)!}{(2sm + 2m)!}\left(\frac{1}{2}\right)^{2m(k-s-1)}$$

is increasing, since

$$4m \leq 2sm + 2 \text{ and therefore } 2(s + 1)\, m + 2m \leq 2(2sm + 1),$$

$$\left[2^{s+1}(2(s + 1)\, m + 2m)\right]^{2m} \leq \left[2^{s+2}(2sm + 1)\right]^{2m},$$

$$2^{2m(s+1)}\left[(2(s + 1)\, m + 1)\,(2(s + 1)\, m + 2) \cdots (2(s + 1)\, m + 2m)\right]$$

$$\leq 2^{2m(s+2)}(2sm + 1)\,(2sm + 2) \cdots (2sm + 2m)\times$$

$$\times \frac{1}{(2sm + 1) \cdots (2sm + 2m)}\left(\frac{1}{2}\right)^{2m(k-s-1)}$$

$$\leq \frac{1}{(2(s + 1)\, m + 1) \cdots (2(s + 1)\, m + 2m)}\left(\frac{1}{2}\right)^{2m(k-s-2)}$$

which yields exactly: $\chi(s) \leq \chi(s + 1)$.

Then we have

$$(k + 1)^{2m} \frac{M_{2sm}}{M_{2(s+1)m}}\left(\frac{1}{2}\right)^{2m(k-s-1)} \leq \left(\frac{c_1}{M_1}\right)^{2m} \frac{(2(k - 1)\, m)!}{(2km)!}\,(k + 1)^{2m}$$

$$\leq \left(\frac{c_1}{M_1}\right)^{2m}\left(\frac{k + 1}{2(k - 1)\, m + 1}\right)^{2m} \leq G_2.$$

Consequently we may write the inequality

$$I_6 \leq \frac{G_2}{\lambda} \sum_{s=-1}^{k-1}\left(\frac{\gamma(\varepsilon)}{\lambda}\right)^{k-s-1} \sigma^s(u, \lambda, R). \tag{2.53}$$

Using (2.48), ..., (2.53), it thus follows that

$$\sigma^k(u, \lambda, R) \leq \varepsilon G^* \sigma^k(u, \lambda\sigma, R) + \frac{G^*}{\lambda} \frac{M_{2(k-1)m}}{M_{2km}} \sigma_0^k(\mathscr{A} u, \lambda, R) + \tag{2.54}$$

$$+ \frac{G^*}{\lambda} \psi^k(u, \lambda, R) + \frac{G^*}{\lambda \varepsilon^{2m}} \sigma_0^k(u, \lambda, R) +$$

$$+ (1 + \gamma(\varepsilon))\, \frac{G^*}{\lambda} \sum_{s=-1}^{k-1}\left(\frac{\gamma(\varepsilon)}{\lambda}\right)^{k-s-1} \sigma^s(u, \lambda, R),$$

with a suitable constant G^* depending on C_1 and on the G_i's.

But $\sigma_0^k(u, \lambda, R) \leq \sigma^{k-1}(u, \lambda, R)$ and therefore we have

$$\sigma^k(u, \lambda, R) \leq \varepsilon G^* \sigma^k(u, \lambda, R) + \frac{G^*}{\lambda} \frac{M_{2(k-1)m}}{M_{2km}} \sigma^{k-1}(\mathscr{A}u, \lambda, R) + \frac{G^*}{\lambda} \psi^k(u, \lambda, R) + \frac{G^*}{\lambda \varepsilon^{2m}} \sigma^{k-1}(u, \lambda, R) + \frac{G^*(1+\gamma(\varepsilon))}{\lambda} \sum_{s=-1}^{k-1} \left(\frac{\gamma(\varepsilon)}{\lambda}\right)^{k-s-1} \sigma^s(u, \lambda, R). \tag{2.55}$$

Let us now fix $\varepsilon = 1/2G^*$; then $\gamma(\varepsilon)$ is fixed and we can determine λ_1 such that for $\lambda \geq \lambda_1$ (2.47) is satisfied for $k \geq 1$.

Finally let us show (2.47) for $k = 0$. Using (2.37) and (2.42) (which also hold for $k = 0$) we deduce from (2.17) that

$$\|u\|_{2m,0,\varrho} \leq C_1^* \left\{ \|\mathscr{A}u\|_{0,0,\varrho+\delta} + \frac{1}{\delta^{2m}} |||\mathscr{B}u|||_{0,\varrho+\delta} + \varepsilon \|u\|_{2m,0,\varrho+\delta} + \frac{1}{\delta^{2m}\varepsilon^{2m}} \|u\|_{0,0,\varrho+\delta} \right\}.$$

We multiply this inequality with $(R-\varrho)^{2m}/(M_0\lambda)$ and take the upper bound for $R/2 \leq \varrho < R$ choosing $\delta = (R-\varrho)/2$. We obtain

$$\sigma^0(u, \lambda, R) \leq C_1^* \left\{ \frac{R}{2\lambda M_0} \|\mathscr{A}u\|_{0,0,R} + \frac{2^{2m}}{\lambda M_0} |||\mathscr{B}u|||_{0,0R} + \varepsilon \sup\left(\frac{(R-\varrho)^{2m}}{(R-\varrho-\delta)^{2m}} \cdot \frac{(R-\varrho-\delta)^{2m}}{M_0 \lambda} \|u\|_{2m,0,\varrho+\delta} \right) + \frac{2^m}{\varepsilon^{2m}} \frac{1}{\lambda M_0} \|u\|_{0,0,R} \right\} \leq$$

$$\leq \frac{G^*}{\lambda} \sigma_0^0(\mathscr{A}u, \lambda, R) + \frac{G^*}{\lambda} \psi^0(u, \lambda, R) + \varepsilon G^* \sigma^0(u, \lambda, R) + \frac{G^*}{\lambda \varepsilon^{2m}} \sigma_0^0(u, \lambda, R)$$

with G^* a suitable constant; thus we have

$$\sigma^0(u, \lambda, R) \leq \varepsilon G^* \sigma^0(u, \lambda, R) + \frac{G^*}{\lambda} \sigma^{-1}(\mathscr{A}u, \lambda, R) + \frac{G^*}{\lambda} \psi^0(u, \lambda, R) + \frac{G^*}{\lambda \varepsilon^{2m}} \sigma^{-1}(u, \lambda, R).$$

Choosing $\varepsilon = 1/(2G^*)$, we can determine λ_1 such that for $\lambda \geq \lambda_1$ we have exactly (2.47) for $k = 0$ (recall that $M_0 = M_{-1} = \cdots = M_{-2m}$). □

We are now ready to prove

Theorem 2.2 *Under the hypotheses of Theorem* 1.3, *if* λ *and* R *are fixed as in Lemma* 2.7, *for every* $u \in \mathscr{D}(\overline{\Omega}_{\varrho_0})$ *satisfying* (1.21) *and* (1.22), *we have*

$$\text{(2.56)} \quad \sigma^k(\mathscr{A}^i u, \lambda, R) \leq \frac{M_{(k+i+1)2m}}{M_{2km}} c_0 (L_* + 2)^{k+i+1} \quad \forall k \geq -1, \ i \geq 0,$$

with $L_* = L_0 \, d^{2m}$.

Proof. If $k = -1$ and i is arbitrary, (2.56) holds thanks to (1.21) and to the definition of $\sigma^{-1}(u, \lambda, R)$. We prove the theorem by induction on k; thus we assume (2.56) to hold for $k - 1$ (and arbitrary i) and apply (2.47) to $\mathscr{A}^i u$ with $i \geq 0$, $k \geq 0$; we have

$$\text{(2.57)} \quad \sigma^k(\mathscr{A}^i u, \lambda, R) \leq \frac{1}{4} \frac{M_{2(k-1)m}}{M_{2km}} \sigma^{k-1}(\mathscr{A}^{i+1} u, \lambda, R) + $$
$$+ \frac{1}{4M_{2m}} \psi^k(\mathscr{A}^i u, \lambda, R) + \frac{1}{4} \sum_{s=-1}^{k-1} \sigma^s(\mathscr{A}^i u, \lambda, R).$$

According to (1.22), we have

$$\psi^k(\mathscr{A}^i u, \lambda, R) \leq \frac{1}{M_{2(k-1)m}} |||\mathscr{B}(\mathscr{A}^i u)|||_{2km,R} \leq c_0 L_0^{k+i+1} \frac{M_{(k+i+1)2m}}{M_{2m(k-1)}}$$
$$\leq c_0 L_0^{k+i+1} \frac{M_{(k+i+1)2m}}{M_{2km}} \frac{M_{2km}}{M_{2m(k-1)}} \leq \big(\text{thanks to (1.10)}\big)$$
$$\leq c_0 L_0^{k+i+1} \frac{M_{(2mk+i+1)}}{M_{2mk}} d^{2mk} M_{2m} \leq c_0 (L_* + 2)^{k+i+1} \frac{M_{2m(k+i+1)}}{M_{2mk}} M_{2m}$$

and therefore

$$\text{(2.58)} \quad \sigma^k(\mathscr{A}^i u, \lambda, R) \leq \frac{1}{4} \frac{M_{2(k-1)m}}{M_{2km}} \sigma^{k-1}(\mathscr{A}^{i+1} u, \lambda, R) + $$
$$+ \frac{1}{4} c_0 (L_* + 2)^{k+i+1} \frac{M_{2m(k+i+1)}}{M_{2mk}} + \frac{1}{4} \sum_{s=-1}^{k-1} \sigma^s(\mathscr{A}^i u, \lambda, R).$$

From which, by induction, it follows that

$$\sigma^k(\mathscr{A}^i u, \lambda, R) \leq \frac{1}{4} \frac{M_{2m(k+i+1)}}{M_{2km}} c_0 (L_* + 2)^{k+i+1} + \frac{1}{4} c_0 (L_* + 2)^{k+i+1} \times$$
$$\times \frac{M_{2m(k+i+1)}}{M_{2mk}} + \frac{1}{4} \sum_{s=-1}^{k-1} \frac{M_{(s+i+1)2m}}{M_{2ms}} C_0 (L_* + 2)^{s+i+1}$$

$$\leq \big(\text{thanks to } (1.13)\big) \leq \frac{1}{2} \frac{M_{2m(k+i+1)}}{M_{2mk}} c_0(L_* + 2)^{k+i+1} +$$

$$+ \frac{1}{4} c_0 \frac{M_{2m(k+i+1)}}{M_{2mk}} (L_* + 2)^{k+i+1} \sum_{s=-1}^{k-1} \frac{1}{(L_* + 2)^{k-s}} \leq$$

$$\leq c_0(L_* + 2)^{k+i+1} \frac{M_{2m(k+i+1)}}{M_{2mk}}$$

and the theorem is proved.

2.4 Bounds for the Normal Derivatives

First of all we note that, according to the Leibniz formula,

$$\sum_{|\alpha|=s} |D^\alpha_{x'} D^q_y(uv)| \leq \sum_{l=0}^{s} \sum_{r=0}^{q} \binom{s}{l}\binom{q}{r} \sum_{|\eta|=l} |D^r_y D^\eta_{x'} u| \times \tag{2.59}$$

$$\times \sum_{|\gamma|=s-l} |D^{q-r}_y D^\gamma_{x'} v|$$

for s and q integers ≥ 0 and every couple of functions u and v. □

Next, for k and q integers ≥ 0, $\lambda > 0$, $\theta > 0$, $0 < R < \varrho_1$, set

$$\sigma_0^{k,q}(u, \lambda, \theta, R) = \frac{1}{M_{(k+q-1)2m} \lambda^{k+q} \theta^k} \sup_{R/2 \leq \varrho < R} (R - \varrho)^{(k+q)2m} \| D^{2mq}_y u \|_{0,2mk,\varrho}$$

and note that

$$\sigma_0^{k,0}(u, \lambda, \theta, R) = \sigma_0^k(u, \theta\lambda, R) \leq \sigma^{k-1}(u, \lambda\theta, R),\ k \geq 0. \tag{2.60}$$ □

We prove

Theorem 2.3. *Under the hypotheses of Theorem* 1.3, *let* $u \in \mathscr{D}(\overline{\Omega}_{\varrho_0})$ *and satisfy* (1.21) *and* (1.22); *then there exist* λ_0 *and* θ_0, *independent of* u, *such that for* $R \leq \varrho_1/2$ *we have*

$$\sigma_0^{k,q}(\mathscr{A}^i u, \lambda_0, \theta_0, R) \leq \frac{M_{(k+q+i)2m}}{M_{(k+q-1)2m}} c_0(L_* + 2)^{q+k+i}, \tag{2.61}$$

$$\forall q, k, i \geq 0.$$

Proof. Thanks to Remark 1.1, we may assume that

$$\mathscr{A} u = D^{2m}_y u + \sum_{t=1}^{2m} \sum_{j=0}^{t} \sum_{|\beta|=j} a_{t,j,\beta}(x', y)\, D^{2m-t}_y D^\beta_{x'} u;$$

and therefore, for $q \geq 0$ and $|\alpha| = 2mk$, we obtain

$$D_y^{2m+2mq} D_{x'}^\alpha u = D_y^{2mq} D_{x'}^\alpha (\mathscr{A} u) - \\ - \sum_{t=1}^{2m} \sum_{j=0}^{t} \sum_{|\beta|=j} D_y^{2mq} D_{x'}^\alpha \big(a_{t,j,\beta}(x', y)\, D_y^{2m-t} D_{x'}^\beta u\big).$$

Applying (2.59) we have

$$(2.62) \qquad \sum_{|\alpha|=mk} |D_y^{2m+2mq} D_{x'}^\alpha u| \leq \sum_{|\alpha|=2mk} |D_y^{2mq} D_{x'}^\alpha (\mathscr{A} u)| + \\ + \sum_{l=0}^{2km} \sum_{r=0}^{2qm} \sum_{t=1}^{2m} \sum_{j=0}^{t} \sum_{|\beta|=j} \binom{2km}{l} \binom{2qm}{r} \sum_{|\eta|=l} |D_y^r D_{x'}^\eta a_{t,j,\beta}(x', y)| \times \\ \times \sum_{|\gamma|=2km-l} |D_y^{2mq-r+2m-t} D_{x'}^{\gamma+\beta} u| \leq$$

(thanks to the fact that $\sum_{|\eta|=l} |D_y^r D_{x'}^\eta a_{t,j,\beta}(x', y)| \leq c L^{l+r} M_{l+r}$)

$$\leq \sum_{|\alpha|=2km} |D_y^{2mq} D_{x'}^\alpha (\mathscr{A} u)| + \sum_{l=0}^{2km} \sum_{r=0}^{2qm} \sum_{t=1}^{2m} \sum_{j=0}^{t} \binom{2km}{l} \binom{2qm}{r} c L^{l+r} \times \\ \times M_{l+r} \sum_{|\beta|=j} \sum_{|\gamma|=2km-l} |D_y^{2qm-r+2m-t} D_{x'}^{\gamma+\beta} u| \leq \sum_{|\alpha|=2km} |D_y^{2qm} D_{x'}^\alpha (\mathscr{A} u)| + \\ + \sum_{l=0}^{km} \sum_{r=0}^{qm} \sum_{t=1}^{m} \sum_{j=0}^{t} \binom{2km}{l} \binom{2qm}{r} L^{l+r} M_{l+r} \sum_{|\mu|=2km-l+j} |D_{x'}^\mu D_y^{2qm-r+2m-t} u|.$$

It follows that

$$\| D_y^{2qm+2m} u \|_{0,2km,\varrho} \leq \| D_y^{2qm} (\mathscr{A} u) \|_{0,2km,\varrho} + \\ + c \sum_{l=0}^{2km} \sum_{r=0}^{2qm} \sum_{t=1}^{2m} \sum_{j=0}^{t} \binom{2km}{l} \binom{2qm}{r} L^{l+r} M_{l+r} \| D_y^{2qm+2m-t-r} u \|_{0,2km-l+j,\varrho}.$$

Applying Lemma 2.2 (and Remark 2.1) with $\varepsilon = 1$, we have

$$\sum_{j=0}^{t} \| D_y^{2qm+2m-t-r} u \|_{0,2km-l+j,\varrho} \leq t \| D_y^{2qm+2m-r-t} u \|_{0,2km-l+t,\varrho} + \\ + t c_m \| D_y^{2qm+2m-r-t} u \|_{0,2km-l,\varrho}.$$

Again by Lemmas 2.2 and 2.3, we have for arbitrary $\varepsilon > 0$:

$$\sum_{t=1}^{2m} \sum_{j=0}^{t} \| D_y^{2qm+2m-r-t} u \|_{0,2km-l+j,\varrho} \leq \varepsilon \gamma_m \| D_y^{2qm+2m-r} u \|_{0,2km-l,\varrho} + \\ + \gamma_m c_m(\varepsilon) \| D_y^{2qm-r} u \|_{0,2(k+1)m-l,\varrho} + \gamma_m c'_m(\varepsilon) \| D_y^{2qm-r} u \|_{0,2km-l,\varrho}$$

where γ_m depends only on m, and $c'_m(\varepsilon)$ depends only on ε and m (ϱ_0 being fixed). Thus we have

$$\begin{aligned}(2.63)\qquad & \| D_y^{2qm+2m} u \|_{0,2km,\varrho} \leq \| D_y^{2qm} \mathscr{A} u \|_{0,2km,\varrho} + \\ & + c \sum_{l=0}^{2km} \sum_{r=0}^{2qm} \binom{2km}{l} \binom{2qm}{r} L^{l+r} M_{l+r} \gamma_m \{ \varepsilon \| D_y^{2(q+1)m-r} u \|_{0,2km-l,\varrho} + \\ & + c_m(\varepsilon) \| D_y^{2qm-r} u \|_{0,2(k+1)m-l,\varrho} + c'_m(\varepsilon) \| D_y^{2qm-r} u \|_{0,2km-l,\varrho} \} \\ & \leq \big(\text{by } (1.10)\big) \leq \| D_y^{2qm} \mathscr{A} u \|_{0,2km,\varrho} + \\ & + \sum_{l=0}^{2km} \sum_{r=0}^{2qm} \binom{2km}{l} \binom{2qm}{r} L_1^{2+r} M_l M_r \{ \varepsilon \| D_y^{2(q+1)m-r} u \|_{0,2km-l,\varrho} + \\ & + c_m(\varepsilon) \| D_y^{2qm-r} u \|_{0,2(k+1)m-l,\varrho} + c'_m(\varepsilon) \| D_y^{2qm-r} u \|_{0,2km-l,\varrho} \}\end{aligned}$$

where $L_1 \geq 1$ depends on c, L and d.

Let us now seek bounds for the terms of type $\| \ \|_{0,2km-l,\varrho}$, $l = 0, 1, \ldots, 2km$, by an expression containing only terms of type $\| \ \|_{0,2sm,\varrho}$ with $s = 0, \ldots, k$. In fact we have

$$\begin{aligned}(2.64)\qquad & \sum_{l=0}^{2km} \binom{2km}{l} L_1^l M_l \| v \|_{0,2km-l,\varrho} = M_0 \| v \|_{0,2km,\varrho} + \\ & + \sum_{i=0}^{2km-1} \binom{2km}{2km-i} L_1^{2km-i} M_{2km-i} \| v \|_{0,i,\varrho} \leq \big(\text{using } (1.8)\big) \\ & \leq M_0 \| v \|_{0,2km,\varrho} + \sum_{i=0}^{2km-1} \frac{c_1}{M_1} M_{2km} L_1^{2km-i} \| v \|_{0,i,\varrho}.\end{aligned}$$

Now let $i = 2sm + \sigma$, with $0 \leq s \leq k-1$, $0 \leq \sigma < 2m$; by Lemma 2.2, we have

$$\| v \|_{0,2sm+\sigma,\varrho} \leq \varepsilon' \| v \|_{0,2(s+1)m,\varrho} + c_m (\varepsilon')^{-\sigma/(2m-\sigma)} \| v \|_{0,2sm,\varrho}$$

and, choosing $\varepsilon' = \dfrac{M_{2sm+\sigma}}{M_{2(s+1)m} L_1^{2m-\sigma}}$ for each fixed s and σ, we obtain, thanks to (1.11),

$$(\varepsilon')^{-\sigma/(2m-\sigma)} \leq d \frac{M_{2sm+\sigma}}{M_{2sm}} L_1^\sigma$$

and therefore:

$$\begin{aligned}(2.65)\qquad \frac{M_{2km}}{M_i} L_1^{2km-i} \| v \|_{0,i,\varrho} \leq & \frac{M_{2mk}}{M_{2(s+1)m}} L_1^{2km-2m-2sm} \| v \|_{0,2(s+1)m,\varrho} + \\ & + c_m d L_1^{2(k-s)m} \frac{M_{2km}}{M_{2sm}} \| v \|_{0,2sm,\varrho}.\end{aligned}$$

Then from (2.63), we deduce

$$\sum_{l=0}^{2km} \binom{2km}{l} L_1^l M_l \|v\|_{0,2km-l,\varrho} \leq c''_m \sum_{s=0}^{k} \frac{M_{2km}}{M_{2sm}} L_1^{2(k-s)m} \|v\|_{0,2sm,\varrho} \tag{2.66}$$

with c''_m depending on m and on M_0.

In analogous fashion,

$$\begin{aligned} &\sum_{r=0}^{2qm} \binom{2qm}{r} L_1^r M_r \|\mathrm{D}_y^{2qm-r} u\|_{0\,0,\varrho} \\ &\leq c'''_m \sum_{s=0}^{q} \frac{M_{2qm}}{M_{2sm}} L_1^{2qm-2sm} \|\mathrm{D}_y^{2sm} u\|_{0,0,\varrho}, \end{aligned} \tag{2.67}$$

with c'''_m depending on m and on M_0.

Using (2.66) and (2.67), we deduce from (2.63):

$$\begin{aligned} &\|\mathrm{D}_y^{2qm+2m} u\|_{0,2km,\varrho} \leq \|\mathrm{D}_y^{2qm} \mathscr{A} u\|_{0,2km,\varrho} + \\ &+ \sum_{p=0}^{k} \sum_{s=0}^{q} \frac{M_{2mk} M_{2qm}}{M_{2pm} M_{2sm}} L_2^{2(k-p+q-s)m} \{\varepsilon \|\mathrm{D}_y^{2sm+m} u\|_{0,2pm,\varrho} + \\ &+ c_m(\varepsilon) \|\mathrm{D}_y^{2sm} u\|_{0,2(p+1)m,\varrho} + c'_m(\varepsilon) \|\mathrm{D}_y^{2sm} u\|_{0,2pm,\varrho}\} \end{aligned} \tag{2.68}$$

with L_2 depending on L_1, m and M_0.

Now multiply (2.68) with

$$\frac{(R-\varrho)^{2(k+q+1)m}}{M_{2(k+q)m} \lambda^{k+q+1} \theta^k}$$

and take the upper bound for $R/2 \leq \varrho < R$; then, R being < 1, we obtain

$$\begin{aligned} &\sigma_0^{k,q+1}(u, \lambda, \theta, R) \leq \frac{M_{2(k+q-1)m}}{\lambda M_{2(k+q)m}} \sigma_0^{k,q}(\mathscr{A} u, \lambda, \theta, R) + \\ &+ \sum_{p=0}^{k} \sum_{s=0}^{q} L_2^{2(k-p+q-s)m} \Big\{ \varepsilon \frac{\sigma_0^{p,s+1}(u, \lambda, \theta, R)}{\lambda^{k-p+q-s} \theta^{k-p}} \frac{M_{2(p+s)m} M_{2km} M_{2qm}}{M_{2(k+q)m} M_{2pm} M_{2sm}} + \\ &+ c_m(\varepsilon) \frac{\sigma_0^{p+1,s}(u, \lambda, \theta, R)}{\lambda^{k-p+q-s} \theta^{k-p-1}} \frac{M_{2(p+s)m} M_{2km} M_{2qm}}{M_{2(k+q)m} M_{2pm} M_{2sm}} + \\ &+ c'_m(\varepsilon) \frac{\sigma_0^{p,s}(u, \lambda, \theta, R)}{\lambda^{k-p+q-s+1} \theta^{k-p}} \frac{M_{2(p+s-1)m} M_{2km} M_{2qm}}{M_{2(k+q)m} M_{2pm} M_{2sm}} \Big\} \\ &\leq \big(\text{using (1.14) and (1.9)}\big) \leq \frac{M_{2(k+q-1)m}}{\lambda M_{2(k+q)m}} \sigma_0^{k,q}(\mathscr{A} u, \lambda, \theta, R) + \\ &+ \sum_{p=0}^{k} \sum_{s=0}^{q} L_2^{2(k-p)m+2(q-s)m} \Big\{ \varepsilon \frac{\sigma_0^{p,s+1}(u, \lambda, \theta, R)}{\lambda^{k-p+q-s} \theta^{k-p}} + \\ &+ c_m(\varepsilon) \frac{\sigma_0^{p+1,s}(u, \lambda, \theta, R)}{\lambda^{k-p+q-s} \theta^{k-p-1}} + \frac{c'_m(\varepsilon)}{\lambda} \frac{\sigma_0^{p,s}(u, \lambda, \theta, R)}{\lambda^{k-p+q-s} \theta^{k-p}} \Big\}. \end{aligned}$$

Finally, setting $L_2^{2m}/(\lambda\theta) = \xi$ and $L_2^{2m}/\lambda = \varphi$, we obtain

$$\sigma_0^{k,q+1}(u, \lambda, \theta, R) \leq \frac{M_{2(k+p-1)m}}{\lambda M_{2(k+q)m}} \sigma_0^{k,q}(\mathscr{A}u, \lambda, \theta, R) + \tag{2.69}$$

$$+ \sum_{p=0}^{k} \sum_{s=0}^{q} \xi^{k-p} \varphi^{q-s} \Big\{ \varepsilon \sigma_0^{p,s+1}(u, \lambda, \theta, R) + \theta c_m(\varepsilon)\, \sigma_0^{p+1,s}(u, \lambda, \theta, R) +$$

$$+ \frac{c'_m(\varepsilon)}{\lambda} \sigma_0^{p,s}(u, \lambda, \theta, R) \Big\}.$$

Now choose $\varepsilon = 1/40$ and, for this choice of ε, take λ_0 and θ_0 such that

$$\begin{cases} \lambda_0\theta_0 \geq \lambda_1,\ \theta_0 c_m(\varepsilon) \leq \dfrac{1}{40}, & \dfrac{c'_m(\varepsilon)}{\lambda_0} \leq \dfrac{1}{40}, \\[2mm] \dfrac{L_2^m}{\lambda_0\theta_0} < \dfrac{1}{2}, \quad \dfrac{L_2^m}{\lambda_0} < \dfrac{1}{2}, & \dfrac{1}{\lambda_0} < \dfrac{1}{10}. \end{cases} \tag{2.70}$$

Then we have

$$\frac{1}{2} \sigma_0^{k,q+1}(u, \lambda_0, \theta_0, R) \leq \frac{1}{10} \frac{M_{2(k+q-1)m}}{M_{2(k+q)m}} \sigma_0^{k,q}(\mathscr{A}u, \lambda_0, \theta_0, R) + \tag{2.71}$$

$$+ \frac{1}{40} \sum_{p=0}^{k-1} \xi^{k-p} \sigma_0^{p,q+1}(u, \lambda_0, \theta_0, R) + \frac{1}{40} \sum_{p=0}^{k} \sum_{s=0}^{q-1} \xi^{k-p} \varphi^{q-s} \times$$

$$\times \sigma_0^{p,s+1}(u, \lambda_0, \theta_0, R) + \frac{1}{40} \sum_{p=0}^{k} \sum_{s=0}^{q} \xi^{k-p} \varphi^{q-s} \{\sigma_0^{p+1,s}(u, \lambda_0, \theta_0, R) +$$

$$+ \sigma_0^{p,s}(u, \lambda_0, \theta_0, R)\}$$

where, if $k = 0$, the sum $\sum_{p=0}^{k-1}$ must be deleted.

Let us apply (2.71) to $\mathscr{A}^i u$; we obtain

$$\sigma_0^{k,q+1}(\mathscr{A}^i u, \lambda_0, \theta_0, R) \leq \frac{1}{5} \frac{M_{2(k+q-1)m}}{M_{2(k+q)m}} \sigma_0^{k,q}(\mathscr{A}^{i+1} u, \lambda_0, \theta_0, R) + \tag{2.72}$$

$$+ \frac{1}{20} \sum_{p=0}^{k-1} \xi^{k-p} \sigma_0^{p,q+1}(\mathscr{A}^i u, \lambda_0, \theta_0, R) +$$

$$+ \frac{1}{20} \sum_{p=0}^{k} \sum_{s=0}^{q} \xi^{k-p} \varphi^{q-s} \sigma_0^{p,s+1}(\mathscr{A}^i u, \lambda_0, \theta_0, R) +$$

$$+ \frac{1}{20} \sum_{p=0}^{k} \sum_{s=0}^{q} \xi^{k-p} \varphi^{q-s} \{\sigma_0^{p+1,s}(\mathscr{A}^i u, \lambda_0, \theta_0, R) + \sigma_0^{p,s}(\mathscr{A}^i u, \lambda_0, \theta_0, R)\}$$

with the same convention as before for the sum $\sum_{p=0}^{k-1}$.

We denote by $p(q, i, k)$ the property (2.61) for the values q, i, k of the parameters and by $P(q, i, k)$ the property $p(q', i', k')$, $0 \leq q' \leq q$, $0 \leq i' \leq i$, $0 \leq k' \leqq k$. From (2.72), we shall verify the following implications:

$$(\alpha) \qquad P(q, i+1, 0) \cup P(q, i, 1) \Rightarrow p(q+1, i, 0)$$

$$(\beta) \qquad \begin{cases} \text{if } k \geq 1, P(q, i+1, k) \cup P(q+1, i, k-1) \cup P(q, i, k+1) \Rightarrow \\ \Rightarrow p(q+1, i, k). \end{cases}$$

Indeed, according to (2.72), we have

$$\begin{aligned}
\sigma_0^{k,q+1}(\mathscr{A}^i u, \lambda_0, \theta_0, R) &\leq \frac{1}{5} \frac{M_{2(k+q-1)m}}{M_{2(k+q)m}} \cdot \frac{M_{2(k+q+i+1)m}}{M_{2(k+q-1)m}} c_0 (L_* + 2)^{k+q+i+1} + \\
&+ \frac{1}{20} \sum_{p=0}^{k-1} \xi^{k-p} \frac{M_{2(p+q+i+1)m}}{M_{2(p+q)m}} c_0 (L_* + 2)^{p+q+i+1} + \\
&+ \frac{1}{20} \sum_{p=0}^{k} \sum_{s=0}^{q-1} \xi^{k-p} \varphi^{q-s} \frac{M_{2(p+s+i+1)m}}{M_{2(p+1)m}} c_0 (L_* + 2)^{p+s+i+1} + \\
&+ \frac{1}{20} \sum_{p=0}^{k} \sum_{s=0}^{q} \xi^{k-p} \varphi^{q-s} \left\{ \frac{M_{2(p+s+i+1)m}}{M_{2(p+s)m}} c_0 (L_* + 2)^{p+s+i+1} + \right. \\
&\left. + \frac{M_{2(p+s+i)m}}{M_{2(p+s-1)m}} c_0 (L_* + 2)^{p+s+i} \right\} \leq \text{(using (1.13))} \\
&\leq \frac{M_{2(k+q+i+1)m}}{M_{2(k+q)m}} c_0 (L_* + 2)^{k+q+i+1} \left\{ \frac{1}{5} + \frac{1}{20} \sum_{p=0}^{k-1} \left(\frac{\xi}{L_* + 2} \right)^{k-p} + \right. \\
&+ \frac{1}{20} \sum_{p=0}^{k} \sum_{s=0}^{q-1} \left(\frac{\xi}{L_* + 2} \right)^{k-p} \left(\frac{\varphi}{L_* + 2} \right)^{q-s} + \\
&\left. + \frac{1}{10} \sum_{p=0}^{k} \sum_{s=0}^{q} \left(\frac{\xi}{L_* + 2} \right)^{k+p} \left(\frac{\varphi}{L_* + 2} \right)^{q-s} \right\} \\
&\leq \frac{M_{2(k+q+i+1)m}}{M_{2(k+\;)m}} c_0 (L_* + 2)^{k+q+i+1}.
\end{aligned}$$

Therefore (2.61) also holds for $\sigma_0^{k,q+1}(\mathscr{A}^i u, \lambda_0, \theta_0, R)$.

We shall now prove the theorem by induction. According to (2.60) and (2.56), since $\lambda_0 \theta_0 \geq \lambda_1$, we know that $P(0, i, k)$ holds, $\forall i$, $k \geq 0$. We assume $P(q', i, k)$, $q' \leq q$, $\forall i$ and k and show $P(q+1, i, k)$, $\forall i$ and k.

By induction and (α), we have $P(q+1, i, 0)$, $\forall i$. Let k be fixed arbitrarily; assume $P(q+1, i, k-1)$; this, together with the induction hypothesis and (β), implies $p(q+1, i, k)$; this in turn shows $P(q+1, i, k)$ and ends the proof of Theorem 2.3.

2.5 Proof of Theorem 1.3

From Theorem 2.3, we can deduce

Proposition 2.1. *Under the hypotheses of Theorem 2.3, there exists an N depending only on $\mathscr{A}$, $\mathscr{B}_j$ and L_0 such that, if $\varrho' = \varrho_1/4$, we have*

$$\sum_{|\alpha|=s} \| \mathrm{D}^\alpha u \|_{L^2(\Omega_{\varrho'})} \leq c_0 N^s M_s \quad \forall s \geq 0. \tag{2.73}$$

Proof. We use (2.61) with $i = 0$ and $R = \varrho_1/2$; then

$$\sigma_0^{k,q}\left(u, \lambda_0, \theta_0, \frac{\varrho_1}{2}\right) \leq \frac{M_{2(k+q)m}}{M_{2(k+q-1)m}} c_0 (L_* + 2)^{k+q}$$

from which, thanks to the definition of $\sigma_0^{k,q}(u, \lambda, \theta, R)$, we deduce

$$\begin{aligned} \| \mathrm{D}_y^{2qm} u \|_{0,2km,\varrho'} &\leq M_{2(k+q)m}\, \lambda_0^{k+q} \theta_0^k (\varrho')^{-(k+q)2m}\, c_0 (L_* + 2)^{k+q} \\ &\leq M_{2(k+q)m} c_0 N_1^{k+q} \end{aligned} \tag{2.74}$$

where N_1 depends on λ_0, θ_0, ϱ', L_* and therefore on L_0, $\mathscr{A}$, $\mathscr{B}_j$.

Let t and r be arbitrary integers ≥ 0 and assume that $t = 2qm + l$ and $r = 2km + j$ with $0 \leq l < 2m$, $0 \leq j < 2m$; then, applying Lemma 2.2 with $\varepsilon = 1$, we have

$$\begin{aligned} \| \mathrm{D}_y^t u \|_{0,r,\varrho'} &\leq \| \mathrm{D}_y^{2(q+1)m} u \|_{0,r,\varrho'} + c_m \| \mathrm{D}_y^{2qm} u \|_{0,r,\varrho'} \\ &\leq \| \mathrm{D}_y^{2(q+1)m} u \|_{0,2(k+1)m,\varrho'} + c_m \| \mathrm{D}^{2(q+1)m} u \|_{0,2km,\varrho'} + \\ &\quad + c_m \| \mathrm{D}_y^{2qm} u \|_{0,2km,\varrho'} + c_m^2 \| \mathrm{D}_y^{2qm} u \|_{0,2km,\varrho'} \leq \big(\text{by (1.9) and (2.74)}\big) \leq \\ &\leq M_{2(k+q+2)m} c_0 (1 + 2c_m + c_m^2)\, N^{k+q+2} \leq \big(\text{again by (1.9)}\big) \leq \\ &\leq M_{(t+r+2m)} c_0 (1 + 2c_m + c_m^2)\, N_1^{k+q+2} \leq \big(\text{using (1.10)}\big) \leq \\ &\leq c_0 N_2^{t+r} M_{t+r}, \end{aligned} \tag{2.75}$$

where N_2 depends on N_1, d, c_m, m.

Finally, we have

$$\begin{aligned} \sum_{|\alpha|=s} \| \mathrm{D}^\alpha u \|_{L^2(\Omega_{\varrho'})} &= \sum_{t=0}^{s} \sum_{|\beta|=s-t} \| \mathrm{D}_y^t \mathrm{D}_x^\beta u \|_{L^2(\Omega_{\varrho'})} = \sum_{t=0}^{s} \| \mathrm{D}_y^t u \|_{0,s-t,\varrho'} \\ &\leq c_0 N_s^2 (s+1)\, M_s \leq c_0 (2N_2)^s\, M_s \end{aligned}$$

whence (2.73), with $N = 2N_2$. ☐

We are now able to prove Theorem 1.3; in fact, applying the Sobolev theorem (see Theorem 9.8, Chapter 1), from (2.73) we obtain

$$\sup_{\overline{\Omega}_{\varrho'}} \sup_{|\alpha|=s} |D^\alpha u| \leq c^* c_0 N_*^s M_s$$

with c^* depending on the dimension n of $\mathbf{R}^n$ and N_* depending on ϱ' and on N. Therefore the theorem is proved.

2.6 Complements and Remarks

The proof of Theorem 1.3 also yields the *theorem on "elliptic iterates" in the interior of the domain, and more precisely*:

Theorem 2.4. *Let $M_k = (k!)^\beta$, with real $\beta \geq 1$ (or more generally $\{M_k\}$ satisfying ((1.6), ..., (1.11)). Let Ω be an arbitrary bounded open set in $\mathbf{R}^n$, A a linear elliptic differential operator on Ω given by (1.1), with $a_{pq} \in \mathscr{D}_{M_k}(\overline{\Omega})$. Then, if $u \in \mathscr{E}(\Omega)$ and if there exist two constants c_0 and L_0 (depending on u) such that*

$$\| A^i u \|_{L^2(\Omega)} \leq c_0 L_0^i M_{2mi}, \ \forall i \geq 0, \tag{2.76}$$

we have $u \in \mathscr{E}_{M_k}(\Omega)$ $(\mathscr{E}_{M_k}(\Omega) = \mathscr{H}(\Omega)$ *if* $M_k = k!)$.

Indeed, we reduce the problem to the case in which Ω is a ball in $\mathbf{R}^n$ and use the same arguments as for Theorem 1.3 (bounds for the tangential derivatives, Sections 2.2 and 2.3), starting from the a priori estimate (3.6) of Chapter 2 (with $l = r = 2m$), instead of (4.39) of Chapter 2. The proof is also simpler than the proof of Theorem 1.3, not only because we no longer need to distinguish between the (tangential and normal) variables, but also because we do not have to study the estimates of the boundary data $\mathscr{B}_j u$ ☐

Of course, we also have a corollary to Theorem 2.4 as an analogue to Corollary 1.1:

Corollary 2.1. *Under the hypotheses of Theorem 2.4 on M_k, Ω and A, if $u \in \mathscr{E}(\Omega)$ and if $Au = f$, with $f \in \mathscr{E}_{M_k}(\Omega)$, then $u \in \mathscr{E}_{M_k}(\Omega)$ $(\mathscr{E}_{M_k}(\Omega) = \mathscr{H}(\Omega)$, if $M_k = k!)$.*

Remark 2.6. It may be possible, with the same type of demonstration, to eliminate conditions (1.9), (1.10), (1.11) on the sequence M_k, keeping (1.6), (1.7), (1.8). This is evident for (1.9), since at least for sufficiently large k, (1.9) follows from (1.15) (which is a consequence of (1.7)).

We have not done any further work on this point, since hypotheses (1.6), ..., (1.11), as we have seen, cover the most important case of Gevrey sequences, $M_k = (k!)^\beta$, $\beta \geq 1$. ☐

Remark 2.7. The analogue to Theorem 2.4 and to its Corollary 2.1 for Beurling-type spaces (see Section 2.3, Chapter 7) has recently been given by O. John [1] using similar techniques. Here we note one of these results of which we shall make use later on:

$$(2.77)\quad \begin{cases} \textit{let } \Omega \textit{ be a bounded open set in } \mathbf{R}^n \textit{ and } A \textit{ an} \\ \textit{elliptic differential operator on } \Omega \textit{ given by (1.1) with } a_{pq} \in \mathscr{H}(\overline{\Omega}); \\ \textit{let } M_k = (k!)^\beta, \textit{ with } \beta > 1. \textit{ Then, if } u \in \mathscr{D}(\overline{\Omega}) \\ \textit{and satisfies:} \\ \qquad \forall L > 0, \textit{ there exists a } c > 0 \textit{ such that} \\ \qquad\qquad \|A^i u\|_{L^2(\Omega)} \leq c L^i M_{2mi} \ \forall i, \\ \textit{it follows that } u \textit{ is of Beurling type in } \Omega, \text{ i.e.} \\ \textit{for every compact set } \mathscr{K} \subset \Omega \textit{ and every } L > 0, \textit{there exists} \\ \textit{a } c' > 0 \textit{ such that} \\ \qquad \sup_{x \in \mathscr{K}} |D^\alpha u(x)| \leq c L^k M_k, \ |\alpha| = k, \ \forall k. \quad \square \end{cases}$$

Remark 2.8. We shall now indicate a procedure which, at least in certain cases, allows the reduction of "elliptic iterates" type theorems to regularity results of the type given by Corollaries 1.1 and 2.1 for a new operator constructed from A.

At first, we assume $M_k = k!$ (analytic case) and, in order to simplify, we further assume homogeneous Dirichlet boundary conditions; in other words, under the hypotheses of Theorem 1.2, for $M_k = k!$, we let $u \in \mathscr{D}(\overline{\Omega})$ satisfy

$$(2.78)\quad \begin{cases} \|A^i u\|_{L^2(\Omega)} \leq c_0 L_0^i (2mi)! \ \ \forall i \geq 0, \\ \gamma_j(A^i u) = 0 \ \ \forall i \geq 0, \ j = 0, \ldots, m-1. \end{cases}$$

Let us introduce an *additional variable* t ($t \in \mathbf{R}^1$) and the (vector) function

$$(2.79)\quad w(t) = \sum_{i=0}^{\infty} (-1)^{i(m+1)} \frac{t^{2mi} A^i u}{(2mi)!},$$

for which we shall verify the following properties:

$$(2.80)\quad \begin{cases} t \to w(t) \text{ is an analytic function on }]-t_0, t_0[\\ \text{(suitable } t_0\text{) with values in } H^{2m}(\Omega); \end{cases}$$

$$(2.81)\quad \begin{cases} \text{in the cylinder } Q_0 = \Omega \times]-t_0, t_0[, \text{ we have} \\ (-1) \dfrac{\partial^{2m} w}{\partial t^{2m}} + A w = 0, \end{cases}$$

$$(2.82)\quad \gamma_j w(t) = 0 \text{ on } \Gamma \text{ and for } t \in]-t_0, t_0[, \ j = 0, \ldots, m-1.$$

Indeed, let us first recall that, for every $v \in H^{2m}(\Omega)$ with $\gamma_j v = 0$, $j = 0, \ldots, m-1$, we have (use the estimate (5.3) of Chapter 2 and Theorem 16.3 of Chapter 1)

(2.83) $$\|v\|_{H^{2m}(\Omega)} \le c_*(\|Av\|_{L^2(\Omega)} + \|v\|_{L^2(\Omega)}) \ (c_* \text{ constant}).$$

Applying (2.83) to $v = A^i u$, it follows that

$$\|A^i u\|_{H^{2m}(\Omega)} \le c_* c_0 [L^i (2im)! + L^{i+1}(2(i+1)\,m)!] \quad \forall i,$$

whence

(2.84) $$\|A^i u\|_{H^{2m}(\Omega)} \le \tilde{c}\tilde{L}^i (2im)! \ (\tilde{c}, \tilde{L} \text{ constants}) \ \forall i.$$

Then (2.80) follows immediately from (2.79) and (2.84).

The verification of (2.81) is as immediate calculation. Finally, since γ_j is a continuous linear operator of $H^{2m}(\Omega)$ into $H^{2m-j-1/2}(\Gamma)$, we have

$$\gamma_j w(t) = \sum_{i=0}^{\infty} (-1)^{i(m+1)} \frac{t^{2mi}}{(2mi)!} \gamma_j(A^i u)$$

(convergence in $H^{2m-j-1/2}(\Gamma)$ for every $t \in]-t_0, t_0[$) and therefore according to (2.78) we obtain (2.82).

Let us now assume that A is not only properly elliptic in Ω, but furthermore satisfies the relation

(2.85) $$(-1)^m A_0(x, \xi) \neq -|A_0(x, \xi)| \quad \forall x \in \overline{\Omega} \text{ and } \forall \xi \in \mathbf{R}^n, \ \xi \neq 0;$$

(note that in particular this hypothesis is satisfied if A is *strongly elliptic* in $\overline{\Omega}$).

Then, in the cylinder Q_0, equation (2.81) is *properly elliptic* in the variables $x_1, \ldots, x_n$ and t.

Thus applying Corollary 1.1 with $M_k = k!$ for the cylinder Q_0 and equation (2.81), (we shall obtain a *global* result on x, but local in t), we see that $w(x, t)$ is analytic in (x, t) in a cylinder $Q_1 = \Omega \times]-t_1, t_1[$, with $t_1 < t_0$; and therefore $u(x) = w(x, 0)$ is analytic on $\overline{\Omega}$.

Of course, we must have demonstrated Corollary 1.1, which can be done directly, without using Theorem 1.2, by the same type of arguments as for the proof of Theorem 1.2, but much simplified.

The same idea can also be developed in the the case of non-analytic Gevrey classes ($M_k \neq k!$), but then the problem is reduced to a regularity problem of the type of Corollaries 1.1 and 2.1, for more complicated operators than elliptic operators, for example for quasi-elliptic operators (and then the advantage of such a method is almost non-existent!). Here is a simple example; take $M_k = (k!)^\beta$ with rational $\beta > 1$, $\beta = p/q$,

p and q integers, still with homogeneous Dirichlet conditions; thus

$$(2.86)\qquad \begin{cases} \|A^i u\|_{L^2(\Omega)} \leq c_0 L_0^i\big((2mi)!\big)^\beta \ \ \forall i \geq 0 \\ \gamma_j(A^i u) = 0 \ \ \forall i \geq 0, \ j = 0, \ldots, m-1. \end{cases}$$

Let us try to prove that $u \in \mathscr{D}_{(k!)^\beta}(\overline{\Omega})$. We introduce

$$(2.87)\qquad w(t) = \sum_{i=0}^{\infty} (-1)^{i(pm+1)} \frac{t^{2mp}}{(2mip)!} A^{qi} u$$

and verify, as for (2.84), that

$$\|A^{iq} u\|_{H^{2m}(\Omega)} \leq \tilde{c}\tilde{L}^i (2pmi)!,$$

so that we still have (2.80). Next, we verify that w is a solution of

$$(2.88)\qquad (-1)^{pm} \frac{\partial^{2mp} w}{\partial t^{2mp}} + A^q w = 0$$

in $Q_0 = \Omega \times]-t_0, t_0[$, and that

$$\gamma_j w(t) = \gamma_j\big(A w(t)\big) = \cdots = \gamma_j\big(A^{q-1} w(t)\big) = 0, \quad t \in]-t_0, t[,$$
$$j = 0, \ldots, m-1.$$

Now if A is *strongly elliptic* or if A *is not only properly elliptic, but also satisfies*

$$(2.89)\qquad (-1)^{mq} A_0^q(x, \xi) \neq -|A_0^q(x, \xi)| \ \forall x \in \Omega, \ \ \forall \xi \in \mathbf{R}^n, \ \xi \neq 0,$$

then equation (2.87) is *quasi-elliptic* in Q_0; by using a theorem of Cavallucci [1], which is a generalization of Corollary 1.1 to quasi-elliptic operators (see also Matzusawa [2]), it follows that $w(x, t)$ is of Gevrey type of order β in x and analytic in t on $\overline{\Omega} \times]-t_1, t_1[$, $t_1 < t_0$; therefore $w(x, 0) = u(x) \in \mathscr{D}_{(k!)^\beta}(\overline{\Omega})$.

Remark 2.9. C. Goulaouic [2] has shown that the spaces $\mathscr{D}_{M_k}(\overline{\Omega})$ *are not* in general "*interpolation spaces*" between the space of analytic functions on $\overline{\Omega}$ and the space $\mathscr{D}(\overline{\Omega})$; Here, we have an essential difference with Volumes 1 and 2, where interpolation was directly applicable. (Still according to Goulaouic [1], [2], interpolation could be used for certain homogeneous problems.) □

3. Application of Transposition; Existence of Solutions in the Space $\mathscr{D}'(\Omega)$ of Distributions

3.1 Generalities

Following the ideas developed systematically in this book, we shall now see how the regularity results of the preceding sections can be used, by applying the method of *transposition*.

Our aim is the study of elliptic boundary value problems in the spaces $\mathscr{D}'(\Omega)$ and $\mathscr{D}'_{M_k}(\Omega)$; we shall first examine the case of distributions, $\mathscr{D}'(\Omega)$. □

Thus, let us again consider the boundary value problem (1.3), adding, to hypotheses (I), (II), (III) of Theorem 1.1, the hypothesis of *normality* of the system $\{B_j\}_{j=0}^{m-1}$ (this hypothesis is essential to the method of transposition, as we have seen in Section 6 of Chapter 2; see also Section 8.3 of Chapter 2) and the hypothesis that the *data of the problem are analytic*; more precisely, we assume that:

i) *Ω, A, $\{B_j\}_{j=0}^{m-1}$ satisfy hypotheses* (I), (II), (III) *of Theorem* 1.1;

ii) *Γ is an analytic variety*;

iii) *the coefficients a_{pq} belong to $\mathscr{H}(\overline{\Omega})$ and the coefficients b_{jh} belong to $\mathscr{H}(\Gamma)$*;

iv) *the system $\{B_j\}_{j=0}^{m-1}$ is normal on Γ.*

Under these hypotheses we can introduce a (formal) *adjoint problem* of (1.3), as we have seen in Chapter 2, by applying Theorem 2.1 of Chapter 2; we choose the system $\{S_j\}_{j=0}^{m-1}$ of "boundary" operators so as to have *Green's formula*:

$$\int_\Omega (Au)\,\bar{v}\,\mathrm{d}x - \int_\Omega u\overline{A^*v}\,\mathrm{d}x = \sum_{j=0}^{m-1}\int_\Gamma S_ju\overline{C_jv}\,\mathrm{d}\sigma - \sum_{j=0}^{m-1}\int_\Gamma B_ju\overline{T_jv}\,\mathrm{d}\sigma \quad \forall u, v \in \mathscr{D}(\overline{\Omega}), \tag{3.1}$$

the operators C_j and T_j, $j = 0, \ldots, m-1$, depending on A, $\{B_j\}_{j=0}^{m-1}$ and $\{S_j\}_{j=0}^{m-1}$, according to Theorem 2.1 of Chapter 2.

We also recall that N and N^* denote the spaces

$$N = \{w \mid w \in \mathscr{D}(\overline{\Omega}), B_jw = 0, j = 0, \ldots, m-1, Aw = 0\}, \tag{3.2}$$

$$N^* = \{w \mid w \in \mathscr{D}(\overline{\Omega}), C_jw = 0, j = 0, \ldots, m-1, A^*w = 0\}. \tag{3.3}$$

We know that N and N^* are finite-dimensional subspaces of $\mathscr{D}(\overline{\Omega})$ (see Chapter 2, Section 5).

Thanks to Corollary 1.1 to Theorem 1.2 of the present chapter, we may state that, under hypotheses i), ..., iv), N and N^* are made up of *analytic functions on $\overline{\Omega}$.* □

Let us now consider the homogeneous adjoint problem:

$$\begin{cases} A^*v = \varphi, \text{ with } \varphi \in \mathscr{D}(\Omega), \\ C_jv = 0, j = 0, 1, \ldots, m-1. \end{cases} \tag{3.4}$$

We can apply Theorem 1.1 of this chapter, of course changing A to A^* and B_j to C_j, which is permissible (see Theorem 2.2, Chapter 2); we obtain the following result.

We introduce the space:

$$X = \{v \mid v \in \mathscr{D}(\overline{\Omega}),; C_j v = 0, j = 0, \ldots, m-1; A^* v \in \mathscr{D}(\Omega)\} \tag{3.5}$$

provided with the inductive limit topology of the spaces

$$X^{(\nu)} = \{v \mid v \in \mathscr{D}(\overline{\Omega}), C_j v = 0, j = 0, \ldots, m-1, A^* v \in \mathscr{D}(\mathscr{K}_\nu; \Omega)\}, \tag{3.6}$$

where $\{\mathscr{K}_\nu\}$ is an increasing sequence of compact sets contained in Ω whose union is Ω and $\mathscr{D}(\mathscr{K}_\nu; \Omega)$ is the Fréchet space of functions of $\mathscr{D}(\Omega)$ with support contained in $\mathscr{K}_\nu$, (see Chapter 7), $X^{(\nu)}$ being provided with the natural Fréchet space topology; thus X is a strict $(\mathscr{L}\mathscr{F})$-space.

We further denote by $\{\mathscr{D}(\Omega); N\}$ the closed subspace of $\mathscr{D}(\Omega)$, defined by

$$\{\mathscr{D}(\Omega); N\} = \{\varphi \mid \varphi \in \mathscr{D}(\Omega); \int_\Omega \varphi \overline{w} \, dx = 0 \quad \forall w \in N\}. \tag{3.7}$$

Then, by application of Theorem 1.1 and by the definition of X, we have (still denoting by A^* the operator deduced from A^* by passage to the quotient by N^*):

Proposition 3.1. *The operator A^* defines an (algebraic and topological) isomorphism of X/N^* onto $\{\mathscr{D}(\Omega); N\}$.* □

Let us now transpose the isomorphism A^* obtained by Proposition 3.1; we obtain

Proposition 3.2. *For every continuous antilinear form $\dot{v} \to L(\dot{v})$ on X/N^*, there exists one and only one element $\dot{u}$ belonging to the space $\{\mathscr{D}(\Omega); N\}'$ (strong dual of $\{\mathscr{D}(\Omega); N\}$) such that*

$$\langle \dot{u}, \overline{A^* \dot{v}} \rangle = L(\dot{v}), \quad \forall \dot{v} \in X/N^*, \tag{3.8}$$

where the brackets denote the duality between $\{\mathscr{D}(\Omega); N\}'$ and $\{\mathscr{D}(\Omega); N\}$; $\dot{u}$ depends continuously on L (for the strong dual topologies). □

Note that

$$\{\mathscr{D}(\Omega); N\}' = \mathscr{D}'(\Omega)/N \tag{3.9}$$

Indeed it is sufficient to note that for every non-empty subset P of a separated locally convex space E, the bipolar set P^{00} is identical to the smallest balanced (circled) and closed convex set for the weak topology which contains P.

We also note that (use (3.1))

$$\int_\Omega w A^* v \, dx = 0, \ \forall w \in N \text{ and } \forall v \in X.$$

Thus we may also express Proposition 3.2 as follows:

(3.10) $\begin{cases} \textit{given a continuous antilinear form } L \textit{ on } X/N^*, \textit{ there} \\ \textit{exists } u \in \mathcal{D}'(\Omega), \textit{ determined up to addition of a function} \\ \textit{of } N, \textit{ such that} \\ \qquad \langle u, \overline{A^*v}\rangle = L(v^\cdot), \forall v^\cdot \in X/N^* \textit{ and } \forall v \in v^\cdot, \\ \textit{the brackets denoting the duality between } \mathcal{D}'(\Omega) \textit{ and } \mathcal{D}(\Omega). \end{cases}$ □

3.2 Choice of the Form L; the Space $\Xi(\Omega)$ and its Dual

Now we have to "separate" the equation from the boundary conditions in (3.10), by choosing L in an appropriate way, according to the procedure followed in this text (see for example Section 6 of Chapter 2.) Formally we choose L to be in the form

$$L(v) = \langle f, \bar{v}\rangle + \sum_{j=0}^{m-1} \langle g_j, \overline{T_j v}\rangle, \tag{3.11}$$

where f and g_j are suitable given "functions" on Ω and on Γ respectively: then we have the equation

$$\langle u, \overline{A^*v}\rangle = \langle f, \bar{v}\rangle + \sum_{j=0}^{m-1} \langle g_j, \overline{T_j v}\rangle, \ \forall v \in X,$$

from which, still formally, thanks to the Green formula, we deduce that

$$Au = f \text{ in } \Omega, \ B_j u = g_j \text{ on } \Gamma.$$

Now the problem is to choose f and g_j and to justify the preceding formal considerations. □

Let us first study the problem as far as the equation $Au = f$ in Ω is concerned. For this purpose, we introduce a space $K(\Omega)$ of distributions on Ω such that

(3.12) $\begin{cases} X \subset K(\Omega) \subset L^2(\Omega) \textit{ with continuous injections}; \\ K(\Omega) \textit{ is reflexive}; \\ \mathcal{D}(\Omega) \textit{ is dense in } K(\Omega) \textit{ (and therefore } K(\Omega) \textit{ is} \\ \textit{a normal space of distributions on } \Omega). \end{cases}$

Spaces $K(\Omega)$ with the properties (3.12) exist: for example $K(\Omega) = L^2(\Omega)$.

But (as we have already seen in Chapter 2) in (3.11) we shall take f in $K'(\Omega)$, dual of $K(\Omega)$. The theory will therefore be the more general, the "greater" $K'(\Omega)$ is, i.e. the "smaller" $K(\Omega)$ is.

We do not know whether there exists a "smallest possible" space $K(\Omega)$ with properties (3.12). If it exists, the optimal space $K(\Omega)$ depends on the boundary conditions (since $X \subset K(\Omega)$). Here, we shall (conforming to the setting of Chapter 2, Section 6.3) construct a particular space $K(\Omega) = \Xi(\Omega)$, which is independent of the boundary conditions, but already seems to be "sufficiently small".

Let $\varrho(x)$ be a function of $\mathscr{D}(\overline{\Omega})$, positive in Ω, vanishing on Γ of the same order as the distance $\mathrm{d}(x, \Gamma)$ of x to Γ (i.e. $\lim\limits_{x \to x_0 \in \Gamma} \dfrac{\varrho(x)}{\mathrm{d}(x, \Gamma)} = \mathrm{d} \neq 0$); we have already introduced and used a function of this type (Chapter 1, Section 11.2, Chapter 2, Section 6.3, ...).

Definition 3.1. $\Xi(\Omega)$ *denotes the space*

$$\Xi(\Omega) = \{u \mid \varrho^{|\alpha|} \, \mathrm{D}^\alpha u \in L^2(\Omega), \, \forall \alpha\},$$

provided with the topology defined by the family of semi-norms

$$\| \varrho^{|\alpha|} \, \mathrm{D}^\alpha u \|_{L^2(\Omega)}, \; \forall \alpha .$$

It is easily verified that $\Xi(\Omega)$ is a complete space; therefore it is a *Fréchet space.*

This space is made up of infinitely differentiable functions in Ω (but not necessarily in $\overline{\Omega}$); of course, we also have

$$X \subset \mathscr{D}(\overline{\Omega}) \subset \Xi(\Omega) \subset L^2(\Omega). \tag{3.13}$$

Proposition 3.3. $\mathscr{D}(\Omega)$ *is dense in* $\Xi(\Omega)$.

Proof. Let $\delta_\nu(x)$ be a sequence of functions of $\mathscr{D}(\Omega)$, $\nu = 1, 2, \ldots,$ such that $\delta_\nu(x) = 1$ for $\mathrm{d}(x, \Gamma) \geq 2/\nu$ and $\delta_\nu(x) = 0$ for $\mathrm{d}(x, \Gamma) \leq 1/\nu$ and

$$|\mathrm{d}(x, \Gamma)|^{|\alpha|} \, |\mathrm{D}^\alpha \, \delta_\nu(x)| \leq c_\alpha \; (c_\alpha \text{ depending only on } \alpha).$$

Now let $u \in \Xi(\Omega)$; then $\delta_\nu \, u \in \mathscr{D}(\Omega)$ and the proposition will be proved if we can verify that

$$\delta_\nu u \to u \text{ in } \Xi(\Omega) \text{ , as } \nu \to +\infty. \tag{3.14}$$

To this end, we must show that $\varrho^{|\alpha|} \, \mathrm{D}^\alpha(\delta_\nu u) \to \varrho^{|\alpha|} \, \mathrm{D}^\alpha u$ in $L^2(\Omega)$; but $\delta_\nu \varrho^{|\alpha|} \, \mathrm{D}^\alpha u \to \varrho^{|\alpha|} \, \mathrm{D}^\alpha u$ in $L^2(\Omega)$; therefore it is sufficient to show that

$$\varrho^{|\alpha|} (\mathrm{D}^\beta \delta_\nu) \, (\mathrm{D}^\gamma u) \to 0 \text{ in } L^2(\Omega), \; |\beta| \geq 1, \; |\beta| + |\gamma| = |\alpha|. \tag{3.15}$$

Now $\varrho^{|\alpha|}(\mathrm{D}^\beta \delta_\nu) \, (\mathrm{D}^\gamma u)$ vanishes for $\mathrm{d}(x, \Gamma) \geq 2\nu$ (since $|\beta| \geq 1$), so that, according to the Lebesgue theorem, we have (3.15), noting that

$$|\varrho^{|\alpha|}(\mathrm{D}^\beta \delta_\nu) \, (\mathrm{D}^\gamma u)| = \left(|\varrho^{|\beta|}(\mathrm{D}^\beta \delta_\nu)\right) (\varrho^{|\gamma|} \, \mathrm{D}^\gamma u) \, | \leq c_\beta \, \varrho^{|\gamma|} \mathrm{D}^\gamma u \in L^2(\Omega). \quad \square$$

We denote the (strong) dual of $\Xi(\Omega)$ by $\Xi'(\Omega)$. Thanks to Proposition 3.3, $\Xi'(\Omega)$ can be identified with a space of distributions on Ω.

We obtain

Proposition 3.4. *Every element f of the space $\Xi'(\Omega)$ may be represented (non-uniquely) by the form*

$$f = \sum_{\text{finite}} D^\alpha(\varrho^{|\alpha|} f_\alpha), \text{ with } f_\alpha \in L^2(\Omega). \tag{3.16}$$

Indeed, we only need to note that every continuous linear form on $\Xi(\Omega)$ may be written

$$M(\varphi) = \sum_{\text{finite}} \int_\Omega g_\alpha \varrho^{|\alpha|} D^\alpha \varphi \, dx, \text{ with } g_\alpha \in L^2(\Omega),$$

and, that according to Proposition 3.3, it is determined by its values for each $\varphi \in \mathscr{D}(\Omega)$ (from which we obtain (3.16) by setting $f_\alpha = (-1)^{|\alpha|} g_\alpha$).

Proposition 3.5. *The space $\Xi(\Omega)$ is reflexive.*

Proof. It is sufficient to show that every bounded and weakly closed subset in $\Xi(\Omega)$ is weakly compact. Now if $\{u_i\}$ is a bounded sequence in $\Xi(\Omega)$, there exists an M_α such that

$$\|\varrho^{|\alpha|} D^\alpha u_i\|_{L^2(\Omega)} \leq M_\alpha, \ \forall i.$$

Then using the "diagonal" process, we can extract a subsequence $\{u_\nu\}$ from $\{u_i\}$ such that $\varrho^{|\alpha|} D^\alpha u_\nu$ converges weakly in $L^2(\Omega)$, for every fixed α, to a function ψ_α which depends on α. But, since u_ν converges weakly to a function u in the sense of $L^2(\Omega)$, $\varrho^{|\alpha|} D^\alpha u_\nu$ converges weakly to $\varrho^{|\alpha|} D^\alpha u$ in the sense of $\mathscr{D}'(\Omega)$. If $f \in \Xi'(\Omega)$, we have, thanks to Proposition 3.4,

$$f = \sum_{|\alpha| \leq l} D^\alpha(\varrho^{|\alpha|} f_\alpha),$$

with a suitable integer l and $f_\alpha \in L^2(\Omega)$; and therefore

$$\langle f, u_\nu \rangle = \sum_{|\alpha| \leq l} \langle D^\alpha(\varrho^{|\alpha|} f_\alpha), u_\nu \rangle = \sum_{|\alpha| \leq 1} (-1)^{|\alpha|} \langle f_\alpha, \varrho^{|\alpha|} D^\alpha u_\nu \rangle$$

and finally

$$\lim_{\nu \to \infty} \langle f, u_\nu \rangle = \sum_{|\alpha| \leq l} (-1)^{|\alpha|} \langle f_\alpha, \varrho^{|\alpha|} D^\alpha u \rangle$$

i.e. u_ν converges weakly to u in $\Xi(\Omega)$. ▯

Conclusion.

Therefore we see that, thanks to (3.13) and to Propositions 3.5 and 3.3, $\Xi(\Omega)$ may be chosen as the space $K(\Omega)$. And we can take f in $\Xi'(\Omega)$ for the choice of L in (3.11).

3.3 Final Choice of the Form L; the Space Y

For the choice of the boundary data g_j, we must first note

Proposition 3.2.

$$v \to \mathscr{T}v = \{T_0 v, \ldots, T_{m-1} v\}$$

is a continuous linear mapping of X into the space $[\mathscr{H}(\Gamma)]^m$.

Proof. If $v \in X$, then by definition, A^*v vanishes in a neighborhood of Γ and $C_j v = 0$ on Γ, $j = 0, \ldots, m-1$. Therefore from Corollary 1.1 to Theorem 1.2 (the local character of this result is evidently a consequence of Theorem 1.3), it follows that v is *analytic* in a neighborhood of Γ and therefore $T_j v \in \mathscr{H}(\Gamma)$, so that $\mathscr{T}(v) \in [\mathscr{H}(\Gamma)]^m$ if $v \in X$.

Furthermore, the spaces X and $\mathscr{H}(\Gamma)$ being of $(\mathscr{L}\mathscr{F})$-type, we may apply the closed graph theorem to $\mathscr{T}$ (see Grothendieck [1]); now $v \to \mathscr{T}v$ is, for example, a continuous mapping of X into $[C^\circ(\Gamma)]^m$ and therefore also of X into $[\mathscr{H}(\Gamma)]^m$. ☐

Remark 3.1. In Section 3.5, we shall see that $\mathscr{T}$ is *surjective.* ☐

We can now choose $g_j \in \mathscr{H}'(\Gamma)$ in (3.11) and thus make our final choice of L; more precisely, we note that

(3.17) $\begin{cases} \text{if } f \in \Xi'(\Omega), \text{ the form } v \to \langle f, \bar{v}\rangle \text{ (the bracket} \\ \text{denoting the duality between } \Xi'(\Omega) \text{ and } \Xi(\Omega)\text{) is} \\ \text{antilinear and continuous on } X; \end{cases}$

(3.18) $\begin{cases} \text{if } g_j \in \mathscr{H}'(\Gamma), \text{ the form } v \to \sum_{j=0}^{m-1} \langle g_j, \overline{T_j v}\rangle \text{ (the} \\ \text{bracket denoting the duality between } \mathscr{H}'(\Gamma) \text{ and } \mathscr{H}(\Gamma)\text{)} \\ \text{is antilinear and continuous on } X. \end{cases}$

Then considering the form

$$L(v) = \langle f, \bar{v}\rangle + \sum_{j=0}^{m-1} \langle g_j, \overline{T_j v}\rangle, \tag{3.19}$$

making the convention that $L(v^\bullet) = L(v)$ for every $v \in X$, element of the class $v^\bullet$ of X/N^*, we see that (3.19) defines a continuous antilinear form $v^\bullet \to L(v^\bullet)$ on X/N^* if and only if

$$\langle f, \bar{v}\rangle + \sum_{j=0}^{m-1} \langle g_j, \overline{T_j v}\rangle = 0, \forall v \in N^*. \tag{3.20}$$

In this way we finally obtain

Theorem 3.1. *Let hypotheses* i), ii), iii), iv) *of Section* 3.1 *be satisfied; let* $f \in \Xi'(\Omega)$ *and* $g_j \in \mathscr{H}'(\Gamma)$, $j = 0, \ldots, m-1$, *with* (3.20). *Then there exists* $u \in \mathscr{D}'(\Omega)$, *determined up to addition of a function in* N, *such that*

$$\langle u, \overline{A^* v} \rangle = \langle f, \bar{v} \rangle + \sum_{j=0}^{m-1} \langle g_j, \overline{T_j v} \rangle, \quad \forall v \in X. \tag{3.21}$$

Furthermore $\{f; g_0, \ldots, g_{m-1}\} \to u$ *is a continuous linear mapping of the closed subspace of* $\Xi'(\Omega) \times [\mathscr{H}'(\Gamma)]^m$, *made up of the elements satisfying* (3.20), *into* $\mathscr{D}'(\Omega)/N$. □

Writing (3.21) for every $v \in \mathscr{D}(\Omega)$ (which is contained in X), it follows that

$$\langle u, \overline{A^* v} \rangle = \langle f, \bar{v} \rangle, \quad \forall v \in \mathscr{D}(\Omega),$$

and therefore u satisfies the equation

$$A u = f \tag{3.22}$$

in the sense of distributions on Ω.

There remains to interpret the boundary conditions which are "contained" in (3.21); and for this purpose we must have a trace theorem for the solution u.

Thanks to (3.22), we see that u belongs to a space Y defined in the following way:

$$Y = \{u \mid u \in \mathscr{D}'(\Omega),\ A u \in \Xi'(\Omega)\}. \tag{3.23}$$

We shall provide Y with the coarsest locally convex topology which makes the mappings $u \to u$ and $u \to A u$ of Y into $\mathscr{D}'(\Omega)$ and $\Xi'(\Omega)$ respectively, continuous.

We must therefore give *trace theorems* for the elements u of Y, as we shall do in the following sections. □

Remark 3.2. Theorem 3.1 still holds if, instead of $\Xi'(\Omega)$, we take the dual $K'(\Omega)$ of a space $K(\Omega)$ satisfying conditions (3.12). Indeed, in order to prove Theorem 3.1, we only used the first and third "abstract" properties (3.12) of the space $\Xi(\Omega)$. □

3.4 Density Theorem

Let us first show

Theorem 3.2. *Let* Ω *be an open set satisfying hypothesis* (I) *of Section* 1.1 *and let* A, *given by* (1.1), *be properly elliptic in* $\bar{\Omega}$ *with* $a_{pq} \in \mathscr{H}(\bar{\Omega})$; *then* $\mathscr{D}(\bar{\Omega})$ *is dense in the space* Y, *defined by* (3.23).

Proof. Let $u \to M(u)$ be a continuous antilinear form on Y; it may be written, the intervening spaces being reflexive:

$$M(u) = \langle f, \bar{u} \rangle + \langle g, \overline{Au} \rangle, \text{ with } f \in \mathscr{D}(\Omega),\ g \in \Xi(\Omega). \tag{3.24}$$

Assume that we have $M(\varphi) = 0$, $\forall \varphi \in \mathscr{D}(\bar{\Omega})$; and let us show that it then follows that we have $M(u) = 0$, $\forall u \in Y$.

Denote by $\tilde{f}$ and $\tilde{g}$ the extensions of f and g to $\mathbf{R}^n$ by zero outside Ω; and let $\mathscr{A}$ be a linear operator of order $2m$ with infinitely differentiable coefficients in $\mathbf{R}^n$ and analytic in a neighborhood $\mathscr{O}$ of $\bar{\Omega}$, which coincides with the operator A in $\bar{\Omega}$ and which is *properly elliptic* in $\bar{\mathscr{O}}$. Then noting that $\tilde{f} \in \mathscr{D}(\mathbf{R}^n)$ and $\tilde{g} \in L^2(\mathbf{R}^n)$, (3.24) may be written

$$M(\varphi) = \langle \tilde{f}, \bar{\Phi} \rangle + \langle \tilde{g}, \overline{\mathscr{A}\Phi} \rangle = 0, \ \forall \Phi \in \mathscr{D}(\mathbf{R}^n), \tag{3.25}$$

where the brackets are taken in the sense of distributions on $\mathbf{R}^n$ and $\varphi =$ restriction of Φ to $\bar{\Omega}$ (and therefore $\varphi \in \mathscr{D}(\bar{\Omega})$); therefore, if $\mathscr{A}^*$ denotes the (formal) adjoint of $\mathscr{A}$, we have, in the sense of distributions on $\mathbf{R}^n$,

$$\mathscr{A}^* \tilde{g} = -\tilde{f}. \tag{3.26}$$

But then, according to the hypoellipticity of $\mathscr{A}^*$ in $\mathscr{O}$ (see Theorem 3.2 of Chapter 2), $\tilde{g}$ is infinitely differentiable in $\mathscr{O}$; and since f is analytic (since zero!) in the complement of its support, which is a compact set $\mathscr{K}$ in Ω, $\tilde{g}$ is also analytic in $\mathscr{O} - \mathscr{K}$, thanks to (3.26) and to Corollary 1.1 of Section 1; but since $\tilde{g}$ vanishes in $\mathscr{O} - \Omega$, we have: $\tilde{g} = 0$ in a neighborhood of Γ, therefore $g \in \mathscr{D}(\Omega)$.

Furthermore, by restriction of (3.26) to Ω, we have

$$A^* g = -f \text{ in } \Omega. \tag{3.27}$$

But then, in (3.24), we have $\langle g, \overline{Au} \rangle = \langle A^* g, \bar{u} \rangle$, since $g \in \mathscr{D}(\Omega)$ and

$$M(u) = \langle f + A^* g, \bar{u} \rangle \ \big(\text{duality between } \mathscr{D}(\Omega) \text{ and } \mathscr{D}'(\Omega)\big)$$

According to (3.27) we therefore have $M(u) = 0$. ▯

3.5 Trace Theorem and Green's Formula in Y

Let us return to the study of the mapping $v \to \mathscr{T} v$ and first show that it is surjective from X onto $[\mathscr{H}(\Gamma)]^m$:

Lemma 3.1. *Let Ω be an open set satisfying* (I) *and* ii), *A be given by* (1.1), *properly elliptic in $\bar{\Omega}$, with $a_{pq} \in \mathscr{H}(\bar{\Omega})$, B_j be given by* (1.2), *with $b_{jh} \in \mathscr{H}(\Gamma)$, the system $\{B_j\}_{j=0}^{m-1}$ being normal on Γ; let L be a fixed positive constant. Consider the space $\mathscr{H}_L(\Gamma)$ (defined in Section* 3.2 *of Chapter* 7). *There exists a continuous linear mapping $\varphi = \{\varphi_0, \ldots, \varphi_{m-1}\} \to v_L(\varphi)$ of $[\mathscr{H}_L(\Gamma)]^m$ into X, such that*

$$T_j v_L(\varphi) = \varphi_j, \; j = 0, \ldots, m-1, \tag{3.28}$$

(continuous "right-inverse" of $\mathscr{C}$).

Proof. 1) Let $\varphi = \{\varphi_0, \ldots, \varphi_{m-1}\}$ be given in $[\mathscr{H}_L(\Gamma)]^m$; we need to construct $v \in \mathscr{D}(\bar{\Omega})$ such that

$$C_j v = 0, \; T_j v = \varphi_j, \; j = 0, \ldots, m-1, \tag{3.29}$$

$$A^* v \in \mathscr{D}(\Omega). \tag{3.30}$$

Applying Lemma 2.1 and the arguments of Lemma 2.2 of Chapter 2, using the fact that $\{C_j, T_j\}$ is a Dirichlet system of order $2m$ on Γ with analytic coefficients on Γ, it all comes down to constructing $v \in \mathscr{D}(\bar{\Omega})$ satisfying (3.30) and

$$\gamma_j v = \psi_j, \; j = 0, \ldots, m-1, \tag{3.31}$$

where ψ_j belongs to $\mathscr{H}_M(\Gamma)$ (M depending on L).

2) Let us then consider the *Cauchy problem*

$$\begin{cases} A^* u = 0 \text{ in a neighborhood of } \Gamma, \\ \gamma_j u = \psi_j \text{ on } \Gamma, \, j = 0, \ldots, 2m-1. \end{cases} \tag{3.32}$$

By the Cauchy-Kovalewska theorem (Γ being compact and all coefficients being analytic), problem (3.32) admits a unique solution in a suitable neighborhood of Γ, this neighborhood depending on M, and therefore on L, but *not* on the choice of φ_j in $\mathscr{H}_L(\Gamma)$. Thus, if we denote by I_ϱ the set of points of Ω such that $d(x, \Gamma) < \varrho$, ϱ sufficiently small, we see that there exists a $\varrho(L)$, such that (3.32) admits a unique solution in $I_{\varrho(L)}$.

Let α be a function of $\mathscr{D}(\bar{\Omega})$ such that $\alpha = 0$ in $\bar{\Omega} - I_{\varrho(L)}$ and $\alpha(x) = 1$ in $I_{\varrho(L)/2}$, for example; this function obviously exists. Finally, set

$$v = \alpha u \text{ in } I_{\varrho(L)}, \; v = 0 \;\text{ in }\; \bar{\Omega} - I_{\varrho(L)}. \tag{3.33}$$

Then (3.31) holds and therefore (3.29) holds as well; furthermore $v \in \mathscr{D}(\bar{\Omega})$ and, in $I_{\varrho(L)/2}$, $A^* v = \alpha A^* u = 0$, therefore $A^* v \in \mathscr{D}(\Omega)$.

3) The construction of v given in 2) defines, L being fixed, a linear mapping $\varphi \to v_L(\varphi)$ of $[\mathscr{H}_L(\Gamma)]^m$ into X ($\varrho(L)$ and α being determined by L).

There only remains to show the continuity, but the spaces $[\mathscr{H}_L(\Gamma)]^m$ and X being in particular ($\mathscr{L}\mathscr{F}$) spaces, we may apply the closed graph theorem. And then it suffices, for example, to note that, by the Cauchy-Kovalewska theorem, $\psi = \{\psi_0, \ldots, \psi_{2m-1}\} \to u =$ solution of (3.32) is a continuous mapping of $[\mathscr{H}_M(\Gamma)]^{2m}$ into $C^\circ(I_{\varrho(L)})$. ☐

We can now prove the trace theorem:

Theorem 3.3. *Under the hypotheses of Lemma 3.1, the mapping $u \to Bu = \{B_0 u, \ldots, B_{m-1} u\}$ of $\mathscr{D}(\bar{\Omega})$ into $[\mathscr{D}(\Gamma)]^m$ extends by continuity to a continuous linear mapping, still denoted by $u \to Bu$, of Y into $[\mathscr{H}'(\Gamma)]^m$; furthermore, for $u \in Y$ and $v \in X$, we have the "Green formula":*

$$\langle Au, \bar{v}\rangle - \langle u, \overline{A^* v}\rangle = - \sum_{j=0}^{m-1} \langle B_j u, \overline{T_j v}\rangle, \tag{3.34}$$

the first bracket denoting the duality between $\Xi'(\Omega)$ and $\Xi(\Omega)$, the second between $\mathscr{D}'(\Omega)$ and $\mathscr{D}(\Omega)$ and $\langle B_j u, T_j v\rangle$ the duality between $\mathscr{H}'(\Gamma)$ and $\mathscr{H}(\Gamma)$.

Proof. 1) Let u be given in Y and let $\varphi = \{\varphi_0, \ldots, \varphi_{m-1}\}$ be given in $[\mathscr{H}(\Gamma)]^m$. Then $\varphi \in [\mathscr{H}_L(\Gamma)]^m$ for a suitable L; we choose $v = v_L(\varphi)$ as in Lemma 3.1 and introduce

$$Z(v_L(\varphi)) = \langle u, \overline{A^* v_L(\varphi)}\rangle - \langle Au, \overline{v_L(\varphi)}\rangle, \tag{3.35}$$

the first bracket denoting the duality between $\mathscr{D}'(\Omega)$ and $\mathscr{D}(\Omega)$ and the second between $\Xi'(\Omega)$ and $\Xi(\Omega)$ (note that $v_L(\varphi) \in X \subset \Xi(\Omega)$).

Let us first verify that $Z(v_L(\varphi))$ does not depend on the "right-inverse" used, but *only* on φ. Indeed, if v_1 and v_2 are two such "right-inverses", then $\chi = v_1 - v_2$ satisfies the conditions

$$C_j \chi = 0,\ T_j \chi = 0,\ j = 0, \ldots, m-1,$$

and therefore, by Lemma 2.1 of Chapter 2:

$$\gamma_j \chi = 0,\ j = 0, \ldots, 2m-1,$$

and since $A^*\chi \in \mathscr{D}(\Omega)$, by the uniqueness of the Cauchy problem (3.32), it follows that $\chi \in \mathscr{D}(\Omega)$. But then we have

$$\langle u, \overline{A^*\chi}\rangle = \langle Au, \bar{\chi}\rangle \text{ and therefore}$$

$$Z(v_1) = Z(v_2).$$

Thus (3.35) depends only on φ and we may write $Z(\varphi)$ instead of $Z(v_L(\varphi))$.

2) The form $\varphi \to Z(\varphi)$ is antilinear on $[\mathscr{H}(\Gamma)]^m$. Indeed, let us verify that $Z(a\varphi_1 + b\varphi_2) = aZ(\varphi_1) + bZ(\varphi_2)$; now, we can find an L such that φ_1 and $\varphi_2 \in \mathscr{H}_L(\Gamma)$ and then use expression (3.35) for $Z(\varphi)$ and the antilinearity of $Z(\varphi)$ follows from the linearity of the mapping $\varphi \to v_L(\varphi)$.

Let us now show that $\varphi \to Z(\varphi)$ is *continuous on* $[\mathscr{H}(\Gamma)]^m$. It suffices to verify that $\varphi \to Z(\varphi)$ is continuous on $[\mathscr{H}_L(\Gamma)]^m$, for every fixed L. But then we use expression (3.35) for $Z(\varphi)$ and the continuity follows from Lemma 3.1.

3) Consequently, we have

$$Z(\varphi) = \sum_{j=0}^{m-1} \langle \tau_j u, \bar{\varphi}_j \rangle, \tag{3.36}$$

where $\tau_j u \in \mathscr{H}'(\Gamma)$.

Still using expression (3.35) for $Z(\varphi)$, it is easily seen that $u \to \tau u = \{\tau_0 u \ \dots \ \tau_{m-1} u\}$ is a linear mapping of Y into $[\mathscr{H}'(\Gamma)]^m$.

Let us now show the continuity of τ. It is sufficient to show that, given a bounded set $\mathscr{B}$ in $[\mathscr{H}(\Gamma)]^m$, there exists a neighborhood of zero in Y, say $\mathscr{V}$, such that

$$|\langle \tau u, \bar{\varphi} \rangle| = \left| \sum_{j=0}^{m-1} \langle \tau_j u, \bar{\varphi}_j \rangle \right| \leq 1, \ \forall u \in \mathscr{V} \text{ and } \varphi \in \mathscr{B}.$$

But $\mathscr{B}$ is necessarily a bounded set in $[\mathscr{H}_L(\Gamma)]^m$ for a suitable L (see Chapter 7, Section 1.2 and 3.2); choose $v_L(\varphi)$ for this L as in Lemma 3.1; then

$$\langle \tau u, \bar{\varphi} \rangle = \langle u, \overline{A^* v_L(\varphi)} \rangle - \langle Au, \overline{v_L(\varphi)} \rangle \tag{3.37}$$

and $v_L(\varphi)$ belongs to a bounded set of X, therefore of $X^{(\nu)}$ with suitably fixed ν and therefore, in particular, $v_L(\varphi)$ belongs to a bounded set $\mathscr{B}_1$ in $\mathscr{D}(\bar{\Omega})$ and $A^* v_L(\varphi)$ to a bounded set $\mathscr{B}_2$ in $\mathscr{D}(\Omega)$; note that $\mathscr{B}_1$ is also bounded in $\Xi(\Omega)$. Then, E° denoting the polar set of E, we can take

$$\mathscr{V} = \{u \mid u \in \tfrac{1}{2} \mathscr{B}_2^0, Au \in \tfrac{1}{2} \mathscr{B}_1^0\},$$

noting that $\mathscr{B}_1^0 \subset \Xi'(\Omega)$. And the desired result follows.

4) Now, taking $u \in \mathscr{D}(\bar{\Omega})$, we deduce from (3.37) and the Green formula (3.1) that

$$\langle \tau u, \bar{\varphi} \rangle = \sum_{j=0}^{m-1} \int_\Gamma B_j u \bar{\varphi}_j \, d\sigma, \forall \varphi \in [\mathscr{H}_L(\Gamma)]^m \text{ and } \forall L,$$

whence

$$\tau u = Bu.$$

Also note that, thanks to Theorem 3.2, τ is the extension by continuity of B.

5) Finally the Green formula (3.34) results from the preceding considerations: for, if $v \in X$, then $\varphi_j = T_j v \in \mathcal{H}(\Gamma)$ and we can take $v_L(\varphi) = v$ in (3.35) and therefore (3.34) follows from (3.35) and (3.36). □

3.6 The Existence of Solutions in the Space Y

We are now ready to give a more precise interpration of Theorem 3.1. Indeed, for the solution u obtained in Theorem 3.1, we can write either formula (3.21) or formula (3.34); but we already know that $Au = f$ (see (3.22)), therefore it follows that

$$\sum_{j=0}^{m-1} \langle B_j u - g_j, \overline{T_j v} \rangle = 0, \forall v \in X,$$

and also, thanks to Lemma 3.1,

$$\sum_{j=0}^{m-1} \langle B_j u - g_j, \bar{\varphi}_j \rangle = 0, \forall \varphi_j \in \mathcal{H}(\Gamma),$$

and therefore $B_j u = g_j$, $j = 0, \ldots, m-1$.

We have therefore shown

Theorem 3.4. *Under hypotheses* i), ii), iii), iv) *of Section* 3.1, *the operator*

$$\mathcal{P}: u \to \mathcal{P}u = \{Au, B_0 u, \ldots, B_{m-1} u\} \tag{3.38}$$

defines an (algebraic and topological) isomorphism of Y/N *onto the space* $\{\Xi'(\Omega) \times [\mathcal{H}'(\Gamma)]^m; N^*, \mathcal{C}\}$ *of elements of* $\Xi'(\Omega) \times [\mathcal{H}'(\Gamma)]^m$, *satisfying* (3.20). □

In other words, the boundary value problem

$$Au = f \text{ in the sense of } \mathcal{D}'(\Omega), \tag{3.39}$$

$$B_j u = g_j \text{ in the sense of the (trace) Theorem 3.3,} \quad j = 0, \ldots, m-1, \tag{3.40}$$

admits a solution $u \in Y$, determined up to addition of a function of N, for every $f \in \Xi'(\Omega)$ and every $g_j \in \mathcal{H}'(\Gamma)$, satisfying the compatibility relations (3.20).

In particular, P is an indexed operator in Y and its index is given by $\chi(P) = \dim N - \dim N^*$.

3.7 Continuity of Traces on Surfaces Neighbouring Γ

In this section we add a complement to the trace theorem of Section 3.5, along the same lines as Theorem 8.1 of Chapter 2.

Thus, let $\{\Gamma_\varrho\}$, $0 \leq \varrho \leq \varrho_0 < 1$, be the family of parallel surfaces to Γ, *tending towards* Γ as $\varrho \to 0$, which was introduced in Section 8.1 of Chapter 2. In the same notation, we may now assume the homeomorphism θ_i and J_{ij} $(i, j = 1, \ldots, N)$ interverning in (8.1), Chapter 2, to be *analytic* (since now Γ is an *analytic* variety, hypothesis ii) of Section 3.1).

Consequently, we may also assume that the homeomorphism

$$x \to \psi(x, \varrho) \text{ of } \Gamma_\varrho \text{ onto } \Gamma, \tag{3.41}$$

defined with the help of θ_i by (8.2) of Section 8.1, Chapter 2, is *analytic*, as well as its inverse ψ^{-1}; also recall that ψ and ψ^{-1} are bounded, together with each of their derivatives, by constants which are independent of ϱ (but depending on the order of the derivative).

Still denoting by Ω_ϱ the open set with boundary Γ_ϱ contained in Ω, we may consider the spaces $X(\Omega_\varrho)$ and $Y(\Omega_\varrho)$ to be constructed from Ω_ϱ in the same way as the spaces X and Y. For this purpose, we assume, as we are allowed to do, that the coefficients of the operators $\{B_j\}$, $\{C_j\}$, $\{S_j\}$, $\{T_j\}$ are defined and analytic not only on Γ, but in $\bar{\Omega} - \Omega_{\varrho_0}$, so that the systems $\{B_j\}$, $\{C_j\}$, $\{S_j\}$, $\{T_j\}$ are *normal* on Γ_ϱ for every ϱ such that $0 < \varrho \leq \varrho_0$.

And we shall set

$$X(\Omega_\varrho) = \{v \mid v \in \mathscr{D}(\bar{\Omega}_\varrho),\ \ C_j v\,|_{\Gamma\varrho} = 0,\ j = 0, \ldots, m-1,\ \ A^* v \in \mathscr{D}(\Omega_\varrho)\},$$
$$Y(\Omega_\varrho) = \{u \mid u \in \mathscr{D}'(\Omega_\varrho),\ \ Au \in \Xi'(\Omega_\varrho)\},$$

with topologies analogous to those of X and Y.

The problem is to define $B_j u\,|_{\Gamma\varrho}$, if possible, for u given in Y, and then to see if $B_j u\,|_{\Gamma\varrho}$ converges (in a topology which remains to be defined) to $B_j u$.

But in fact it is not possible to define $B_j u\,|_{\Gamma\varrho}$ for *arbitrary* u in Y; indeed, the mapping "restriction of Ω to Ω_ϱ" *does not map* $\Xi'(\Omega)$ into $\Xi'(\Omega_\varrho)$ (the transposed mapping is

$$\varphi \to \tilde{\varphi} = \text{extension of } \varphi \text{ from } \Omega_\varrho \text{ to } \Omega \text{ by } 0 \text{ outside } \Omega_\varrho,$$

and this mapping does not send $\Xi(\Omega_\varrho)$ into $\Xi(\Omega)$).

This leads us to considerably restrict the class of u's for which we shall solve the problem. We introduce:

$$Y_1 = Y_1(\Omega) = \{u \mid u \in \mathscr{D}'(\Omega),\ Au \in L^2(\Omega) + \mathscr{E}'(\Omega)\} \tag{3.42}$$

(where $\mathscr{E}'(\Omega)$ = dual of $\mathscr{E}(\Omega)$ = space of distributions having *compact support* in Ω); in definition (3.42), we have provided $L^2(\Omega) + \mathscr{E}'(\Omega)$ with the dual topology of $L^2(\Omega) \cap \mathscr{E}(\Omega)$ and Y_1 with the coarsest topology which makes the mappings $u \to u$ and $u \to Au$ of Y_1 into $\mathscr{D}'(\Omega)$ and Y_1 into $L^2(\Omega) + \mathscr{E}'(\Omega)$ respectively, continuous.

By the same type of proof as for Theorems 3.2 and 3.3, we verify that

$$\left\{\begin{array}{l}\mathscr{D}(\bar{\Omega}) \text{ is dense in } Y_1 \text{ and we have, for } Y_1, \text{ a trace} \\ \text{theorem analogous to Theorem 3.3.}\end{array}\right. \tag{3.43}$$

We shall now solve the problem *for the elements of* Y_1.

First of all, if $u \in Y_1$, then $(Au)_{\Omega\varrho}$ = restriction of Au to Ω_ϱ, satisfies

$$(Au)_{\Omega\varrho} \in L^2(\Omega_\varrho) + \mathscr{E}'(\Omega_\varrho) \text{ for } \textit{sufficiently small } \varrho \tag{3.44}$$

and *therefore*

$$(Au)_{\Omega\varrho} = A(u_{\Omega\varrho}) \in \Xi'(\Omega_\varrho). \tag{3.45}$$

Therefore, in particular, we have $u_{\Omega\varrho} \in Y(\Omega_\varrho)$.

We can therefore define $B_j u|_{\Gamma\varrho}$ (element of $\mathscr{H}'(\Gamma_\varrho)$), by the trace theorem of Section 3.5 applied to $Y(\Omega_\varrho)$, and then, by transfer of structure with (3.41), we define

$$B_j^{(\varrho)} u = \text{image of } B_j u\,|_{\Gamma\varrho} \text{ under (3.41)}. \tag{3.46}$$

We have: $B_j^{(\varrho)} u \in \mathscr{H}'(\Gamma)$. We shall now show:

$$B_j^{(\varrho)} u \to B_j u \textit{ in } \mathscr{H}'(\Gamma) \textit{ as } \varrho \to 0,\ \forall u \in Y_1. \tag{3.47}$$

For this purpose we must show that

$$\langle B_j^{(\varrho)} u \to B_j u, \varphi_j\rangle \to 0 \textit{ as } \varrho \to 0, \tag{3.48}$$

uniformly for φ_j in a bounded set $\mathscr{B}$ of $\mathscr{H}(\Gamma)$, and therefore (see Chapter 7) in a bounded set $\mathscr{B}$ of $\mathscr{H}_L(\Gamma)$ for suitable L.

Thanks to (3.43), we can approximate u in Y_1 with a sequence $\{u_\nu\}$ in $\mathscr{D}(\bar{\Omega})$. Then we have

$$\begin{aligned}\langle B_j^{(\varrho)} u - B_j u, \varphi_j\rangle &= \langle B_j^{(\varrho)} u - B_j^{(\varrho)} u_\nu, \varphi_j\rangle + \langle B_j^{(\varrho)} u_\nu - B_j u_\nu, \varphi_j\rangle + \\ &+ \langle B_j u_\nu - B_j u, \varphi_j\rangle.\end{aligned} \tag{3.49}$$

Thanks to (3.43), the last bracket in (3.49) goes to zero as $\nu \to +\infty$, uniformly for φ_j in $\mathscr{B}$.

The second bracket on the right-hand side of equation (3.49) goes to zero as $\varrho \to 0$, uniformly for φ_j in $\mathscr{B}$, for each *fixed* ν.

We must therefore show that

$$\langle B_j^{(\varrho)} u - B_j^{(\varrho)} u_0, \varphi_j\rangle \to 0 \text{ as } \nu \to +\infty, \tag{3.50}$$

uniformly for φ_j in $\mathscr{B}$ and for ϱ in $[0, \varrho_0[$, $j = 0, \ldots, m-1$. After transfer of structure by (3.41), this amounts to showing that

$$\langle B_j(u - u_\nu)\,|_{\Gamma\varrho}, \varphi_{\varrho,j}\rangle \to 0 \text{ as } \nu \to +\infty, \tag{3.51}$$

uniformly for $\varphi_{\varrho,j}$ in the bounded set $\mathscr{B}_\varrho$ of $\mathscr{H}(\Gamma_\varrho)$, the set corresponding to $\mathscr{B}$ by (3.41), and for ϱ in $[0, \varrho_0[$, $j = 0, \ldots, m-1$.

But we set

$$\varphi_\varrho = \{\varphi_{\varrho,0}, \ldots, \varphi_{\varrho,m-1}\}$$

and let $v(\varphi_\varrho) \in X(\Omega_\varrho)$ be the right-inverse of φ_ϱ, analogue to the one of Lemma 3.1, but passing from Γ_ϱ to Ω_ϱ.

Green's formula (3.34) applied to Ω_ϱ yields

$$\sum_{j=0}^{m-1} \langle B_j(u-u_\nu)\,|_{\Gamma_\varrho}, \bar{\varphi}_{\varrho,j}\rangle = \langle (u-u_\nu), \overline{A^*v(\varphi_\varrho)}\rangle - \tag{3.52}$$
$$- \langle A(u-u_\nu)_{\Omega_\varrho}, \overline{v(\varphi_\varrho)}\rangle,$$

where on the right-hand side, the first (resp. second) bracket denotes the antiduality between $\mathscr{D}'(\Omega_\varrho)$ and $\mathscr{D}(\Omega_\varrho)$ (resp. $\Xi'(\Omega_\varrho)$ and $\Xi(\Omega_\varrho)$).

Let w_ϱ = extension of $A^*v(\varphi_\varrho)$ to Ω by 0 outside Ω_ϱ; then $w_\varrho \in \mathscr{D}(\Omega)$ and remains in a bounded set of $\mathscr{D}(\Omega)$ as $\varrho \to 0$, therefore

$$\langle (u-u_\nu)_{\Omega_\varrho}, \overline{A^*v(\varphi_\varrho)}\rangle = \langle u-u_\nu, \bar{w}_\varrho\rangle \to 0.$$

On the other hand:

$$A(u-u_\nu)_{\Omega_\varrho} = (g_\nu)_{\Omega_\varrho} + (f_\nu)_{\Omega_\varrho},$$

where $g_\nu \to 0$ in $\mathscr{E}'(\Omega)$, $f_\nu \to 0$ in $L^2(\Omega)$.

The distributions g_ν have their support in a fixed compact set $\mathscr{K}$ in Ω; let $\theta \in \mathscr{D}(\Omega)$, $\theta = 1$ in the neighborhood of $\mathscr{K}$; then, *for sufficiently small* ϱ:

$$\langle (g_\nu)_{\Omega_\varrho}, \overline{v(\varphi_\varrho)}\rangle = \langle g_\nu, \overline{\theta \widetilde{v(\varphi_\varrho)}}\rangle,$$

where:

$$\widetilde{v(\varphi_\varrho)} = \text{extension of } v(\varphi_\varrho) \text{ by 0 outside } \Omega.$$

Therefore

$$\langle A(u-u_\nu)_{\Omega_\varrho}, \overline{v(\varphi_\varrho)}\rangle = \langle g_\nu, \overline{\theta \widetilde{v(\varphi_\varrho)}}\rangle + \int_{\Omega_\varrho} f_\nu \overline{v(\varphi_\varrho)}\, dx \tag{3.53}$$

and it can easily be seen that each term on the right-hand side of (3.53) goes to zero; and this completes the proof of (3.47). ☐

Remark 3.3. Of course, the condition "$u \in Y_1$", Y_1 defined by (3.42), is not the "optimal" condition which guarantees the correctness of (3.47).

We note in fact that the results of this chapter can be extended to the spaces $\Xi_p(\Omega)$, $1 < p < \infty$ (constructed as $\Xi(\Omega)$ but replacing L^2 with L^p) and to their duals $\Xi'_{p'}(\Omega) = (\Xi_p(\Omega))'$.

Taking Au in $L^p(\Omega) + \mathscr{E}'(\Omega)$, $1 < p < \infty$, we obtain a result which is analogous to (3.47) (and for $1 < p < 2$, this yields a less restrictive condition than $u \in Y_1$). ☐

4. Existence of Solutions in the Space $\mathscr{D}'_{M_k}(\Omega)$ of Ultra-Distributions

4.1 Generalities

The method used in Section 3 also applies to an even more general setting, if we consider the equation $Au = f$ in the sense of *ultra-distributions*.

Let us again assume that the open set Ω and the operators A and B_j, $j = 0, \ldots, m - 1$, satisfy hypotheses i), ii), iii), iv) of Section 3.1.

Further, let

$$M_k = (k!)^\beta, \text{ fixed real } \beta > 1 \tag{4.1}$$

(or more generally, let $\{M_k\}$ satisfy (1.6), ..., (1.11) and (1.2) of Section 1.1, Chapter 7; then thanks to (1.10), $\{M_k\}$ also satisfies (1.3) of Section 1.1, Chapter 7).

The idea is to use $\mathscr{D}_{M_k}(\Omega)$ instead of $\mathscr{D}(\Omega)$ (and therefore $\mathscr{D}'_{M_k}(\Omega)$ instead of $\mathscr{D}'(\Omega)$) in the theory developed in Section 3.

First, we introduce the space

$$\begin{aligned} X_{M_k} &= \{v \mid v \in \mathscr{D}_{M_k}(\bar{\Omega});\ C_j v = 0,\ j = 0, \ldots, m-1, \\ &\qquad A^* v \in \mathscr{D}_{M_k}(\Omega)\}, \end{aligned} \tag{4.2}$$

where $\mathscr{D}_{M_k}(\bar{\Omega})$ and $\mathscr{D}_{M_k}(\Omega)$ are the spaces defined in Chapter 7 (Definitions 1.1 and 1.2), the space X_{M_k} is provided with the inductive limit topology of the spaces $X^{(\nu)}_{M_k}$ defined as follows:

$$\begin{aligned} X^{(\nu)}_{M_k} &= \{v \mid v \in \mathscr{D}_{M_k}(\bar{\Omega}; \chi(L_\nu)),\ C_j v = 0,\ j = 0, \ldots, m-1, \\ &\qquad A^* v \in \mathscr{D}_{M_k}(\Omega; \mathscr{K}_\nu, L_\nu)\} \end{aligned} \tag{4.3}$$

where $\{L_\nu\}$ is an increasing sequence of numbers which tends towards $+\infty$, $\{\mathscr{K}_\nu\}$ is an increasing sequence of compact sets contained in Ω and whose union is Ω, $\chi(L)$ is the function introduced in Corollary 1.1 to Theorem 1.2, A^* and C_j replacing A and B_j, respectively, and where the spaces $\mathscr{D}_{M_k}(\bar{\Omega}; \chi(L_\nu))$, $\mathscr{D}_{M_k}(\Omega; \mathscr{K}_\nu, L_\nu)$ are the Banach spaces defined in Sections 1.3 and 1.2 of Chapter 7, respectively.

It is then natural to provide $X^{(\nu)}_{M_k}$ with the norm of the graph, which makes it a Banach space.

Thus X_{M_k} is an $(\mathscr{L}\mathscr{F})$-space (and even better, according to the properties of the spaces $\mathscr{D}_{M_k}(\bar{\Omega})$ and $\mathscr{D}_{M_k}(\Omega)$; cf. Chapter 7, Sections 1.2 and 1.3).

We also introduce the space

$$\{\mathscr{D}_{M_k}(\Omega); N\} = \left\{\varphi \mid \varphi \in \mathscr{D}_{M_k}(\Omega),\ \int_\Omega \varphi \bar{w}\, dx = 0 \quad \forall w \in N\right\}$$

closed subspace of $\mathscr{D}_{M_k}(\Omega)$.

Applying Theorem 1.1 and Corollary 1.1 to Theorem 1.2, we obtain

$$(4.4) \qquad \begin{cases} A^* \textit{ defines an (algebraic and topological)} \\ \textit{isomorphism of } X_{M_k}/N^* \textit{ onto } \{\mathscr{D}_{M_k}(\Omega); N\}. \end{cases}$$

If, as we have done in Section 3.1, we transpose (4.4) and if we note that

$$(4.5) \qquad \{\mathscr{D}_{M_k}(\Omega); N\} = \mathscr{D}'_{M_k}(\Omega)/N,$$

we obtain the following Proposition (see (3.10)):

$$(4.6) \qquad \begin{cases} \textit{given a continuous antilinear form } L \textit{ on } X_{M_k}/N, \textit{ there exists} \\ u \in \mathscr{D}'_{M_k}(\Omega), \textit{ determined up to addition of a function of } N, \\ \textit{such that} \\ \qquad \langle u, \overline{A^* v}\rangle = L(v^{\cdot}) \ \forall v^{\cdot} \in X_{M_k}/N^* \textit{ and } v \in v^{\cdot}, \\ \textit{the bracket denoting the duality between } \mathscr{D}'_{M_k}(\Omega) \textit{ and } \mathscr{D}_{M_k}(\Omega). \\ \textit{Furthermore } u \textit{ depends continuously on } L. \end{cases}$$

As in Section 3.2, we thus arrive at the problem of choosing the form L; in particular, we are led to introduce the spaces $K_{M_k}(\Omega)$ of functions such that (see (3.11))

$$(4.7) \qquad \begin{cases} X_{M_k} \subset K_{M_k}(\Omega) \subset L^2(\Omega), \textit{ with continuous injections,} \\ K_{M_k}(\Omega) \textit{ is reflexive,} \\ \mathscr{D}_{M_k}(\Omega) \textit{ is dense in } K_{M_k}(\Omega). \end{cases}$$

Such spaces always exist, for example $L^2(\Omega)$, since $\mathscr{D}(\Omega)$ is dense in $L^2(\Omega)$ and since $\mathscr{D}_{M_k}(\Omega)$ is dense in $\mathscr{D}(\Omega)$ (see Section 1.2 of Chapter 7). Our goal is to introduce "the smallest possible" space $K_{M_k}(\Omega)$ (see the comments at the beginning of Section 3.2 which are still valid in the present setting). For this purpose, the next Section introduces a space $\Xi_{M_k}(\Omega)$ which is analogous to $\Xi(\Omega)$. ☐

4.2 The Space $\Xi_{M_k}(\Omega)$ and its Dual

Again let $\varrho(x)$ be the function introduced in Section 3.2 for the Definition 3.1 of $\Xi(\Omega)$. For fixed $L > 0$, we first define the space

$$(4.8) \qquad \Xi^L_{M_k}(\Omega) = \Big\{u \mid u \in \mathscr{E}(\Omega) \text{ such that there exists a } c, \text{ depending on } u, \text{ such that } \sum_{|\alpha|=k} \|\varrho^k D^\alpha u\|_{L^2(\Omega)} \leq c L^k M_k, \ \forall k \geq 0\Big\};$$

provided with the norm

$$\|u\|_L = \sup_k \sum_{|\alpha|=k} \frac{\|\varrho^k \mathrm{D}^\alpha u\|_{L^2(\Omega)}}{L^k M_k}$$

it is a Banach space.

Next we define

$$\Xi_{M_k}(\Omega) = \operatorname{ind\,lim}_{L\to+\infty} \Xi^L_{M_k}(\Omega), \tag{4.9}$$

L increasing monotonically to $+\infty$.

Proposition 4.1. *The space $\Xi_{M_k}(\Omega)$ is reflexive.*

Proof. It is sufficient to note that $\Xi_{M_k}(\Omega)$ may be equivalently defined as the inductive limit of a monotonically increasing sequence of Hilbert spaces and to apply the results of Chapter 7, Section 1.2, Remark 1.3.

Indeed, let us consider the space

$$*\Xi^L_{M_k}(\Omega) = \left\{u \mid u \in \mathscr{E}(\Omega), \sum_{k=0}^{\infty} \frac{1}{(L^k M_k)^2} \sum_{|\alpha|=k} \|\varrho^k \mathrm{D}^\alpha u\|^2_{L^2(\Omega)} < +\infty\right\} \tag{4.10}$$

for fixed $L > 0$.

Provided with the norm

$$\left(\sum_{k=0}^{\infty} \frac{1}{(L^k M_k)^2} \sum_{|\alpha|=k} \|\varrho^k \mathrm{D}^\alpha u\|^2_{L^2(\Omega)}\right)^{1/2}, \tag{4.11}$$

it is a Hilbert space.

Thanks to the properties of the sequence $\{M_k\}$, it is easily seen that

$$\Xi_{M_k}(\Omega) = \operatorname{ind\,lim}_{n\to\infty} *\Xi^{L_n}_{M_k}(\Omega), \tag{4.12}$$

where L_n is a positive, monotonically increasing sequence which tends to $+\infty$. □

It follows directly from the definitions that

$$\mathscr{D}_{M_k}(\overline{\Omega}) \subset \Xi_{M_k}(\Omega) \tag{4.13}$$

and therefore also that

$$X_{M_k} \subset \Xi_{M_k}(\Omega). \quad □ \tag{4.14}$$

Therefore, in order to verify that $\Xi_{M_k}(\Omega)$ satisfies conditions (4.7), there only remains to show

Proposition 4.2. *The space $\mathscr{D}_{M_k}(\Omega)$ is dense in $\Xi_{M_k}(\Omega)$.*

Proof. Let u be given in $\Xi_{M_k}(\Omega)$, therefore

(4.15) $u \in \Xi^L_{M_k}(\Omega)$ for a suitable L (and we may assume $L \geq 1$).

For the time being, let us assume that

$$(4.16)\quad \begin{cases} \textit{there exists a sequence } \delta_\nu \textit{ of functions of } \mathscr{D}(\Omega), \\ \nu = 1, 2, \ldots, \textit{ such that} \\ \delta_\nu(x) = 1 \textit{ if } \mathrm{d}(x, \Gamma) \geq \dfrac{2}{\nu}, \; \delta_\nu(x) = 0 \textit{ if } \mathrm{d}(x, \Gamma) \leq \dfrac{1}{\nu}, \\ |\mathrm{d}(x, \Gamma)|^k \, |\mathrm{D}^\alpha \, \delta_\nu(x)| \leq c_2 L^2 M_k \text{ for } |\alpha| = k \text{ and } k > 0, \\ c_2 = \text{constant}. \end{cases}$$

Then the function $\delta_\nu u \in \mathscr{D}_{M_k}(\Omega)$ and Proposition 4.2 will be demonstrated if we can show that $\delta_\nu u \to u$ in $\Xi_{M_k}(\Omega)$.

More specifically, we shall prove that

$$(4.17)\qquad \delta_\nu u \to u \text{ in } \Xi_{M_k}^{L+\eta}(\Omega), \text{ with arbitrary } \eta > 0.$$

Indeed, let

$$Y_{\nu,k} = \frac{1}{(L+\eta)^k M_k} \sum_{|\alpha|=k} \| \varrho^k (\mathrm{D}^\alpha(\delta_\nu u) - \mathrm{D}^\alpha u) \|_{L^2(\Omega)}.$$

We have

$$Y_{\nu,k} \leq z_{\nu,k} + \mathscr{L}_{\nu,k},$$

where

$$z_{\nu,k} = \frac{1}{(L+\eta)^k M_k} \sum_{|\alpha|=k} \| (\delta_\nu - 1) \, \varrho^k \, \mathrm{D}^\alpha u \|_{L^2(\Omega)},$$

$$\mathscr{L}_{\nu,k} = \frac{1}{(L+\eta)^k M_k} \sum_{j=1}^{k} \binom{k}{j} \sum_{\substack{|\lambda|=j \\ |\mu|=k-j}} \| (\varrho^j \, \mathrm{D}^\lambda \, \delta_\nu) \, (\varrho^{k-j} \, \mathrm{D}^\mu u) \|_{L^2(\Omega)}.$$

Next, set

$$Y_\nu = \sup_k Y_{\nu,k}, \; z_\nu = \sup_k z_{\nu,k}, \; \mathscr{L}_\nu = \sup_k \mathscr{L}_{\nu,k}.$$

Then (4.17) is equivalent to $Y_\nu \to 0$, which will be shown if we can show that

$$(4.18)\qquad z_\nu \to 0,$$

$$(4.19)\qquad \mathscr{L}_\nu \to 0.$$

But thanks to the fact that $u \in \Xi_{M_k}^L(\Omega)$, we have

$$z_{\nu,k} \leq \frac{1}{(L+\eta)^k M_k} c L^k M_k,$$

therefore

$$z_\nu \leq \sup_{0 \leq k \leq N} z_{\nu,k} + c \left(\frac{L}{L+\eta} \right)^N$$

and this proves (4.18) if we can verify that

$$z_{\nu,k} \to 0, \text{ for } \nu \to +\infty, \textit{ fixed } k,$$

which is immediate.

Let us now show (4.19). We have

$$\mathscr{L}_{\nu,k} \leq \frac{c'}{(L+\eta)^k M_k} \sum_{j=1}^{k} \binom{k}{j} L^j M_j L^{k-j} M_{k-j} \leq \frac{c''}{(L+\eta)^k} \sum_{j=1}^{k} L^j L^{k-j} \leq$$

$$\leq c''' k \frac{L^k}{(L+\eta)^k} \quad (c', c'', c''' \text{ suitable constants}).$$

Therefore, for sufficiently great N:

$$\mathscr{L}_\nu \leq \sup_{0 \leq k \leq N} \mathscr{L}_{\nu,k} + c''' N \left(\frac{L}{L+\eta}\right)^N$$

and we have the desired result if we show that

$$\mathscr{L}_{\nu,k} \to 0, \text{ for } \nu \to +\infty, \textit{ fixed } k.$$

Now for fixed k this is immediate, since, for $|\lambda| \geq 1$, $\varrho^{|\lambda|} \mathrm{D}^\lambda \delta_\nu$ is bounded on $\bar{\Omega}$ and vanishes except on a set Ω_ν whose measure tends to zero as $\nu \to +\infty$.

Thus there remains to show (4.16). Via "local maps" we are led to show the existence of a sequence $\theta_\nu(t)$ of functions of one variable t in $[0, 1]$ such that

$$\theta_\nu(t) = 1 \text{ if } t \geq 2/\nu,\ \theta_\nu(t) = 0 \text{ if } t \leq \frac{1}{\nu},\ 0 \leq \theta_\nu(t) \leq 1,$$
$$|t^k \theta_\nu^{(k)}(t)| \leq c_2 L^k M_k,\ \forall k,$$

and again, at the risk of having to translate and change ν to 2ν later, it is sufficient to show the existence of χ_ν with

$$\text{(4.20)} \qquad \begin{cases} \chi_\nu(t) = 1 \text{ if } t \geq \dfrac{2}{\nu},\ 0 \leq \chi_\nu(t) \leq 1, \\ \chi_\nu^{(j)}(0) = 0 \quad \forall j, \end{cases}$$

$$\text{(4.21)} \qquad |t^k \chi_\nu^{(k)}(t)| \leq c L^k M_k \quad \forall k.$$

To this end, we start with a function $\varrho \in \mathscr{D}_{M_k}(\mathbf{R})$ with compact support in $[0, 1]$,

$$\varrho \geq 0,\ \int_0^1 \varrho\, \mathrm{d}t = 1,\ |\varrho^{(k)}(t)| \leq c' \left(\frac{L}{2}\right)^k M_k \quad \forall k.$$

Next we set

$$\varrho_\nu(t) = \nu\varrho(\nu t),\ \psi_\nu(t) = \begin{cases} 0 \text{ if } t \leq 0, \\ \nu t \text{ if } 0 \leq t \leq \dfrac{1}{\nu} \\ 1 \text{ if } t > \dfrac{1}{\nu}, \end{cases}$$

and define

$$\chi_\nu = \varrho_\nu * \psi_\nu .$$

Conditions (4.20) hold; and for (4.21) we have

$$|\chi_\nu^{(k)}(t)| = \left| \int_0^t \psi_\nu(t-\sigma)\, \varrho_\nu^{(k)}(\sigma)\, d\sigma \right| \leq \nu^k \int_0^{1/\nu} |\varrho^{(k)}(\nu\sigma)|\, \nu\, d\sigma =$$

$$= \nu^k \int_0^1 |\varrho^{(k)}(\sigma)|\, d\sigma \leq c'\nu^k \left(\frac{L}{2}\right)^k M_k,$$

therefore, if $0 \leq t \leq 2/\nu$:

$$|t^k \chi_\nu^{(k)}(t)| \leq \left(\frac{2}{\nu}\right)^k \nu^k \left(\frac{L}{2}\right)^k M_k, \text{ whence } (4.21). \ \square$$

$\Xi'_{M_k}(\Omega)$ shall denote the (strong) dual of $\Xi_{M_k}(\Omega)$. Thanks to Proposition 4.2, $\Xi'_{M_k}(\Omega)$ can be identified with a subspace of $\mathcal{D}'_{M_k}(\Omega)$. More precisely, we have:

Proposition 4.3. *Every element f of $\Xi'_{M_k}(\Omega)$ may be represented (non-uniquely) in the form*

$$f = \sum_{k=0}^{\infty} \sum_{|\alpha|=k} D^\alpha(\varrho^k f_\alpha), \tag{4.22}$$

with $f_\alpha \in L^2(\Omega)$ and $\sum_{k=0}^{\infty} \sum_{|\alpha|=k} L^k M_k \|f_\alpha\|_{L^2(\Omega)} < +\infty$, $\forall L > 0$.

4.3 The Space Y_{M_k} and the Existence of Solutions in Y_{M_k}

Still in analogy with Section 3, we now introduce the space

$$Y_{M_k} = \{u \mid u \in \mathcal{D}'_{M_k}(\Omega),\ Au \in \Xi'_{M_k}(\Omega)\}, \tag{4.23}$$

provided with the coarsest, locally convex topology such that the mappings $u \to u$ and $u \to Au$ of Y_{M_k} into $\mathcal{D}'_{M_k}(\Omega)$ and $\Xi'_{M_k}(\Omega)$ respectively, are continuous.

We shall study a trace theorem for the space Y_{M_k} and show that the *space of traces* $\{B_j u\}_{j=0}^{m-1}$ *of the* u's $\in Y_{M_k}$ *is again* $[\mathcal{H}'(\Gamma)]^m$ (as for Y). $\square$

Indeed, we first show, as for Proposition 3.6, that $v \to \mathscr{C}v$ is a continuous linear mapping of X_{Mk} into $[\mathscr{H}(\Gamma)]^m$ (note that the hypotheses on A and B_j are the same as in Section 3 and that if $v \in X_{Mk}$, A^*v still vanishes in a neighborhood of Γ).

This and the properties of $\Xi_{Mk}(\Omega)$ allow us to choose the form L in (4.6) in the following manner:

$$L(v) = \langle f, \bar{v} \rangle + \sum_{j=0}^{m-1} \langle g_j, \overline{T_j v} \rangle, \; f \in \Xi'_{Mk}(\Omega), \; g_j \in \mathscr{H}'(\Gamma), \tag{4.24}$$

the first bracket denoting the duality between $\Xi'_{Mk}(\Omega)$ and $\Xi_{Mk}(\Omega)$ and the others between $\mathscr{H}'(\Gamma)$ and $\mathscr{H}(\Gamma)$.

As for (3.22), it then follows that the solution u of (4.6) with L given by (4.24) satisfies

$$Au = f \textit{ in the sense of } \mathscr{D}'_{Mk}(\Omega). \tag{4.25}$$

It can again be shown that (analogue to Theorem 3.2)

$$\mathscr{D}(\bar{\Omega}) \textit{ is dense in } Y_{Mk}. \tag{4.26}$$

Indeed, it is sufficient to reproduce the proof of Theorem 3.2 with the obvious formal modifications: in (3.24), $f \in \mathscr{D}_{Mk}(\Omega)$, $g \in \Xi_{Mk}(\Omega)$, but (3.25)—(3.27) are again understood in the sense of distributions; we only have to add that, thanks to (3.27) and to Corollary 1.1 to Theorem 1.2 (note that since the coefficients a_{pq} of A are analytic, they also belong to $\mathscr{D}_{Mk}(\bar{\Omega})$), g belongs not only to $\mathscr{D}(\Omega)$ but also to $\mathscr{D}_{Mk}(\Omega)$, and then we have

$$M(u) = \langle f + A^*g, \bar{u} \rangle \text{ (duality between } \mathscr{D}'_{Mk}(\Omega) \text{ and } \mathscr{D}_{Mk}(\Omega)\text{)}. \quad \square$$

Remark 4.1. With the same type of proof, it can be shown that $\mathscr{D}_{Mk}(\bar{\Omega})$ is dense in Y_{Mk}. □

Next, we again have a lemma analogous to Lemma 3.1, with X_{Mk} replacing X; indeed, we again apply the Cauchy-Kovalewska Theorem in the same way as in Section 3.5, and the only thing to modify in the proof is the choice of the function α in (3.33), now taken in $\mathscr{D}_{Mk}(\bar{\Omega})$, which is always possible according to Section 1.2 of Chapter 7.

We are thus led to the following trace theorem.

Theorem 4.1. *Under the hypotheses of Lemma* 3.1, *the mapping* $u \to Bu = \{B_0 u, \ldots, B_{m-1} u\}$ *of* $\mathscr{D}(\bar{\Omega})$ *into* $[\mathscr{D}(\Gamma)]^m$ *extends by continuity to a continuous linear mapping, still denoted by* $u \to Bu$, *of* Y_{Mk} *into* $[\mathscr{H}'(\Gamma)]^m$; *furthermore, for* $u \in Y_{Mk}$ *and* $v \in X_{Mk}$, *we have*:

$$\langle Au, \bar{v} \rangle - \langle u, \overline{A^*v} \rangle = - \sum_{j=0}^{m-1} \langle B_j u, \overline{T_j v} \rangle, \tag{4.27}$$

where the first bracket denotes the duality between $\Xi'_{M_k}(\Omega)$ *and* $\Xi_{M_k}(\Omega)$, *the second between* $\mathscr{D}'_{M_k}(\Omega)$ *and* $\mathscr{D}_{M_k}(\Omega)$, *and* $\langle B_j u, \overline{T_j v}\rangle$ *the duality between* $\mathscr{H}'(\Gamma)$ *and* $\mathscr{H}(\Gamma)$.

The proof is completely analogous to the one of Theorem 3.3. □

Finally, still with the same arguments as in Section 3.6, we obtain the following existence theorem.

Theorem 4.2. *Under the hypotheses* i), ii), iii), iv) *of Section* 3.1, *the operator* $\mathscr{P}$, *given by* (3.38), *defines an isomorphism of* Y_{M_k}/N *onto the space* $\{\Xi'_{M_k}(\Omega)\times[\mathscr{H}'(\Gamma)]^m;\ N^*, \mathscr{C}\}$ *of elements of* $\Xi'_{M_k}(\Omega)\times[\mathscr{H}'(\Gamma)]^m$ *satisfying* (3.20). □

Remark 4.2. For all results obtained in this Section, we can take, instead of $\Xi'_{M_k}(\Omega)$, a "general" space $K'_{M_k}(\Omega)$, dual of the space $K_{M_k}(\Omega)$ satisfying (4.7); indeed only properties (4.7) intervened in the proofs; the space Y_{M_k} is then defined by

$$Y_{M_k} = \{u \mid u \in \mathscr{D}'_{M_k}(\Omega),\ Au \in K'_{M_k}(\Omega)\}. \quad □$$

Remark 4.3. Assume that we do not have the *optimal* regularity theorem in the classes $\{M_k\}$ up to the boundary of Ω (Theorem 1.2), but only[1] regularity in the interior of Ω in the classes $\{M_k\}$ and *analytic regularity* up to the boundary of Ω: more precisely, we only have Theorem 1.1, Corollary 1.1 with $M_k = k!$ and Corollary 2.1 with arbitrary M_k. Then it is still possible to develop the preceding theory; we only need to adapt the definition of the space X_{M_k} by taking:

$$X^*_{M_k} = \{v \mid v \in \mathscr{D}(\overline{\Omega}),\ C_j v = 0,\ j = 0, \ldots, m-1,\ A^* v \in \mathscr{D}_{M_k}(\Omega)\}.$$

If $K^*_{M_k}(\Omega)$ is a space satisfying (4.7) (but with $X^*_{M_k}$ instead of the space X_{M_k}), we replace the space Y_{M_k} with

$$Y^*_{M_k} = \{u \mid u \in \mathscr{D}'_{M_k}(\Omega),\ Au \in K^{*\prime}_{M_k}(\Omega)\}\ \big(K^{*\prime}_{M_k}(\Omega) = \text{dual of } K^*_{M_k}(\Omega)\big);$$

to obtain the following result:

$\mathscr{P}$ is *an isomorphism* of $Y^*_{M_k}/N$ *onto the space* $\{K^{*\prime}_{M_k}(\Omega)\times[\mathscr{H}'(\Gamma)]^m;\ N^*, \mathscr{C}\}$ *of elements of* $K^{*\prime}_{M_k}(\Omega)\times[\mathscr{H}'(\Gamma)]^m$ *satisfying* (3.20). □

Remark 4.4. The preceding Remark and Remark 2.8 of Section 2.6 show the possibility of a new generalization of the results of this Section to spaces of ultra-distributions $\mathscr{B}'_{M_k}(\Omega)$ of Beurling type (see Section 2.3, Chapter 7).

Still under hypotheses i), ii), iii), iv), set

$$\tilde{X}_{M_k} = \{v \mid v \in \mathscr{D}(\overline{\Omega}),\ C_j v = 0,\ j = 0, \ldots, m-1,\ A^* v \in \mathscr{B}_{M_k}(\Omega)\}$$

[1] A situation of this type appears in Remark 4.4.

and

$$\tilde{Y}_{Mk} = \{u \mid u \in \mathscr{B}'_{Mk}(\Omega),\ Au \in \tilde{K}'_{Mk}(\Omega)\},$$

where $\tilde{K}'_{Mk}(\Omega)$ is the dual of a space $\tilde{K}_{Mk}(\Omega)$ satisfying the relations (see (4.7))

$$\begin{cases} \tilde{X}_{Mk} \subset \tilde{K}_{Mk}(\Omega) \subset L^2(\Omega) \text{ with continuous injections,} \\ \tilde{K}_{Mk}(\Omega) \text{ is reflexive,} \\ \mathscr{B}_{Mk}(\Omega) \text{ is dense in } \tilde{K}_{Mk}(\Omega). \end{cases}$$

Then we still have a *trace theorem* analogous to Theorems 3.3 and 4.1 with the *same trace space* and we still have the result:

$$\begin{cases} \textit{the operator } \mathscr{P} \textit{ is an isomorphism of } \tilde{Y}_{Mk}/N \\ \textit{onto the space } \{\tilde{K}'_{Mk}(\Omega) \times [\mathscr{H}'(\Gamma)]^m;\ N^*, \Phi\} \\ \textit{of elements of } \tilde{K}'_{M,k}(\Omega) \times [\mathscr{H}'(\Gamma)]^m \textit{ satisfying } (3.20). \end{cases} \quad (4.28)$$

□

4.4 Application to the Regularity in the Interior of Ultra-Distribution Solutions of the Equation $Au = f$

In the preceding Sections, we have seen that the different spaces Y, Y_{Mk}, $\tilde{Y}_{Mk}$ *have the same space "of traces"* (see Theorems 3.3 and 4.1, Remark 4.4). Now this leads to an interesting consequence on the regularity in the interior of Ω of solutions of the equation $Au = f$, in the sense of ultra-distributions. Indeed, we have

Corollary 4.1. *Let Ω be a bounded open set in $\mathbf{R}^n$ with boundary Γ, an $(n-1)$-dimensional variety, Ω being locally on one side of Γ, and A a properly elliptic linear operator on Ω, with analytic coefficients in $\bar{\Omega}$; then every ultra-distribution u of $\mathscr{D}'_{Mk}(\Omega)$ or of $\mathscr{B}'_{Mk}(\Omega)$ which is a solution of $Au = f$, with (for example) $f \in L^2(\Omega)$, is a distribution of $\mathscr{D}'(\Omega)$.*

Proof. Taking into account the fact that the Dirichlet conditions $\{\gamma_j u\}_{j=0}^{m-1}$ cover *all* properly elliptic operators (Remark 1.3, Chapter 2), we may apply the (trace) Theorem 4.1 (or Remark 4.4) and we see that u admits a trace $\gamma u = \{g_0, \ldots, g_{m-1}\}$ with $g_j \in \mathscr{H}'(\Gamma)$; and of course f and $g_0, \ldots, g_{m-1}$ satisfy the compatibility conditions (3.20) of the Dirichlet problem

$$Au = f,\ \gamma_j u = g_j,\ j = 0, \ldots, m-1. \quad (4.29)$$

Therefore thanks to Theorem 3.4, problem (4.29) admits at least one solution $w \in \mathscr{D}'(\Omega)$; then the difference $u - w$, also taking into account the fact that $\mathscr{D}'(\Omega) \subset \mathscr{D}'_{M_k}(\Omega)$ (or $\mathscr{B}'_{M_k}(\Omega)$), satisfies the conditions:

$$u - w \in \mathscr{D}'_{M_k}(\Omega) \ \left(\text{or } \mathscr{B}'_{M_k}(\Omega)\right),$$

$$A(u - w) = 0, \ \gamma_j(u - w) = 0, \ j = 0, \ldots, m - 1.,$$

and therefore, thanks to Theorem 4.2 (or to (4.28)), $u - w \in N$ and therefore $u \in \mathscr{D}'(\Omega)$. □

This Corollary thus reduces the problem of the regularity in the interior of Ω, Ω *now being an arbitrary open set in* $\mathbf{R}^n$, *of solutions of a properly elliptic equation in the sense of ultra-distributions of Gevrey or of Beurling type to that of ordinary distributions; it follows for example, also applying Theorem* 3.2 *of Chapter* 2 *and Corollary* 1.1 *of this Chapter, that if* f *is analytic in* Ω, *then* u *is also analytic in* Ω. □

5. Comments

Analyticity "*in the interior*" of solutions of linear elliptic equations with analytic coefficients (Corollary 2.1 with $\beta = 1$) is a classical result, shown by E. Picard and S. Bernstein for second-order equations and by Petrowski [1] for equations of arbitrary order; for other proofs, see also F. John [1], Morrey-Nirenberg [1]. For operators with constant coefficients this property is *characteristic* for elliptic operators (see Petrowski [1]).

Analyticity *up to the boundary* of solutions of elliptic boundary value problems (Corollary 1.1 with $\beta = 1$) is due to Morrey-Nirenberg [1].

For the extensions to Gevrey spaces, either "in the interior" or "at the boundary" (Corollaries 1.1 and 2.1 with $\beta = 1$ or $\{M_k\}$ arbitrary), see A. Friedman [2, 3], Murthy [1]. For Beurling spaces, regularity in the interior has been shown by Bjorck [1] and O. John [1].

The theorem on *elliptic iterates* was first considered for the analytic case and *in the interior* of the domain (Theorem 2.4 with $\beta = 1$); particular cases have been studied by Aronszajn [1], Nelson [1], Komatsu [1, 3], the general result being given by Kotake-Narashiman [1]. The extension to spaces of class M_k and *up to the boundary* in the general form of Theorem 1.2 is shown here for the first time; the method of proof takes its inspiration from that of Morrey-Nirenberg [1] and we also use techniques of Kotake-Narashiman [1]; for the case of Dirichlet boundary conditions, see Lions-Magenes [3,4] (see also Roumieu [2], Theorem 5, where the theorem *in the interior* is shown for the Laplace operator). For result (2.7) (theorem on elliptic iterates in the interior in Beurling spaces), see O. John [1].

For a more abstract point of view ("analytic domination" of operators of a Hilbert or Banach space), see Nelson [1], Goodman [1].

Let us finally call attention to the numerous generalizations of regularity results in analytic or Gevrey classes, which have recently been obtained for operators which are "close" to elliptic operators (but which we have not studied in this text):

a) elliptic differential *systems*, see Morrey-Nirenberg [1], Morrey [1], Petrowski [1], A. Friedman [2];

b) *hypoelliptic*, and in particular *quasi-elliptic*, differential operators and hypoelliptic convolution operators, see Hörmander [1, 2], Friberg [1], Pini [1, 2], Volevic [1], Cavallucci [1, 2], Matsuzawa [1, 2, 3], A. Friedman [1], Shilov [1], ... (see also Chapter 10); in particular, let us call attention to the work of Matsuzawa [1, 2], who generalizes Theorem 1.2 on elliptic iterates to quasi-elliptic operators;

c) *pseudo-differential* operators, see Hörmander [3], Boutet de Monvel [1, 2], Boutet de Monvel-Krée [1], Krée [2], ...

The results of Section 3 (application of transposition and non-homogeneous boundary value problems in $\mathscr{D}'(\Omega)$) are given by the authors in [1]; for the generalization to the spaces $\mathscr{D}'_{M_k}(\Omega)$ given in Section 4, see Lions-Magenes [5] (and also Magenes [3, 4]). The property of continuity of traces on surfaces neighboring Γ (Section 3.7) takes its pattern from the formulation of boundary conditions given by Cimmino [1, 2], who characterizes the "traces" on Γ of harmonic functions in Ω by different methods; for the Dirichlet problem for the Laplace operator, see also Johnson [1, 2], Simon [1], J. Douglas [1].

The "traces" of solutions of the homogeneous equation $Au = 0$ and the boundary value problems $Au = 0$, $B_j u = g_j$, have been studied recently, within the framework of the hyperfunctions of Satô, by Komatsu [4] for equations with constant coefficients and by Shapira [4, 5] for equations with variable coefficients.

A generalization of the theory of non-homogeneous boundary value problems in the space $\mathscr{D}'(\Omega)$, when the boundary of Ω is not analytic and may even be highly irregular, is given by D. G. Schaeffer [1].

We also call attention to the work of Krée [1], who, by use of pseudo-differential operators, studies non-homogeneous boundary value problems with *distribution* boundary data on Γ.

Non-homogeneous boundary value problems for analytic pseudo-differential operators have been studied by Boutet de Monvel [1, 2].

Concerning regularity in the interior of *ultra-distribution* solutions of elliptic equations (Section 4.4), see Lions-Magenes [5] (and also Magenes [3, 4]) for Corollary 4.1; by more direct methods, without using the boundary value problems, the problem has been studied for operators with constant coefficients by: Chou [1] for Gevrey ultra-distributions, Björck [1] for Beurling ultra-distributions, Bengel [2, 3], Harvey [1], Komatsu [2, 4] for Satô hyperfunctions, Silva [3] for Silva ultra-distributions; Boutet de Monvel and Krée [1] have solved the problem for operators with variable coefficients and also for certain pseudo-differential operators. Recently Schapira [4], using the boundary value problems, has also shown the regularity in the interior of hyperfunction solutions of elliptic equations with analytic coefficients.

An application of the results of Section 3 to singular perturbations has been given by D. Huet [1].

Let us finally point out a different approach to the problems studied in this Chapter, which is due to Baouendi and Geymonat [1]: viewed as an operator of $\mathscr{D}(\Omega) \to \mathscr{D}(\Omega)$, A is evidently not an isomorphism; but it can be completed with an operator K of $(\mathscr{H}(\Gamma))^m \to \mathscr{D}(\Omega)$ so that the operator $\{A, K\}$ defined by

$$\{A, K\} \{u, \varphi\} = Au + K\varphi$$

is an isomorphism of $\mathscr{D}(\Omega) \times (\mathscr{H}(\Gamma))^m \to \mathscr{D}(\Omega)$.

The isomorphism $\{A, K\}$ is then *transposed* and the problem is to interpret the transposed isomorphism. If one would like to obtain a concrete interpretation (with trace theorems) of this transposed isomorphism, then one must of course introduce spaces analogous to the spaces K of Section 3.2; for details, see Baouendi and Geymonat, *loc. cit.*

6. Problems

6.1. As we have noted in Remark 2.6, is it possible to suppress conditions (1.9)—(1.11) on the sequence $\{M_k\}$ in the hypotheses of Theorem 1.2?

6.2. In what way can Theorem 1.2 on elliptic iterates be extended to the case $M_k = (k!)^\beta$, with $0 < \beta < 1$? This problem is also related to the analytic characterization of the spaces $\mathrm{D}(A^\infty; M_k)$, with for example $M_k = k!$, which we shall discuss in Chapters 9, Section 7 and 10, Section 3; for the one-dimensional case, see Lions-Magenes [2], Section 7.

This problem is also related to polyharmonic functions in the sense of Aronszajn [1, 2].

6.3. Generalization up to the boundary of the result of O. John [1] in Beurling spaces, see Remark 2.7.

6.4. Examples of spaces of the type $K(\Omega)$, $K_{M_k}(\Omega)$ (see (3.12) and (4.7)), different from $\Xi(\Omega)$ and $\Xi_{M_k}(\Omega)$, also as a function of the given boundary operators $\{B_j\}$. It would also be of interest to study the union of the spaces $K'(\Omega)$ for all "admissible" $K(\Omega)$ (see Baouendi and Geymonat [1]).

6.5. Is it possible to interpolate between the results of Section 1 and those of Sections 3 and 4? For interpolation between spaces of infinitely differentiable functions and spaces of analytic functions, see C. Goulaouic [1, 2] and Remark 2.9.

6.6. Non-homogeneous boundary value problems in unbounded open sets, for example in $\Omega = \mathbf{R}^n_+$. This topic offers the opportunity to develop a very interesting entire chapter (see Schapira [5], and Manaresi [1] for a very particular case).

6.7. Non-homogeneous boundary value problems in spaces of ultra-distributions different from those considered in this text, for example, for the hyperfunctions of Satô with $f \neq 0$.

6.8. Can the normality condition on the operators $\{B_j\}$ be eliminated? (certain results are already given in Krée [1], Boutet de Monvel [1] and Komatsu [4]).

6.9. Generalization of this Chapter's theory to elliptic *systems* (see Problem 2.1 of Chapter 2).

6.10. The case of open sets whose boundaries are varieties of dimension less than $n - 1$ (see Problem 2.9 of Chapter 2).

6.11. The study of transmission problems in spaces of ultra-distributions.

6.12. To which classes of elliptic operators which *degenerate or have singularities on the boundary or in the interior* can the theory of this chapter be extended? For results related to this type of operators, in Sobolev spaces, see Baouendi [1], Baouendi-Goulaouic [1], Derridj [1], Oleinik [1], Shimakura [1, 2], Vishik [1] and Zuili [1].

6.13. Let A be an elliptic operator of order $2m$ with C^∞-coefficients but not of class $\{M_k\}$, the boundary Γ being only of class C^∞. What can be said about the functions u such that (1.17), (1.18) hold? If $M_k = k!$, is the class of these functions quasi-analytic? For one-dimensional cases, questions of this type have been studied by M. K. Fage [1] and V. G. Kriptun [1].

Chapter 9

Evolution Equations in Spaces of Distributions and Ultra-Distributions

This Chapter makes use of the *vector*-valued distributions and ultra-distributions of Chapter 7. From the point of view of evolution boundary value problems, we assume the knowledge of the essentials of Chapter 3 and of the beginning of Chapter 4.

The aim of this Chapter is to see whether the solutions of linear evolution equations belong to M_k-classes, if the data belong to M_k-classes. (Once in possession of such results, by transposition and trace theorems, we "automatically" obtain well-posed non-homogeneous problems in spaces of ultra-distributions.)

In order to perform this study, we reconsider the various methods for evolution problems:
variational (or energy) *methods* (Sections 1, 2, 4, 5, 6, 9, 10), *the Laplace transform method* (Section 8), *the semi-group method,* in the usual sense (Section 7) or in the sense of distributions and ultra-distributions (Section 11), *the method of transmutations* (Section 3), which allows for the study of problems *with singularities* in M_k-classes.

1. Regularity Results. Equations of the First Order in t

1.1 Orientation and Notation

We first consider the *variational* setting corresponding to Chapter 3 (Volume 1) and give *regularity in t* results and, in particular, *M_k-regularity in t* results.

Let V and H be two Hilbert spaces on $\mathbf{C}$, with

(1.1) $V \subset H$, V is dense in H, the injection $V \to H$ being continuous.

Identify H with its antidual so that, if V' denotes the antidual of V, we have

$$V \subset H \subset V'. \tag{1.2}$$

We denote by $\| \|$, $| |$ and $\| \|_*$ the norms in V, H and V' respectively. In case of ambiguity $\| \|_X$ shall denote the norm in X. $\square$

Let

(1.3) $$u, v \to a(t; u, v), \quad t \in \mathbf{R},$$

be a family of sesquilinear forms which are continuous on $V \times Y$ and such that we always have

(1.4) $\forall u, v \in V$, the function $t \to a(t; u, v)$ is infinitely differentiable.

We also assume (*coerciveness*):

(1.5) $$\begin{cases} \text{for all } T < \infty, \text{ there exist } \lambda(T) = \lambda \text{ and} \\ \alpha(T) = \alpha > 0, \text{ such that} \\ \operatorname{Re} a(t; v, v) + \lambda |v|^2 \geq \alpha \|v\|^2, \ \forall v \in V \text{ and } t \leq T. \end{cases}$$

Let $A(t) \in \mathscr{L}(V; V')$ be the operator defined by the triplet $\{a(t; u, v), V, H\}$ (see Chapter 2, Section 9); thus

(1.6) $$a(t; u, v) = (A(t) u, v), \quad \forall u, v \in V,$$

where (f, v) denotes the scalar product (linear in f, antilinear in v) between $f \in V'$ and $v \in V$.

We aim to study the t-regularity properties of the equation

(1.7) $$A(t) u + \frac{du}{dt} = f,$$

where the supports of f and u in t are bounded on the left.

1.2 Regularity in the Spaces $\mathscr{D}_+$

In general, X being (for example) a Banach space, we denote by $\mathscr{D}_+(\mathbf{R}; X)$ the space of functions

$$t \to \varphi(t)$$

which are infinitely differentiable mappings of $\mathbf{R} \to X$, and whose support is bounded on the left, i.e.

$$\varphi(t) = 0 \quad \text{for} \quad t \leq t_\varphi, \quad t_\varphi \text{ depending on } \varphi.$$

If we set

(1.8) $$\mathscr{D}_a([a, \infty[; X) = \{\varphi \mid \varphi \in \mathscr{D}_+(\mathbf{R}; X), \ \varphi(t) = 0 \ \text{ if } \ t \leq a\},$$

provided with the topology of uniform convergence in X on every compact set $[a, T]$, variable T, of the functions and each of their derivatives, then

$$\mathscr{D}_+(\mathbf{R}; X) = \operatorname*{ind\,lim}_{a \to -\infty} \mathscr{D}_a([a, \infty[; X) \tag{1.9}$$

(see Remark 4.3, Chapter 7.) □

Our first regularity in t result is

Theorem 1.1. *Assume that hypotheses* (1.4) *and* (1.5) *hold. Then the mapping*

$$u \to A(t)\, u + \frac{\mathrm{d}u}{\mathrm{d}t}$$

is an isomorphism of $\mathscr{D}_+(\mathbf{R}; V) \to \mathscr{D}_+(\mathbf{R}; V')$.

Proof. 1) It is immediate that $u \to A(t)\, u + \dfrac{\mathrm{d}u}{\mathrm{d}t}$ is a continuous linear mapping of $\mathscr{D}_+(\mathbf{R}; V) \to \mathscr{D}_+(\mathbf{R}; V)'$. Therefore we only need to show that if in (1.7) the function f belongs to $\mathscr{D}_+(\mathbf{R}; V')$ then u belongs to $\mathscr{D}_+(\mathbf{R}; V)$ and depends continuously on f.

We assume that $f \in \mathscr{D}_0([0, \infty[; V')$ (which is not restrictive, since 0 plays no special role on $\mathbf{R}_t$). We shall show, and this will be sufficient, that $u \in \mathscr{D}_0([0, \infty[; V)$ and depends continuously on f in this space.

Now f belongs, in particular, to $L^2(0, T; V')$, arbitrary finite T; therefore, according to Chapter 3 (Volume 1), there exists a unique u which is a solution of (1.7) and satisfies

$$u \in L^2(0, T; V), \quad u' = \frac{\mathrm{d}u}{\mathrm{d}t} \in L^2(0, T; V'), \; u(0) = 0, \tag{1.10}$$

and

$$\|u\|_{L^2(0,T;V)} \leq c\, \|f\|_{L^2(0,T;V')} \tag{1.11}$$

(where c depends on T).

Since $u(0) = 0$, we may identify u defined for $t \geq 0$ with its extension by 0 for $t < 0$, equation (1.7) being valid on $t > 0$ or on $\mathbf{R}$.

2) Now we apply the method of *differential quotients* (which we have already used previously on several occasions). In general, we set

$$g_h(t) = \frac{1}{h}\,[g(t) - g(t-h)], \quad h > 0; \tag{1.12}$$

it follows from (1.7) (valid on $\mathbf{R}$) that

$$A(t-h)\, u(t-h) + \frac{\mathrm{d}}{\mathrm{d}t} u(t-h) = f(t-h), \tag{1.13}$$

hence that

$$A(t)\, u_h + u_h' = f_h - A_h(t)\, u(t-h), \tag{1.14}$$

and (1.11) applies to (1.14) (considered as an equation in u_h); it follows that

(1.15) $$\|u_h\|_{L^2(0,T;V)} \leq c\,\|f_h\|_{L^2(0,T;V')} + c\,\|A_h u(t-h)\|_{L^2(0,T;V')}.$$

But the c's denoting various constants, it then follows from the fact that $f \in \mathscr{D}_0([0,\infty[;\, V')$ and from (1.4) that

$$\|f_h\|_{L^2(0,T;V')} \leq c\,\|f'\|_{L^2(0,T;V')},$$

$$\|A_h(t)\|_{\mathscr{L}(V;V')} \leq c\,\|A'(t)\|_{\mathscr{L}(V;V')},$$

so that (1.15) yields:

(1.16) $$\begin{cases} \|u_h\|_{L^2(0,T;V)} \leq c, \\ c = \text{constant, independent of } h. \end{cases}$$

Thus u_h remains in a bounded set of $L^2(0, T;\, V)$ as $h \to 0$, and since $u_h \to u'$ in $\mathscr{D}'(]0, T[;\, V)$, it follows that $u' \in L^2(0, T;\, V)$ and that (passing to the limit in (1.14)):

(1.17) $$A(t)\,u' + A'(t)\,u + u'' = f',$$

equality on $\mathbf{R}$ (or on $\{t > 0\}$, noting that $u'(0) = 0$).

Furthermore

(1.18) $$\|u'\|_{L^2(0,T;V)} \leq c(\|f\|_{L^2(0,T;V')} + \|f'\|_{L^2(0,T;V')}).$$

Of course we can iterate on this procedure. It follows that $u \in \mathscr{D}_0([0,\infty[;\, V)$ and that, for all k, we have

(1.19) $$u^{(k+1)} + \sum_{j=0}^{k} \binom{k}{j} A^{(k-j)}(t)\, u^{(j)}(t) = f^{(k)}(t)$$

$\left(\text{where } \varphi^{(j)} = \dfrac{\mathrm{d}^j}{\mathrm{d}t^j}\varphi\right)$.

Since finally we have

(1.20) $$\|u^{(k)}\|_{L^2(0,T;V)} \leq c_{k,T}(\|f\|_{L^2(0,T;V')} + \cdots + \|f^{(k')}\|_{L^2(0,T;V')}),$$

we see that

$$f \to u$$

is a continuous mapping of $\mathscr{D}_0([0,\infty[;\, V') \to \mathscr{D}_0([0,\infty[;\, V)$ [(1)]. ☐

[(1)] Indeed (see Remark 4.3, Chapter 7), $\mathscr{D}_0([0,\infty[;\, X) = \text{proj lim}\ \mathscr{D}_0([0, T];\, X)$, and we can provide $\mathscr{D}_0([0, T];\, X)$ with the norms $\sum_{j=0}^{k} \|\varphi^{(j)}\|_{L^2(0,T;X)}$.

1.3 Regularity in the Spaces $\mathscr{D}_{+,M_k}$

We now make the hypothesis

(1.21) $\qquad t \to A(t)$ *belongs to the class* $\mathscr{E}_{M_k}(\mathbf{R}; \mathscr{L}(V; V'))$,

$\mathscr{E}_{M_k}(\mathbf{R}; F)$ being defined in Chapter 7.

The sequence M_k satisfies the usual hypotheses (see (1.1), ..., (1.4), Chapter 7), but we recall them here:

(1.22) $\qquad$ the sequence M_k is logarithmically convex,

(1.23) $\qquad$ the sequence M_k is non-quasi-analytic:

$$\sum_{k=1}^{\infty} \frac{M_{k-1}}{M_k} < \infty,$$

(1.24) $$M_{k+1} \leq H^k M_k \quad \forall k,$$

and

(1.25) $$\binom{k}{j} M_{k-j} M_j \leq c_1 M_k \quad \forall k \text{ and } \forall j \text{ with } 0 \leq j \leq k.$$

We shall prove the *theorem on M_k-regularity in t*:

Theorem 1.2. *Assume that $a(t; u, v)$ satisfies* (1.5) *and* (1.21) *and that the sequence $\{M_k\}$ satisfies* (1.22), ..., (1.25). *Then the mapping*

(1.26) $$u \to u' + A(t) u$$

is an isomorphism of $\mathscr{D}_{+,M_k}(\mathbf{R}; V)$ onto $\mathscr{D}_{+,M_k}(\mathbf{R}; V')$.

Proof. 1) We shall first show (and this is the essential point) that for arbitrary finite T, if $f \in \mathscr{D}_{0,M_k}([0, T], \mathscr{L}; V')$ (see (4.6), Chapter 7), then $u \in \mathscr{D}_{0,M_k}([0, T], B; V)$ (for a suitable B which we shall specify), where u is the solution of (1.7).

Let us make this precise.

We are led (by the change of variable $u \to e^{\lambda t} u$, which does not affect the eventual property of belonging to a class $\{M_k\}$) to the case where

(1.27) $$\operatorname{Re} a(t; v, v) \geq \alpha \|v\|^2, \quad \forall v \in V, \ \forall t \in [0, T].$$

Generally, for $\varphi \in L^2(0, T; V)$ (resp. $\varphi \in L^2(0, T; H)$, resp. $\varphi \in L^2(0, T; V')$), we set

(1.28) $$\begin{cases} \nu_1(\varphi) = \left(\int_0^T \|\varphi(t)\|^2 \, dt \right)^{1/2}, \\ \nu_0(\varphi) = \left(\int_0^T |\varphi(t)|^2 \, dt \right)^{1/2}, \\ \nu_{-1}(\varphi) = \left(\int_0^T \|\varphi(t)\|_*^2 \, dt \right)^{1/2}. \end{cases}$$

By hypothesis (and according to Chapter 7, Section 5.5):

$$\nu_{-1}(f^{(k)}) \le d\mathscr{L}^k M_k \quad \forall k. \tag{1.29}$$

According to (1.21):

$$\| A^{(k)}(t) \|_{\mathscr{L}(V;V')} \le cL^k M_k \quad \forall k, \forall t \in [0, T]. \tag{1.30}$$

We shall show that

$$\nu_1(u^{(k)}) \le \delta B^k M_k \quad \forall k, \tag{1.31}$$

with

$$\delta = \frac{2d}{\alpha}, \quad B = \max\left(\left(1 + \frac{2cc_1}{\alpha}\right) L, \mathscr{L}\right). \tag{1.32}$$

It follows from (1.7) that (see Chapter 3)

$$\int_0^T \operatorname{Re} a\left(t; u(t), u(t)\right) \mathrm{d}t + \frac{1}{2} |u(T)|^2 = \operatorname{Re} \int_0^T \left(f(t), u(t)\right) \mathrm{d}t$$

from which

$$\alpha \nu_1(u)^2 \le \nu_1(u)\, \nu_{-1}(f)$$

and therefore

$$\nu_1(u) \le \frac{1}{\alpha} \nu_{-1}(f). \tag{1.33}$$

Thus (1.31) holds for $k = 0$. Let us assume that it holds up to $(k - 1)$. Note that (1.19) may be written

$$\frac{\mathrm{d}}{\mathrm{d}t} u^{(k)} + A(t)\, u^{(k)} = f^{(k)}(t) - \sum_{j=0}^{k-1} \binom{k}{j} A^{(k-j)}(t)\, u^{(j)}(t) \tag{1.34}$$

so that, applying (1.33) to equation (1.34), we obtain:

$$\nu_1(u^{(k)}) \le \frac{1}{\alpha} \nu_{-1}(f^{(k)}) + \frac{1}{\alpha} \sum_{j=0}^{k-1} \binom{k}{j} \nu_{-1}(A^{(k-j)} u^{(j)}). \tag{1.35}$$

According to (1.29), we have

$$\frac{1}{\alpha} \nu_{-1}(f^{(k)}) \le \frac{d}{\alpha} \mathscr{L}^k M_k \le \left(\text{by (1.32)}\right) \frac{\delta}{2} B^k M_k.$$

Therefore in order to obtain (1.31), we must show that

$$\xi = \frac{1}{\alpha} \sum_{j=0}^{k-1} \binom{k}{j} \nu_{-1}(A^{(k-j)} u^{(j)}) \le \frac{\delta}{2} B^k M_k. \tag{1.36}$$

But

$$\nu_{-1}(A^{(k-j)} u^{(j)}) \le \left(\text{according to (1.30)}\right) cL^{k-j} M_{k-j} \nu_{-1}(u^{(j)}) \le$$
$$\le \text{(by the induction hypothesis) } c\,\delta L^{k-j} B^j M_{k-j} M_j,$$

and therefore

$$\xi \leq \frac{c\delta}{\alpha} \sum_{j=0}^{k-1} \binom{k}{j} M_{k-j} M_j L^{k-j} B^j \leq$$

$$\leq \text{(according to (1.25))} \frac{cc_1\delta}{\alpha} M_k \sum_{j=0}^{k-1} L^{k-j} B^j ,$$

so that (1.36) holds if

$$\frac{cc_1}{\alpha} \sum_{j=0}^{k-1} \left(\frac{L}{B}\right)^{k-j} \leq \frac{1}{2}. \tag{1.37}$$

Now

$$\sum_{j=0}^{k-1} \left(\frac{L}{B}\right)^{k-j} \leq \frac{1}{\frac{B}{L} - 1}$$

and (1.37) follows from $B \geq \left(1 + \frac{2cc_1}{\alpha}\right) L$.

2) We recall that the space $\mathscr{D}_{0,M_k}([0, T], B; V)$ (for example) is a *Banach space* with the norm

$$\sup_k \frac{1}{B^k M_k} v_1(u^{(k)}),$$

so that, by the closed graph theorem (or by direct inspection of the estimates in 1)) the mapping

$$f \to u$$

is *continuous* from

$$\mathscr{D}_{0,M_k}([0, T], L; V') \to \mathscr{D}_{0,M_k}([0, T], B; V).$$

By passage to the inductive limit in B and then in L, we see that $f \to u$ is a continuous mapping of

$$\mathscr{D}_{0,M_k}([0, T]; V') \to \mathscr{D}_{0,M_k}([0, T]; V).$$

By passage to the projective limit in T, we obtain the continuity from

$$\mathscr{D}_{0,M_k}([0, +\infty[; V') \to \mathscr{D}_{0,M_k}([0, +\infty[; V);$$

as we have already noted, 0 plays no particular role on $\mathbf{R}_t$; therefore if $f \in \mathscr{D}_{a,M_k}([a, +\infty[; V')$, then $u \in \mathscr{D}_{a,M_k}([a, +\infty[; V)$ and the mapping $f \to u$ is continuous in these spaces. By passage to the inductive limit in a $(\to -\infty)$, we finally obtain the continuity of the mapping $f \to u$ from $\mathscr{D}_{+,M_k}(\mathbf{R}; V') \to \mathscr{D}_{+,M_k}(\mathbf{R}; V)$. ☐

1.4 Regularity in Beurling Spaces

We define (see Chapter 7, Section 5.4) the Beurling space

(1.38)
$$\mathscr{B}_{+,M_k}(\mathbf{R}; X) = \operatorname*{ind\,lim}_{a\to-\infty}\left(\operatorname*{proj\,lim}_{b\to+\infty}\left(\operatorname*{proj\,lim}_{L\to 0} \mathscr{D}_{a,M_k}([a, b], L; X)\right)\right).$$

A natural question to ask is whether Theorem 1.2 holds in Beurling classes; i.e.: *for f belonging to $\mathscr{B}_{+,M_k}(\mathbf{R}; V')$, do we have $u \in \mathscr{B}_{+,M_k}(\mathbf{R}; V)$?* □

We must of course *modify* the hypotheses on $a(t; u, v)$. We do not know the "optimal" hypotheses under which the desired result is true; we shall only give sufficient conditions.

Let us introduce a *second sequence* $\{M_k^*\}$ with:

(1.39) M_k^* satisfies hypotheses analogous to (1.22), (1.23) and (1.24).

Next we assume that the sequences $\{M_k\}$ and $\{M_k^*\}$ satisfy

$$(1.40) \qquad \begin{cases} \binom{k}{j} M_{k-j}^* M_j \leq \varepsilon_k M_k, \quad \forall k, \; 0 \leq j \leq k-1, \\ \varepsilon_k \to 0 \;\text{ if }\; k \to \infty. \end{cases}$$

We shall prove

Theorem 1.3. *Assume that the sequences $\{M_k\}$ and $\{M_k^*\}$ satisfy* (1.22), (1.23), (1.24), (1.39) *and* (1.40). *Let $a(t; u, v)$ satisfy* (1.5) *and*

$$(1.41) \qquad \begin{cases} \textit{for every compact set } K \textit{ in } \mathbf{R}, \forall L, \textit{ there exists} \\ \textit{a } c \textit{ such that} \\ \| A^{(k)}(t) \|_{\mathscr{L}(V; V')} \leq c L^k M_k^*, \quad \forall t \in K, \; \forall k. \end{cases}$$

Then $u \to A(t)u + u'$ is an isomorphism of $\mathscr{B}_{+,M_k}(\mathbf{R}; V)$ onto $\mathscr{B}_{+,M_k}(\mathbf{R}; V')$.

Proof. We assume that we have reduced the problem to the case for which (1.27) holds. We see (as in the proof of Theorem 1.2) that it is sufficient to show: under the hypotheses of the theorem, and *B being a fixed positive number, there exists δ such that*

$$(1.42) \qquad \nu_1(u^{(k)}) \leq \delta B^k M_k, \; \forall k.$$

Now, by hypothesis, there exist c and d such that

$$(1.43) \qquad \nu_{-1}(f^{(k)}) \leq d B^k M_k, \; \forall k,$$

and

$$(1.44) \qquad \| A^{(k)}(t) \|_{\mathscr{L}(V; V')} \leq c \left(\frac{B}{2}\right)^k M_k^* \; \forall k, t \in [0, T].$$

According to (1.40), we can find k_0 such that

$$\binom{k}{j} M^*_{k-j} M_j \leq \frac{\alpha}{2c} M_k \quad \text{for} \quad k \geq k_0,\ 0 \leq j \leq k-1. \tag{1.45}$$

We shall show (1.42) with

$$\delta = \max\left[\max_{0 \leq j \leq k_0} \frac{1}{L^j M_j} \nu_1(u^{(j)}), \frac{2d}{\alpha}\right], L = B/2. \tag{1.46}$$

Of course, according to (1.46), (1.42) holds for $j \leq k_0$. Let us assume that it holds up to $k-1$ ($k-1 > k_0$) and show that it holds for k. As in the proof of Theorem 1.2, we have:

$$\nu_1(u^{(k)}) \leq \frac{1}{\alpha} \nu_{-1}(f^{(k)}) + \frac{1}{\alpha} \sum_{j=0}^{k-1} \binom{k}{j} \nu_{-1}(A^{(k-j)} u^{(j)}). \tag{1.47}$$

But according to (1.43),

$$\frac{1}{\alpha} \nu_{-1}(f^{(k)}) \leq \frac{d}{\alpha} B^k M_k \leq (\text{by } (1.46)) \frac{\delta}{2} B^k M_k,$$

and thus (1.42) holds if

$$\frac{1}{\alpha} \sum_{j=0}^{k-1} \binom{k}{j} \nu_{-1}(A^{(k-j)} u^{(j)}) \leq \frac{\delta}{2} B^k M_k. \tag{1.48}$$

But

$$\begin{aligned} \frac{1}{\alpha} \sum_{j=0}^{k-1} \binom{k}{j} \nu_{-1}(A^{(k-j)} u^{(j)}) &\leq (\text{by } (1.44)) \frac{c}{\alpha} \sum_{j=0}^{k-1} \binom{k}{j} L^{k-j} M^*_{k-j} \nu_{-1}(u^{(j)}) \leq \\ &\leq (\text{by induction}) \frac{c}{\alpha} \delta \sum_{j=0}^{k-1} \binom{k}{j} L^{k-j} B^j M^*_{k-j} M_j \leq \\ &\leq (\text{by } (1.45)) \frac{c\delta}{\alpha} \frac{\alpha}{2c} B^k M_k \sum_{j=0}^{k-1} \left(\frac{L}{B}\right)^{k-j} \leq \\ &\leq \left(\text{since } \frac{L}{B} = \frac{1}{2}\right) \frac{\delta}{2} B^k M_k, \end{aligned}$$

whence (1.48). □

1.5 First Applications

Let $H = L^2(\Omega)$ and let V be a closed vector subspace of $H^m(\Omega)$ (notation of Chapter 1, Volume 1) with

$$H_0^m(\Omega) \subset V \subset H^m(\Omega). \tag{1.49}$$

Let $a_{p,q}(x, t)$ be given functions with $|p|, |q| \leq m$, $x \in \Omega$, $t \in \mathbf{R}$, satisfying

$$t \to a_{p,q}(., t) \text{ belong to } \mathscr{E}_{M_k}(\mathbf{R}; L^\infty(\Omega)). \tag{1.50}$$

For $u, v \in V$, set

$$a(t; u, v) = \sum_{|p|,|q| \leq m} \int_\Omega a_{p,q}(x, t)\, \mathrm{D}^q u\, \overline{\mathrm{D}^p v}\, \mathrm{d}x\,, \tag{1.51}$$

and assume (1.5) to hold.

From (1.50), it follows that for every compact set $[t_0, t_1] \subset \mathbf{R}$, there exist c and L such that

$$\left\| \frac{\partial^k}{\partial t^k} a_{p,q}(.\,, t) \right\|_{L^\infty(\Omega)} \leq c L^k M_k \quad \forall k, \forall t \in [t_0, t_1],$$

and consequently

$$|(A^{(k)}(t)\, u, v)| \leq c L^k M_k \sum_{|p|,|q| \leq m} \|\mathrm{D}^q u\|_{L^2(\Omega)} \|\mathrm{D}^p v\|_{L^2(\Omega)} \leq \tilde{c} L^k M_k \|u\| \, \|v\|$$

(where $\| \ \|$ denotes the norm in V), *so that* (1.21) *holds.*

Remark 1.1. If we assume that

$$t \to a_{p,q}(.\,, t) \text{ belongs to } \mathscr{B}_{M_k^*}(\mathbf{R}; L^\infty(\Omega)), \tag{1.52}$$

then (1.41) holds. ☐

Therefore, under hypothesis (1.50) (*resp.* (1.52)), *Theorem* 1.2 (*resp.* 1.3) *holds.*

Remark 1.2. Theorems 1.2 and 1.3 yield results on regularity (in $\{M_k\}$ classes) *in the variable t.* With hypotheses (1.49), (1.50), alone, *there are no* regularity results in the variable $x \in \Omega$; see Chapter 4 and Remark 1.4 below. ☐

Remark 1.3. Generally, for the applications, one takes "$H = L^2(\Omega)$". We could also consider other spaces, as for example $H = H_0^1(\Omega)$ (see Chapter 3, Section 4.7.5). Thus Theorem 1.2, together with the remarks of Chapter 3, Section 4.7.5, yields the following result: let f be given in $\mathscr{D}_{+,M_k}(H_0^1(\Omega))$ and let u be the solution, with support in t bounded on the left, of

$$\frac{\partial u}{\partial t} + \Delta^2 u = f, \tag{1.53}$$

$$u = 0, \ \frac{\partial \Delta u}{\partial n} = 0 \ \text{ on } \Gamma \times \mathbf{R}_t; \tag{1.54}$$

then

$$u \in \mathscr{D}_{+,M_k}(V), \tag{1.55}$$

where

$$V = \left\{ v \mid v \in H_0^1(\Omega), \frac{\partial}{\partial x_i} \Delta v \in L^2(\Omega), i = 1, \ldots, n \right\}. \quad \square$$

Remark 1.4. Assume that we have a result on the regularity in the variable x of the following type:

$$(1.56) \quad \begin{cases} \text{if } f \in H^s(\Omega), \text{ integer } s \geq 0^{((1))} \text{ and if } w(t) \text{ is the solution in } V \text{ of} \\ a\big(t; w(t), v\big) = (f, v), \ \forall v \in V^{((2))}, \\ \text{then } w(t) \in H^{s+2m}(\Omega) \cap V \text{ and} \\ \|w(t)\|_{H^{s+2m}(\Omega)} \leq \beta_s \|f\|_{H^s(\Omega)}, \ t \in [0, T]. \end{cases}$$

This hypothesis is satisfied if the coefficients $a_{p,q}$ and the boundary Γ of Ω are sufficiently regular and if V is defined by "regular" differential boundary conditions (see Chapter 2, Section 9.7). We also assume that

$$(1.57) \qquad t \to A(t) \text{ belongs to } \mathscr{E}_{M_k}\big(\mathbf{R}; \mathscr{L}(H^{s+2m}(\Omega); H^s(\Omega))\big).$$

Then we have

Theorem 1.4. *Assume that* $A(t)$ *satisfies* (1.5), (1.56) *and* (1.57) *and that the sequence* $\{M_k\}$ *satisfies* (1.22), ..., (1.25). *Then the solution* u, *with support in* t *bounded on the left, of*

$$(1.58) \qquad A(t)\, u + u' = f, \quad f \in \mathscr{D}_{+,M_k}\big(H^s(\Omega)\big),$$

of which we already know that it belongs to $\mathscr{D}_{+,M_k}(V)$ *(according to Theorem* 1.2*), belongs to* $\mathscr{D}_{+,M_k}(H^{s+2m}(\Omega))$. *Furthermore*

$$(1.59) \qquad \begin{gathered} f \to u \text{ is a continuous mapping of} \\ \mathscr{D}_{+,M_k}\big(H^s(\Omega)\big) \to \mathscr{D}_{+,M_k}\big(H^{s+2m}(\Omega)\big). \end{gathered}$$

Proof. 1) Let $g \in \mathscr{D}_{0,M_k}([0, T]; H^s(\Omega))$ and $w(t)$ be the solution of

$$(1.60) \qquad a\big(t; w(t), v\big) = \big(g(t), v\big), \ t \leq T.$$

Then

$$(1.61) \qquad w \in \mathscr{D}_{0,M_k}\big([0, T]; H^{s+2m}(\Omega)\big).$$

Indeed, we first show (by the method of differential quotients, as in the proof of Theorem 1.1) that $t \to w(t)$ is an infinitely differentiable

((1)) One could also take non-integer s, and, in certain cases, $s < 0$ (see Chapter 2).

((2)) We assume that we have reduced the problem to the case (1.27) on the compact set, say $[0, T]$, on which t varies.

mapping of $t \leq T \to V$ (or $H^{s+2m}(\Omega)$) and that:

$$a\big(t;\, w^{(k)}(t),\, v\big) = \big(g^{(k)}(t),\, v\big) - \sum_{j=0}^{k-1} \binom{k}{j} a^{(k-j)}\big(t;\, w^{(j)}(t),\, v\big). \tag{1.62}$$

(1.61) follows via estimates analogous to those made in the proof of Theorem 1.2.

2) Let u, in $\mathscr{D}_{+,M_k}(\mathbf{R};\, V)$, be a solution of (1.58). Since then $u' \in \mathscr{D}_{+,M_k}(\mathbf{R};\, V)$, it follows that

$$A(t)\, u = f - u' \in \mathscr{D}_{+,M_k}\big(\mathbf{R};\, H^{\min(s,m)}(\Omega)\big), \tag{1.63}$$

and applying (1.61) (with s replaced by $\min(s, m)$, as the result still holds for this case) it follows that

$$u \in \mathscr{D}_{0,M_k}\big([0,\, T];\, H^{\min(s,m)+2m}(\Omega)\big),$$

therefore u' belongs to the same space and (1.63) yields

$$f - u' \in \mathscr{D}_{0,M_k}\big([0,\, T];\, H^{\min(s,3m)}(\Omega)\big)$$

and therefore

$$u \in \mathscr{D}_{0,M_k}\big([0,\, T];\, H^{\min(s,3m)+2m}(\Omega)\big)$$

and so on, until we obtain the desired result.

2. Equations of the Second Order in t

2.1 Statement of the main Results

The notation is the same as in the preceding Section. Again we assume that

$$a(t;\, u,\, v) = \overline{a(t;\, u,\, v)} \quad \forall u, v \in V. \tag{2.1}$$

We shall prove the following results:

Theorem 2.1. *Assume that hypotheses* (1.4), (1.5) *and* (2.1) *hold. Then the operator*

$$u \to A(t)\, u + \frac{\mathrm{d}^2 u}{\mathrm{d}t^2}$$

is an isomorphism of $\mathscr{D}_+(\mathbf{R};\, V)$ *onto* $\mathscr{D}_+(\mathbf{R};\, V')$.

Theorem 2.2. *Assume that hypotheses* (1.5), (2.1) *and* (1.21) *hold and that the sequence* $\{M_k\}$ *satisfies* (1.22), ..., (1.25). *Then the operator*

$$u \to A(t)\, u + \frac{\mathrm{d}^2 u}{\mathrm{d}t^2}$$

is an isomorphism of $\mathscr{D}_{+,M_k}(\mathbf{R};\, V)$ *onto* $\mathscr{D}_{+,M_k}(\mathbf{R};\, V')$

These theorems will be proved in Sections 2.2 and 2.3.

Remark 2.1. In each of the above theorems, (2.1) may be replaced by a hypothesis of the type: "the principal part" of $a(t; u, v)$ is hermitian, see Lions [5]. ☐

Remark 2.2. We could also consider the operator given by:

$$A(t)\, u + \frac{d^2 u}{dt^2} + \int_0^t N(t, \sigma)\, u(\sigma)\, d\sigma. \tag{2.2}$$

If we assume that

$$t, \sigma \to N(t, \sigma)$$

is of class M_k in both variables $t, \sigma, \sigma \leq t$, with values in $\mathscr{L}(V; V')$, then *Theorem 2.2 is still valid for the operator defined in* (2.2). Theorem 2.1 holds if $N(t, \sigma)$ is of class C^∞. ☐

2.2 Proof of Theorem 2.1

Let f be given in $\mathscr{D}_+(\mathbf{R}; V')$, we may assume that

$$f(t) = 0 \quad \text{for} \quad t \leq 0;$$

e t u be the solution of

$$A(t)\, u + u'' = f, \tag{2.3}$$

with (see Theorem 9.3, Chapter 3), for arbitrary finite T,

$$\begin{cases} u \in L^\infty(0, T; H), \\ u' \in L^\infty(0, T; V'), \end{cases} \tag{2.4}$$

$$u(0) = 0, \quad u'(0) = 0. \tag{2.5}$$

According to (2.5), we may consider that (2.3) holds for $t > 0$ or on $\mathbf{R}$, by extending u by 0 for $t \leq 0$.

We also have

$$\|u\|_{L^\infty(0,T;H)} + \|u'\|_{L^\infty(0,T;V')} \leq c\, \|f\|_{L^2(0,T;V')}, \tag{2.6}$$

c depending on T.

We need to show that $u \in \mathscr{D}_+(\mathbf{R}; V)$.

For this purpose, we apply the finite difference[1] method, as in the proof of Theorem 1.1. In the notation of this proof, we obtain:

$$A(t)\, u_h + u_h'' = f_h - A_h(t)\, u(t - h). \tag{2.7}$$

[1] In order to avoid technical difficulties for the interpretation of (2.3) and (2.7), we may also approximate, in the sense of $\mathscr{D}_+(\mathbf{R}; V')$, f with $f_n \in \mathscr{D}_+(\mathbf{R}; V)$; since the estimates which follow are then *independent of* n, the result will follow. We could also use the method of Faedo-Galerkin (see Section **4**, for the case of Schroedinger equations).

Considering (2.7) as an equation in u_h, we may apply (2.6). We obtain

$$\| u_h \|_{L^\infty(0,T;H)} + \| u_h' \|_{L^\infty(0,T;V')} \leq c \| f_h \|_{L^2(0,T;V')} + c \| A_h(t)\, u(t-h) \|_{L^2(0,T;V')}$$

from which, thanks to the hypotheses on f and $A(t)$, we obtain that

$$\| u_h \|_{L^\infty(0,T;H)} + \| u_h' \|_{L^\infty(0,T;V')} \leq \text{constant},$$

and since $u_h \to u'$ (resp. $u_h' \to u''$) in $\mathscr{D}'(]0, T[; H)$ (resp. $\mathscr{D}'(]0, T[; V')$) it follows that

$$u' \in L^\infty(0, T; H), \; u'' \in L^\infty(0, T; V')$$

and passing to the limit in (2.7):

$$A(t)\, u' + A'(t)\, u + u^{(3)} = f', \tag{2.8}$$

with

$$u'(0) = u''(0) = 0.$$

But we also have:

$$A(t)\, u = f - u'' \in L^2(0, T; V'),$$

whence

$$u \in L^2(0, T; V).$$

By iterating on this procedure, we obtain the theorem. □

2.3 Proof of Theorem 2.2

2.3.1 Reduction of the Problem

Since "$t = 0$" plays no particular role, and on the other hand, the topological problems being solved as at the end of the proof to Theorem 1.2, it all comes down to showing that if $f \in \mathscr{D}_{0,M_k}([0, T]; V')$, then the solution u of (2.3), which belongs to $\mathscr{D}_+(\mathbf{R}; V)$, satisfies

$$u \in \mathscr{D}_{0,M_k}([0, T]; V).$$

We shall verify that everything then reduces to showing the following: let $f \in \mathscr{D}_{0,M_k}([0, T]; V')$ and let u be the solution in $\mathscr{D}_+(\mathbf{R}; V)$ of

$$A(t)\, u + \beta(t)\, u' + u'' = f, \tag{2.9}$$

where

$$\beta \in \mathscr{E}_{M_k}(\mathbf{R}), \beta \geq 0, \tag{2.10}$$

$$(A(t)\, v, v) \geq \alpha \, \|v\|^2, \; \alpha > 0, \; v \in V, \; t \in [0, T], \tag{2.11}$$

$$(A'(t)\, v, v) \leq -\gamma \, \|v\|^2, \; \gamma > 0, \; v \in V, \; t \in [0, T]; \tag{2.12}$$

then

$$u \in \mathscr{D}_{0,M_k}([0, T]; V). \tag{2.13}$$

Indeed, first setting $u = e^{kt}w$, equation (2.3) becomes

$$\big(A(t) + k^2 I\big)\, w + 2kw' + w'' = e^{-kt} f,$$

or, changing the notation, and denoting the operator $A(t) + k^2 I$ by $A(t)$,

$$A(t)\, u + \eta u' + u'' = f, \tag{2.14}$$

where $\eta \in \mathbf{R}_+$, and where (choosing k in an appropriate manner) (2.11) *holds*. But (2.12) is not necessarily satisfied, and in order to obtain a condition of this type (condition which plays an essential role in the estimates which follow) we must make a *change of variable.*

Set

$$s = \exp(\lambda t),\ \lambda > 0 \text{ chosen further on},$$

$$v(s) = u\left(\frac{1}{\lambda}\log s\right).$$

On $s \geq 1$, equation (2.14) becomes:

(2.15)

$$\frac{d^2}{ds^2}v + \left(\frac{1}{s} + \frac{\alpha}{\lambda s}\right)\frac{d}{ds}v(s) + \frac{1}{\lambda^2 s^2}A\left(\frac{1}{\lambda}\log s\right)v(s) = \frac{1}{\lambda^2 s^2}f\left(\frac{1}{\lambda}\log s\right),$$

$$v(1) = 0,\ \frac{d}{ds}v(1) = 0. \tag{2.16}$$

But

$$\frac{d}{ds}\left(\frac{1}{\lambda^2 s^2}A\left(\frac{1}{\lambda}\log s\right)\right) = \frac{1}{\lambda^2 s^3}\left[-2A\left(\frac{1}{\lambda}\log s\right) + \frac{1}{\lambda}A'\left(\frac{1}{\lambda}\log s\right)\right].$$

But by virtue of (2.11), (1.3) and (1.4) we can always choose λ in such a manner that

$$\left(\left(-2A\left(\frac{1}{\lambda}\log s\right) + \frac{1}{\lambda}A'\left(\frac{1}{\lambda}\log s\right)\right)v, v\right) \leq -\alpha\,\|v\|^2 \quad \forall v \in V \tag{2.17}$$

for

$$\frac{1}{\lambda}\log s \in [0, T].$$

Replacing s with $t + 1$ and still denoting by $A(t)$ the operator

$$\frac{1}{\lambda^2(t+1)^2}A\left(\frac{1}{\lambda}\log(t+1)\right),$$

and changing the notation, we see that we are led back to (2.9) with (2.10) (2.11) and (2.12) (thanks to (2.17)). Since the *classes M_k are invariant* with respect to the changes of functions and of variables we have made we are indeed led to proving (2.13).

2.3.2 Proof of (2.13)

We use the notation introduced in (1.28).

The hypotheses can be interpreted as follows:

$$\nu_{-1}(f^{(k)}) \leq d\mathscr{L}^k M_k \quad \forall k, \tag{2.18}$$

$$\| A^{(k)}(t) \|_{\mathscr{L}(V;V')} \leq cL^k M_k \quad \forall k, t \in [0, T], \tag{2.19}$$

$$|\beta^{(k)}(t)| \leq d_1 \mathscr{L}_1^k M_k \quad \forall k, t \in [0, T]. \tag{2.20}$$

For the proof it is sufficient to show that

$$\nu_1(u^{(k)}) \leq \delta (2B)^k M_k \quad \forall k \tag{2.21}$$

where

$$\delta = 64 \max \left\{ \frac{d}{\sqrt{\alpha\gamma'}}, \frac{\sqrt{T}\, d\mathscr{L}}{\gamma} \frac{cc_1}{\alpha}, c_1 d_1 T, \frac{c_1 c L}{\gamma} \right\} \tag{2.22}$$

$$B = (1 + \delta) \mathscr{L}_2, \text{ with } \mathscr{L}_2 = \max \{\mathscr{L}, \mathscr{L}H, L, \mathscr{L}_1, LH\}. \tag{2.23}$$

Let us prove first that for $k = 0, 1, 2, \ldots$, we have the inequalities

(2.24)

$$\left(\frac{\alpha}{\gamma(2k+1)}\right)^{1/2} \nu_1(u^{(k)}) + \bigl(\gamma(2k+1)\bigr)^{-1/2} \nu_0(u^{(k+1)}) + \nu_1\bigl((T-t)^{1/2} u^{(k)}\bigr) \leq$$

$$\leq 8\bigl(\alpha\gamma(2k+1)\bigr)^{-1/2} \left\{ \nu_{-1}(f^{(k)}) + \sum_{j=0}^{k-1} \binom{k}{j} \nu_{-1}(A^{(k-j)} u^{(j)}) \right\} +$$

$$+ 8\bigl(\gamma(2k+1)\bigr)^{-1/2} \left\{ T \sum_{j=0}^{k-1} \binom{k}{j} \sup_{0 \leq t \leq T} |\beta^{(k-j)}(t)|\, \nu_0(u^{(j+1)}) \right\} +$$

$$+ 8\bigl(\gamma(2k+1)\bigr)^{-1} \left\{ \nu_{-1}\bigl((T-t)^{1/2} f^{(k+1)}\bigr) + \right.$$

$$+ \sum_{j=0}^{k-1} \binom{k}{j} \nu_{-1}\bigl((T-t)^{1/2} A^{(k-j+1)} u^{(j)}\bigr) +$$

$$\left. + \sum_{j=0}^{k-2} \binom{k}{j} \nu_{-1}\bigl((T-t)^{1/2} A^{(k-j)} u^{(j+1)}\bigr) \right\}.$$

We agree to replace by 0 those summations which extend over an empty set of indices.

To begin with, let us assume that, taking the kth order t derivative of (2.9) we obtain that

(2.25)

$$u^{(k+2)} + \beta u^{(k+1)} + \sum_{j=0}^{k-1} \binom{k}{j} \beta^{(k-j)} u^{(j+1)} + \sum_{j=0}^{k-1} \binom{k}{j} A^{(k-j)} u^{(j)} + A u^{(k)} = f^{(k)}.$$

Taking the scalar product of both sides of (2.25) with $u^{(k+1)}$ we obtain

(2.26)

$$\frac{\mathrm{d}}{\mathrm{d}t}\left(\sum_{j=0}^{k-1}\binom{k}{j}(A^{(k-j)}u^{(j)}, u^{(k)})\right) - \sum_{j=0}^{k-1}\binom{k}{j}(A^{(k-j+1)}u^{(j)}, u^{(k)}) -$$

$$-\sum_{j=0}^{k-2}\binom{k}{j}(A^{(k-j)}u^{(j+1)}, u^{(k)}) - \binom{k}{k-1}(A'u^{(k)}, u^{(k)}) + (Au^{(k)}, u^{(k+1)}) +$$

$$+\sum_{j=0}^{k-1}\binom{k}{j}\beta^{(k-j)}(u^{(j+1)}, u^{(k+1)}) + \beta\,|u^{(k+1)}|^2 + (u^{(k+2)}, u^{(k+1)}) =$$

$$= \frac{\mathrm{d}}{\mathrm{d}t}(f^{(k)}, u^{(k)}) - (f^{(k+1)}, u^{(k)}).$$

Let us take twice the real part of (2.26) and let us use the formula

$$\frac{\mathrm{d}}{\mathrm{d}t}\big(A(t)\,v(t), v(t)\big) = 2\mathrm{Re}\,\big(A(t)\,v(t), v'(t)\big) + \big(A'(t)\,v(t), v(t)\big)$$

which follows from (2.1); we obtain

(2.27)

$$\frac{\mathrm{d}}{\mathrm{d}t}\,2\mathrm{Re}\sum_{j=0}^{k-1}\binom{k}{j}(A^{(k-j)}u^{(j)}, u^{(k)}) - 2\mathrm{Re}\sum_{j=0}^{k-1}\binom{k}{j}(A^{(k-j+1)}u^{(j)}, u^{(k)}) -$$

$$-2\mathrm{Re}\sum_{j=0}^{k-2}\binom{k}{j}(A^{(k-j)}u^{(j+1)}, u^{(k)}) - (2k+1)\,(A'u^{(k)}, u^{(k)}) +$$

$$+\frac{\mathrm{d}}{\mathrm{d}t}(Au^{(k)}, u^{(k)}) + 2\beta\,|u^{(k+1)}|^2 + 2\mathrm{Re}\sum_{j=0}^{k-1}\binom{k}{j}\beta^{(k-j)}(u^{(j+1)}, u^{(k+1)}) +$$

$$+\frac{\mathrm{d}}{\mathrm{d}t}|u^{(k+1)}|^2 = \frac{\mathrm{d}}{\mathrm{d}t}\,2\mathrm{Re}(f^{(k)}, u^{(k)}) - 2\mathrm{Re}(f^{(k+1)}, u^{(k)}).$$

We now multiply (2.27) by $T - t$ and integrate from 0 to T; we obtain

(2.28)

$$\int_0^T (Au^{(k)}, u^{(k)})\,\mathrm{d}t + \int_0^T |u^{(k+1)}|^2\,\mathrm{d}t - (2k+1)\int_0^T (T-t)\,(A'u^{(k)}u^{(k)})\,\mathrm{d}t +$$

$$+ 2\int_0^T (T-t)\,\beta\,|u^{(k+1)}|^2\,\mathrm{d}t =$$

$$= 2\mathrm{Re}\int_0^T (f^{(k)}, u^{(k)})\,\mathrm{d}t - 2\mathrm{Re}\int_0^T (T-t)\,(f^{(k+1)}, u^{(k)})\,\mathrm{d}t -$$

$$- 2\mathrm{Re}\sum_{j=0}^{k-1}\binom{k}{j}\int_0^T (A^{(k-j)}u^{(j)}, u^{(k)})\,\mathrm{d}t +$$

$$+ 2\mathrm{Re} \sum_{j=0}^{k-1} \binom{k}{j} \int_0^T (T-t)\,(A^{(k-j+1)}u^{(j)}, u^{(k)})\,dt +$$

$$+ 2\mathrm{Re} \sum_{j=0}^{k-2} \binom{k}{j} \int_0^T (T-t)\,(A^{(k-j)}u^{(j+1)}, u^{(k)})\,dt +$$

$$+ 2\mathrm{Re} \sum_{j=0}^{k-1} \binom{k}{j} \int_0^T (T-t)\,\beta^{(k-j)}(u^{(j+1)}, u^{(k+1)})\,dt.$$

By virtue of (2.10), (2.11), (2.12) we have therefore

(2.29)

$$\alpha[\nu_1(u^{(k)})]^2 + [\nu_0(u^{(k+1)})]^2 + \gamma(2k+1)\left[\nu_1\big((T-t)^{1/2}\,u^{(k)}\big)\right]^2 \le$$

$$\le \left\{2\nu_{-1}(f^{(k)}) + 2\sum_{j=0}^{k-1}\binom{k}{j}\nu_{-1}(A^{(k-j)}u^{(j)})\right\}\nu_1(u^{(k)}) +$$

$$+ \left\{2T\sum_{j=0}^{k-1}\binom{k}{j}\sup_{0\le t\le T}|\beta^{(k-j)}(t)|\,\nu_0(u^{(j+1)})\right\}\nu_0(u^{(k+1)}) +$$

$$+ \left\{2\nu_{-1}\big((T-t)^{1/2}\,f^{(k+1)}\big) + 2\sum_{j=0}^{k-1}\binom{k}{j}\nu_{-1}\big((T-t)^{1/2}A^{(k-j+1)}u^{(j)}\big) +\right.$$

$$\left.+ 2\sum_{j=0}^{k-2}\binom{k}{j}\nu_{-1}\big((T-t)^{1/2}\,A^{(k-j)}u^{(j+1)}\big)\right\}\nu_1\big((T-t)^{1/2}\,u^{(k)}\big).$$

In the same manner, we show (2.29) for $k = 0, 1, \ldots$ Then we obtain (2.24), using the fact that the inequality

$$ax^2 + by^2 + cz^2 \le dx + ey + fz \quad \text{with} \quad a, b, c > 0 \quad \text{and} \quad d, e, f \ge 0$$

implies $\sqrt{\dfrac{a}{c}}\,x + \sqrt{\dfrac{b}{c}}\,y + z \le \dfrac{4d}{\sqrt{ac}} + \dfrac{4e}{\sqrt{bc}} + \dfrac{4}{c}f.$

For $j = 0, 1, 2, \ldots$ we now set

(2.30)

$$X_j = \left(\frac{\alpha}{\gamma(2j+1)}\right)^{1/2}\nu_1(u^{(j)} + \big(\gamma(2j+1)\big)^{-1/2}\,\nu_0(u^{(j+1)}) + \nu_1\big((T-t)^{1/2}\,u^{(j)}\big)$$

and we are going to show

(2.31) $$X_k \le \delta B^k M_k,$$

hence (2.21) and the theorem follows.

For $k = 0$ (2.31) is true, since (2.24) for $k = 0$ implies:

$$X_0 \le 8(\alpha\gamma)^{-1/2}\,\nu_{-1}(f) + 8\gamma^{-1}\nu_{-1}\big((T-t)^{1/2}f'\big);$$

therefore, by virtue of (2.18) and (1.24) we have

$$X_0 \leq 8(\alpha\gamma)^{-1/2}\, dM_0 + 8\gamma^{-1}\sqrt{T}\, d\mathscr{L}M_1 \leq$$
$$\leq 8(\alpha\gamma)^{-1/2}\, dM_0 + 8\gamma^{-1}\sqrt{T}\, d\mathscr{L}M_0 \leq \delta M_0 .$$

Let us admit (2.31) up to $k-1$ and let us verify (2.31) for k; by virtue of (2.18), (2.19) and (2.20), we see that (2.24) implies

$$X_k \leq 8\big(\alpha\gamma(2k+1)\big)^{-1/2}\left\{d\mathscr{L}^k M_k + \sum_{j=0}^{k-1}\binom{k}{j} cL^{k-j} M_{k-j}\nu_1(u^{(j)})\right\} +$$
$$+ 8\big(\gamma(2k+1)\big)^{-1/2}\, T \sum_{j=0}^{k-1}\binom{k}{j} d_1 \mathscr{L}_1^{k-j} M_{k-j}\nu_0(u^{(j+1)}) +$$
$$+ 8\big(\gamma(2k+1)\big)^{-1}\left\{\sqrt{T}\, d\mathscr{L}^{k+1} M_{k+1} +\right.$$
$$+ \sum_{j=0}^{k-1}\binom{k}{j} cL^{k-j+1} M_{k-j+1}\nu_1\big((T-t)^{1/2}\, u^{(j)}\big) +$$
$$\left. + \sum_{j=1}^{k-1}\binom{k}{j-1} cL^{k-j+1} M_{k-j+1}\nu_1\big((T-t)^{1/2}\, u^{(j)}\big)\right\}$$

$$\left(\text{using (1.24) and } \frac{1}{2k+1}\binom{k}{j-1} \leq \binom{k}{j} \text{ for } 1 \leq j \leq k \text{ and } k \geq 1\right) \leq$$
$$\leq 8\big(\alpha\gamma(2k+1)\big)^{-1/2}\, d\mathscr{L}^k M_k + \frac{8\sqrt{T}\, d\mathscr{L}}{\gamma(2k+1)}(\mathscr{L}H)^k M_k +$$
$$+ \sum_{j=0}^{k-1}\binom{k}{j} cL^{k-j} M_{k-j} 8\big(\alpha\gamma(2k+1)\big)^{-1/2}\left(\frac{\gamma(2j+1)}{\alpha}\right)^{1/2} \times$$
$$\times \left[\left(\frac{\alpha}{\gamma(2j+1)}\right)^{1/2} \nu_1(u^{(j)})\right] +$$
$$+ \sum_{j=0}^{k-1}\binom{k}{j} T\, d_1 \mathscr{L}_1^{k-j} M_{k-j}\; 8\big(\gamma(2k+1)\big)^{-1/2}\big(\gamma(2j+1)\big)^{1/2} \times$$
$$\times \left[\big(\gamma(2j+1)\big)^{-1/2}\nu_0(u^{(j+1)})\right] +$$
$$+ 8\big(\gamma(2k+1)\big)^{-1}\sum_{j=0}^{k-1}\binom{k}{j} cL(LH)^{k-j} M_{k-j}\nu_1\big((T-t)^{1/2}\, u^{(j)}\big) +$$
$$+ 8\gamma^{-1}\sum_{j=1}^{k-1}\binom{k}{j} cL(LH)^{k-j} M_{k-j}\nu_1\big((T-t)^{1/2} u^{(j)}\big) \leq \text{using (2.30)} \leq$$

$$\leq 8(\alpha\gamma)^{-1/2}\, d\mathcal{L}^k M_k + 8\gamma^{-1}\sqrt{T}\, d\mathcal{L}(\mathcal{L}H)^k M_k +$$

$$+ \sum_{j=0}^{k-1} \frac{8}{\alpha}\binom{k}{j} cL^{k-j} M_{k-j} X_j \frac{M_j}{M_j} + \sum_{j=0}^{k-1} 8\binom{k}{j} T\, d_1 \mathcal{L}_1^{k-j} M_{k-j} X_j \frac{M_j}{M_j} +$$

$$+ \sum_{j=0}^{k-1} \frac{8}{\gamma}\binom{k}{j} cL(LH)^{k-j} M_{k-j} X_j \frac{M_j}{M_j} +$$

$$+ \sum_{j=1}^{k-1} \frac{8}{\gamma}\binom{k}{j} cL(LH)^{k-j} M_{k-j} X_j \frac{M_j}{M_j} +$$

$$+ \sum_{j=1}^{k-1} \frac{8}{\gamma}\binom{k}{j} cL(LH)^{k-j} M_{k-j} \frac{M_j}{M_j} X_j \leq$$

$\leq$ using hypothesis (1.25), (2.22) and (2.23) $\leq$

$$\leq \frac{\delta}{2} M_k B^k + \sum_{j=0}^{k-1} \frac{8}{\alpha} cc_1 L^{k-j} M_k \frac{X_j}{M_j} + \sum_{j=0}^{k-1} 8 c_1 T d_1 \mathcal{L}_1^{k-j} M_k \frac{X_j}{M_j} +$$

$$+ \sum_{j=0}^{k-1} \frac{16}{\gamma} cc_1 L(LH)^{k-j} M_k \frac{X_j}{M_j} \leq \frac{\delta}{2} B^k M_k + \sum_{j=0}^{k-1} \frac{\delta}{2} \mathcal{L}_2^{k-j} \frac{X_j}{M_j} M_k \leq$$

$$\leq \text{by the induction hypothesis } \leq \frac{\delta}{2} B^k M_k +$$

$$+ \frac{\delta}{2} \sum_{j=0}^{k-1} \mathcal{L}_2^{k-j} \delta B_j M_k \leq \frac{\delta}{2} B^k M_k + \frac{\delta}{2} \delta \mathcal{L}_2^k M_k \sum_{j=0}^{k-1} \frac{B^j}{\mathcal{L}_2^j} =$$

$$= \frac{\delta}{2} B_k M_k + \frac{\delta}{2} \delta \mathcal{L}_2^k M_k \frac{1 - \left(\dfrac{B}{\mathcal{L}_2}\right)^k}{1 - \dfrac{B}{\mathcal{L}_2}} =$$

$$= \frac{\delta}{2} B^k M_k + \frac{\delta}{2} \delta \mathcal{L}_2 M_k \frac{B^k - \mathcal{L}_2^k}{B - \mathcal{L}_2} \leq$$

$$\leq \frac{\delta}{2} B^k M_k + \frac{\delta}{2} \delta \mathcal{L}_2 M_k \frac{B^k}{B - \mathcal{L}_2} = \frac{\delta}{2} B^k M_k \left(1 + \frac{\delta \mathcal{L}_2}{B - \mathcal{L}_2}\right) =$$

$= \text{using (2.23)} = \delta B^k M_k$. □

Remark 2.3. Examples analogous to those in Section 1.5 (this time letting $A^*(t) = A(t)$) can be given. Likewise, there exists a result analogous to Theorem 1,4.

3. Singular Equations of the Second Order in t

3.1 Statement of the Main Results

We shall study the M_k-regularity properties of the solution of the *singular equation*:

$$A(t)\, u + u'' + \frac{2\lambda + 1}{t} u' = f, \quad t > 0, \tag{3.1}$$

for $\lambda \in \mathbf{C}$ with

$$u(0) = 0, \; u'(0) = 0. \quad \square \tag{3.2}$$

We shall prove the following theorem.

Theorem 3.1. *Let the hypotheses of Theorem* 2.2 *hold; thus*: (1.5), (2.1), (1.21) *and* (1.22), ..., (1.25) *hold. Let* $\lambda \in \mathbf{C}$, *with*

$$\lambda \neq -1, -2, -3, \ldots \tag{3.3}$$

Furthermore assume that the function $t \to A(t)$ *is even.*
Let f *be given with*

$$f \in \mathscr{D}_{0,M_k}([0, \infty[; V'). \tag{3.4}$$

Then problem (3.1), (3.2) *admits a unique solution* u *which satisfies*

$$u \in \mathscr{D}_{0,M_k}([0, \infty[; V). \tag{3.5}$$

Furthermore $f \to u$ *is a continuous mapping of*

$$\mathscr{D}_{0,M_k}([0, \infty[; V') \to \mathscr{D}_{0,M_k}([0, \infty[; V).$$

Remark 3.1. The same result is true (the method of proof remaining unchanged) for the equation

$$A(t)\, u + u'' + \frac{2\lambda + 1}{t} u' + M(t)\, u' = f, \tag{3.6}$$

where

$$t \to M(t) \text{ is an even function, } \in \mathscr{E}_{M_k}(\mathbf{R}). \quad \square \tag{3.7}$$

3.2 Proof of Theorem 3.1

We use *the transmutation operators* (see Delsarte [1], Lions [8]). We know that there exist operators B_λ and $\mathscr{B}_\lambda$ with the following properties: if $\mathscr{D}_0$ denotes the space of C^∞ functions on $t \geq 0$ which, together with all their derivatives, vanish at the origin, then:

$$\text{(3.8)}\qquad \begin{cases} B_\lambda \in \mathscr{L}(\mathscr{D}_0; \mathscr{D}_0), \\ \lambda \to B_\lambda \text{ being an entire holomorphic function of} \\ \mathbf{C} \to \mathscr{L}(\mathscr{D}_0; \mathscr{D}_0); \end{cases}$$

$$\text{(3.9)}\qquad \begin{cases} \mathscr{B}_\lambda \in \mathscr{L}(\mathscr{D}_0; \mathscr{D}_0),\ \lambda \neq -1, -2, \ldots, \\ \lambda \to \mathscr{B}_\lambda \text{ being a meromorphic function of} \\ \mathbf{C} \to \mathscr{L}(\mathscr{D}_0; \mathscr{D}_0), \text{ with poles at } -1, -2, \ldots; \end{cases}$$

$$\text{(3.10)}\qquad \begin{cases} \mathrm{D}^2 B_\lambda = B_\lambda L_\lambda \text{ on } \mathscr{D}_0, \text{ if } \mathrm{D} = \dfrac{\mathrm{d}}{\mathrm{d}t} \text{ and if} \\ L_\lambda = \mathrm{D}^2 + \dfrac{2\lambda + 1}{t}\mathrm{D}; \end{cases}$$

$$\text{(3.11)}\qquad B_\lambda \mathscr{B}_\lambda = \mathscr{B}_\lambda B_\lambda = \text{Identity in } \mathscr{D}_0,\ \lambda \neq -1, -2, \ldots$$

Further on, we shall prove the following results:

$$\text{(3.12)}\qquad B_\lambda \in \mathscr{L}(\mathscr{D}_{0,M_k}; \mathscr{D}_{0,M_k});$$

$$\text{(3.13)}\qquad \mathscr{B}_\lambda \in \mathscr{L}(\mathscr{D}_{0,M_k}; \mathscr{D}_{0,M_k}),\ \lambda \neq -1, -2, \ldots;$$

$$\text{(3.14)}\qquad \begin{cases} B_\lambda A(t)\, \mathscr{B}_\lambda \varphi = A(t)\, \varphi + \displaystyle\int_0^t N_\lambda(t, \sigma)\, \varphi(\sigma)\, \mathrm{d}\sigma^{((1))}, \\ \text{where } \varphi \in \mathscr{D}_{0,M_k}([0, \infty[; V) \text{ and } t, \sigma \to N(t, \sigma) \\ \text{is a function of class } M_k \text{ in } t \text{ and } \sigma,\ \sigma \leq t, \\ t \in \mathbf{R}, \text{ with values in } \mathscr{L}(V; V'). \end{cases}$$

Thanks to the transmutation operators B_λ and $\mathscr{B}_\lambda$, Theorem 3.1 reduces to Theorem 2.2. Indeed (3.1) is *equivalent* (if $\lambda \neq -1, \ldots$) to

$$\text{(3.15)}\qquad B_\lambda A(t)\, \mathscr{B}_\lambda(B_\lambda u) + \mathrm{D}^2(B_\lambda u) = B_\lambda f.$$

Set $B_\lambda u = w$; according to (3.14), equation (3.15) is equivalent to

$$\text{(3.16)}\qquad A(t)\, w + \int_0^t N_\lambda(t, \sigma)\, w(\sigma)\, \mathrm{d}\sigma + w''(t) = B_\lambda f.$$

Now, according to (3.12):

$$\text{(3.17)}\qquad B_\lambda f \in \mathscr{D}_{0,M_k}([0, \infty[; V')$$

((1)) In this formula, the operators B_λ and $\mathscr{B}_\lambda$ are extended to vector-valued functions—which is immediate. We assume that $\lambda \neq -1, -2, \ldots$

and according to Theorem 2.2 and Remark 2.2, we obtain:

$$w \in \mathcal{D}_{0,M_k}([0, \infty[; V). \tag{3.18}$$

Then

$$u = \mathcal{B}_\lambda w \in \mathcal{D}_{0,M_k}([0, \infty[; V)$$

and since w depends continuously on $B_\lambda f$, we have the theorem if we can show (3.12), (3.13), (3.14).

Proof of (3.12). We use the *explicit formulas for* B_λ.

For $-1 < \operatorname{Re}\lambda < -1/2$, we have

$$B_\lambda \varphi(x) = b_\lambda \int_0^1 t^{2\lambda+1}(1-t^2)^{-\lambda-3/2}\, \varphi(tx)\, dt \tag{3.19}$$

where

$$b_\lambda = \frac{1}{2\sqrt{\pi}}\, \Gamma(\lambda+1)\, \Gamma\left(-\lambda-\frac{1}{2}\right) \quad (\Gamma = \text{Euler function}). \tag{3.20}$$

Thus

$$(B_\lambda \varphi)^{(k)}(x) = b_\lambda \int_0^1 t^{2\lambda+1+k}(1-t^2)^{-\lambda-3/2}\, \varphi^{(k)}(tx)\, dt,$$

from which, for x belonging to a compact set $[0, x_0]$:

(3.21)

$$|(B_\lambda \varphi)^{(k)}(x)| \le |b_\lambda|\, c L^k M_k \int_0^1 t^{2\operatorname{Re}\lambda+1+k}(1-t^2)^{-\operatorname{Re}\lambda-3/2}\, dt \le \hat{c} L^k M_k \quad \forall k.$$

Next, for

$$-1 < \operatorname{Re}\lambda < n - 1/2 \ (n \text{ positive integer}),$$

B_λ is given, by analytic continuation of (3.19), in the form (see Lions [8], page 69):

$$B_\lambda \varphi(x) = \frac{(-1)^n\, b_\lambda}{(2\lambda+1)(2\lambda-1)\cdots(2\lambda-(2n-3))} \int_0^1 t^{2\lambda-(2n-3)} \times$$
$$\times (1-t^2)^{\lambda-(2n-3)/2}\, T_n\varphi(t, x)\, dt,$$

with

$$T_1\varphi(t, x) = \frac{d}{dt}\big(t\varphi(tx)\big),$$

$$T_2\varphi(t, x) = \frac{d}{dt}\big(t^3 T_1\varphi(t,x)\big),$$

....

$$T_n\varphi(t, x) = \frac{d}{dt}\big(t^3 T_{n-1}\varphi(t, x)\big).$$

Then

$$\frac{\partial^k}{\partial x^k} T_n \varphi(t, x) = \frac{1}{x^k} T_n(x^k \varphi^{(k)})(t, x)$$

and we obtain relations analogous to (3.21) for $-1 < \operatorname{Re} \lambda < n - 1/2$.

For $-1 - n/2 < \operatorname{Re} \lambda < -1/2$, we use another expression for the analytic continuation of B_λ; for $\lambda \neq -3/2, -5/2, \ldots$:

$$B_\lambda \varphi(x) = \frac{(-1)^n b_\lambda}{(2\lambda + 2)(2\lambda + 3) \cdots (2\lambda + n + 1)} \int_0^1 t^{2\lambda+n+1} (1 - t^2)^{\lambda + (n+1)/2} \times$$
$$\times \mathscr{U}_n \varphi(t, x) \, dt,$$

with

$$\mathscr{U}_1 \varphi(t, x) = \frac{d}{dt} \left((1 - t^2)^{1/2} \varphi(tx) \right),$$

$$\mathscr{U}_2 \varphi(t, x) = \frac{d}{dt} \left((1 - t^2)^{3/2} \mathscr{U}_1 \varphi(t, x) \right),$$

$$\cdots\cdots\cdots\cdots\cdots\cdots\cdots\cdots\cdots$$

$$\mathscr{U}_n \varphi(t, x) = \frac{d}{dt} \left((1 - t^2)^{3/2} \mathscr{U}_{n-1} \varphi(t, x) \right).$$

Again, relations analogous to (3.21) follow.

Finally, we verify (as in Lions [8], page 74) that $\lambda = -3/2, -5/2, \ldots$ *are not* exceptional values *when* $\varphi \in \mathscr{D}_0$—and we again obtain estimates analogous to (3.21). ▯

Proof of (3.13). For $\operatorname{Re} \lambda > -1/2$, the operator $\mathscr{B}_\lambda$ is given by

$$\mathscr{B}_\lambda \varphi(x) = \beta \lambda \int_0^1 (1 - t^2)^{\lambda - 1/2} \varphi(tx) \, dt, \tag{3.22}$$

where

$$\beta_\lambda = \frac{2}{\sqrt{\pi}} \frac{\Gamma(\lambda + 1)}{\Gamma(\lambda + 1/2)}. \tag{3.23}$$

For $\varphi \in \mathscr{D}_{0,M_k}$, it follows immediately that

$$|(\mathscr{B}_\lambda \varphi)^{(k)}(x)| \leq |\beta \lambda| \, c L^k M_k \int_0^1 (1 - t^2)^{\operatorname{Re}\lambda - 1/2} t^k \, dt \leq \tilde{c} L^k M_k \quad \forall k. \tag{3.24}$$

We then carry out the analytic continuation of $\mathscr{B}_\lambda$ by formulas (see Lions [8], page 78) which preserve the properties analogous to (3.24), for $\lambda \neq -1, -2, \ldots$ ▯

Proof of (3.14). For $|\operatorname{Re} \lambda| < 1/2$, we verify (see Lions [5], page 265) that

$$B_\lambda A(t) \mathscr{B}_\lambda - A(t) = t \int_0^t \mathscr{C}_\lambda A(t, \tau) \, d\tau, \tag{3.25}$$

where

$$(3.26)\begin{cases} \mathcal{C}_\lambda A(t,\tau) = \gamma_\lambda \int\limits_0^{2\pi} \sin^{-2\lambda+2}\theta \cos^{-2\lambda}\theta A_1\big((t^2\sin^2\theta + \tau^2\cos^2\theta)^{1/2}\big)\, d\theta, \\ A_1(t) = \frac{1}{t} A'(t) \text{ (note that } A \text{ is \textit{even} in } t), \\ \gamma_\lambda = 2\frac{1}{\Gamma(\lambda+1/2)\,\Gamma(-\lambda+1/2)}. \end{cases}$$

Property (3.14), for $|\mathrm{Re}\,\lambda| < 1/2$, follows.

We then pass to the general case, when $\lambda \neq -1, -2, \ldots$, by analytic continuation. □

4. Schroedinger-Type Equations

4.1 Statement of the Main Results

In this Section we shall study the regularity properties in $\mathcal{D}_+$ and in $\mathcal{D}_{+,M_k}$ of the solution of the Schroedinger equation:

$$(4.1)\qquad iA(t)\,u + u' = f, \quad \left(i = \sqrt{-1}\right),$$

where the support of u in t is bounded on the left.

We shall prove the following results:

Theorem 4.1. *Let the hypotheses of Theorem* 2.1 *be satisfied, i.e.* (1.4), (1.5) *and* (2.1) *hold. Then the mapping*

$$(4.2)\qquad u \to iA(t)\,u + u'$$

is an isomorphism of $\mathcal{D}_+(\mathbf{R}; V)$ *onto* $\mathcal{D}_+(\mathbf{R}; V')$.

Theorem 4.2. *Let the hypotheses of Theorem* 2.2 *be satisfied, i.e.* (1.5), (2.1), (1.21) *and* (1.22), ..., (1.25) *hold. Then the mapping* (4.2) *is an isomorphism of* $\mathcal{D}_{+,M_k}(\mathbf{R}; V)$ *onto* $\mathcal{D}_{+,M_k}(\mathbf{R}; V')$.

4.2 Proof of Theorem 4.1

Changing u to $\exp(-i\lambda t)\,u$ and choosing λ appropriately, we reduce the problem to the case for which

$$\big(A(t)\,v, v\big) \geq \alpha\,\|v\|^2, \quad \alpha > 0, \quad v \in V, \quad t \in [0, T]\ [(1)]$$

and by eventual renormalisation of V we are led to the case for which

$$(4.3)\qquad \big(A(t)\,v, v\big) \geq \|v\|^2, \quad v \in V, \quad t \in [0, T].$$

[(1)] It is sufficient to reason on $[0, T]$, arbitrary fixed finite T.

Next, we make the change of variable (as in Section 2.3):

$$s = \exp(\lambda t).$$

Setting

$$v(s) = u\left(\frac{1}{\lambda}\log s\right),$$

we obtain the equation, on $s \geq 1$,

$$\frac{i}{\lambda s} A\left(\frac{1}{\lambda}\log s\right) v(s) + \frac{\mathrm{d}}{\mathrm{d}s} v(s) = \frac{1}{\lambda s} f\left(\frac{1}{\lambda}\log s\right). \tag{4.4}$$

Then

$$\frac{\mathrm{d}}{\mathrm{d}s}\left(\frac{1}{\lambda s} A\left(\frac{1}{\lambda}\log s\right)\right) = \frac{1}{\lambda^2 s^2}\left[-\lambda A\left(\frac{1}{\lambda}\log s\right) + A'\left(\frac{1}{\lambda}\log s\right)\right]$$

and (as in Section 2.3) we are thus led to consider the case for which

$$(A'(t)\, v, v) \leq -\gamma \, \|v\|^2, \ \gamma > 0, \ t \in [0, T], \ v \in V. \quad \square \tag{4.5}$$

We shall now prove Theorem 4.1, assuming that (4.3), (4.5) hold.

We start with the following result[1]: *if* $f \in L^2(0\ T; V')$ *with* $\frac{\mathrm{d}f}{\mathrm{d}t} \in L^2(0\ T; V')$ *then there exists a unique* u *in* $L^2(0\ T; V)$ *solution of* (4.1) *with*

$$u(0) = 0. \tag{4.6}$$

Furthermore (again using the notation introduced in (1.28)):

$$\nu_1(u) \leq 2(\nu_{-1}(f) + T\nu_{-1}(f')). \tag{4.7}$$

Indeed, for example applying the Faedo-Galerkin method (see Chapter 3), it is sufficient to show the estimate (4.7) for regular functions satisfying (4.1) or the equivalent equation

$$A(t)\, u - iu' = -if. \tag{4.1a}$$

Taking the scalar product of the two sides of (4.1a) with $(T - t)\, u'$ and taking twice the real part of both sides, we obtain

(4.8)

$$\int_0^T (T - t)\left[\frac{\mathrm{d}}{\mathrm{d}t} a(t; u, u) - a'(t; u, u)\right] \mathrm{d}t = -2\mathrm{Re}\, i \int_0^T (T - t)\, (f, u')\, \mathrm{d}t.$$

Therefore

$$\int_0^T a(t; u, u)\, \mathrm{d}t - \int_0^T (T - t)\, a'(t; u, u)\, \mathrm{d}t$$
$$= -2\mathrm{Re}\, i\left(\int_0^T (f, u)\, \mathrm{d}t - \int_0^T (T - t)\, (f', u)\, \mathrm{d}t\right)$$

---

[1] See also Pozzi [1] for other results of this type.

from which, since we have (4.3) and (4.5), we obtain

$$\nu_1(u)^2 \leq 2(\nu_{-1}(f) + T\nu_{-1}(f'))\, \nu_1(u)$$

whence (4.7). □

This time, in order to show the regularity, we do not use the method of differential quotients, but rather the Faedo-Galerkin method; the approximate solutions being regular in t, it is sufficient to *formally* differentiate equation (4.1a) with respect to t and then obtain an estimate analogous to (4.7) for $\nu_1(u')$. More precisely, we have, formally

$$A(t)\, u' + A'(t)\, u + iu'' = if. \tag{4.9}$$

We take the scalar product with $(T - t)\, u''$. We obtain

$$\int_0^T (T - t) \left[\frac{\mathrm{d}}{\mathrm{d}t} a(t; u', u') - a'(t; u', u')\right] \mathrm{d}t +$$

$$+ 2\mathrm{Re} \int_0^T (T - t)\, (A'(t)\, u, u'')\, \mathrm{d}t = 2\mathrm{Re}\left(-\mathrm{i} \int_0^T (T - t)\, (f', u'')\, \mathrm{d}t\right)$$

from which since $a'(t; v\ v) \leq 0$:

$$\left\{\begin{aligned} &\int_0^T a(t; u', u')\, \mathrm{d}t + 2\mathrm{Re} \int_0^T A'(t)\, u, u')\, \mathrm{d}t - \\ &- 2\mathrm{Re} \int_0^T (T - t)\, (A'(t)\, u', u')\, \mathrm{d}t - 2\mathrm{Re} \int_0^T (T - t)\, (A''(t), u, u')\mathrm{d}t \\ &\leq 2\mathrm{Re}\left(-\mathrm{i} \int_0^T (f', u')\, \mathrm{d}t - \mathrm{i} \int_0^T (T - t)\, (f'', u')\, \mathrm{d}t\right). \end{aligned}\right. \tag{4.10}$$

Once again applying (4.5), it follows that

$$\nu_1(u')^2 \leq 2(\nu_{-1}(f') + T\nu_{-1}(f''))\, \nu_1(u') + c\nu_1(u)\, \nu_1(u')$$

whence (c denoting a constant):

$$\nu_1(u') \leq c(\nu_{-1}(f) + \nu_{-1}(f') + \nu_{-1}(f'')). \tag{4.11}$$

This justifies the formal differentiation with respect to t. By iterating on this procedure, we obtain the desired result. □

4.3 Proof of Theorem 4.2

We consider the problem on the interval $[0, T]$, on which we assume (4.3) and (4.5) to be satisfied.

The hypotheses may be interpreted as:

$$\nu_{-1}(f^{(k)}) \leq d\mathscr{L}^k M_k,\ \forall k, \tag{4.12}$$

$$\| A^{(k)}(t) \|_{\mathscr{L}(V; V')} \leq cL^k M_k,\ \forall k,\ t \in [0, T]. \tag{4.13}$$

We have to show that

$$\nu_1(u^{(k)}) \leq \delta(2B)^k M_k \quad \forall k, \tag{4.14}$$

where δ and B depend only on α, γ, d, $\mathscr{L}$, c, L, T, c_1, L, H.

The proof is analogous to the proof of formula (2.21) given in Theorem 2.2. We show first the relation similar to (2.24), without the term $\nu_0(u^{(k+1)})$ (we take the kth order t-derivative of (4.1) and we take the scalar product of both sides with $\frac{1}{i} u^{(k+1)}$ and then twice the real part of the result; next, we multiply by $(T - t)$ and integrate from 0 to T; the term $\left(u^{(k+1)}, \frac{1}{i} u^{(k+1)}\right)$ drops out). We then complete the proof as for the proof of (2.21). □

5. Stability Results in M_k-Classes

5.1 Parabolic Regularization

Continuing in the same notation, let us assume that

$$a(t; u, v) = \overline{a(t; v, u)} \quad \forall u, v \in V, \tag{5.1}$$

$$\big(A(t) v, v\big) \geq \alpha \|v\|^2, \ t \in [0, T], \ \alpha > 0, \ v \in V. \tag{5.2}$$

In Chapter 3, Section 8.5, we have seen that we can "approximate" (in the spaces L^2; see Theorem 8.3, Chapter 3) the problem

$$A(t) u + u'' = f, \ u(0) = 0, \ u'(0) = 0, \tag{5.3}$$

with the problem (called *parabolic regularization* of (5.3)):

$$A(t) u_\varepsilon + \varepsilon A(t) u_\varepsilon' + u_\varepsilon'' = f, \ u_\varepsilon(0) = 0, \ u_\varepsilon'(0) = 0, \ \varepsilon > 0. \tag{5.4}$$

A natural question to ask is whether, if f and A are of class $\{M_k\}$, the solution u_ε of (5.4) converges to u as $\varepsilon \to 0$, in the sense of the spaces $\mathscr{D}_{+,M_k}$. □

The answer is yes; we have:

Theorem 5.1. *Let the hypotheses of Theorem 2.2 be satisfied. Let f be given in $\mathscr{D}_{+,M_k}(\mathbf{R}; V')$. Let u (resp. u_ε) be the solution in $\mathscr{D}_{+,M_k}(\mathbf{R}; V)$ of*

$$A(t) u + u'' + = f \tag{5.5}$$

(resp. of

$$A(t) u_\varepsilon + \varepsilon A(t) u_\varepsilon' + u_\varepsilon'' + = f, \ \varepsilon > 0^{((1))}). \tag{5.6}$$

((1)) According to Theorem 1.2, the solution u_ε of (5.6) does belong to $\mathscr{D}_{+,M_k}(\mathbf{R}; V)$, since, according to Chapter 3, Section 8, equation (5.6) reduces to the parabolic case of Section 1.

Then:

$$u_\varepsilon \to u \text{ in } \mathscr{D}_{+,M_k}(\mathbf{R}; V) \text{ as } \varepsilon \to 0. \tag{5.7}$$

Proof. 1) Changing u to $\exp(kt)\, u$, equations (5.5), (5·6) become

$$(A + k^2)\, u + 2ku' + u'' = f, \tag{5.8}$$

$$(A + k^2 + \varepsilon A k)\, u_\varepsilon + (2k + \varepsilon A)\, u_\varepsilon' + u_\varepsilon'' = f, \tag{5.9}$$

respectively.

Assume that $f \in \mathscr{D}_{0,M_k}([0, T]; V')$; then $u_\varepsilon \in \mathscr{D}_{0,M_k}([0, T]; V)$ and is sufficient to show that

$$u_\varepsilon \to u \text{ in } \mathscr{D}_{0,M_k}([0, T]; V). \tag{5.10}$$

Thus we consider the problem on the interval $[0, T]$, and make the change of variable (as in Section 2.3):

$$s = \exp(\lambda t).$$

Setting

$$v(s) = u\left(\frac{1}{\lambda}\log s\right), \quad v_\varepsilon(s) = u_\varepsilon\left(\frac{1}{\lambda}\log s\right),$$

equations (5.8) and (5.9) become respectively, for $s \geq 1$:

$$\frac{d^2 v}{ds^2} + \left(\frac{1}{s} + \frac{2k}{\lambda s}\right)\frac{dv}{ds} + \frac{1}{\lambda^2 s^2}\left[A\left(\frac{1}{\lambda}\log s\right) + k^2\right] v = \frac{1}{\lambda^2 s^2} f\left(\frac{1}{\lambda}\log s\right), \tag{5.11}$$

$$\begin{cases} \dfrac{d^2 v_\varepsilon}{ds^2} + \left\{\dfrac{1}{s} + \dfrac{2k - \varepsilon k^2}{\lambda s} + \dfrac{\varepsilon}{\lambda s}\left[A\left(\dfrac{1}{\lambda}\log s\right) + k^2\right]\right\}\dfrac{dv_\varepsilon}{ds} + \\ + \left\{\dfrac{1}{\lambda^2 s^2}\left[A\left(\dfrac{1}{\lambda}\log s\right) + k^2 - \varepsilon k^3\right] + \right. \\ \left. + \dfrac{\varepsilon k}{\lambda^2 s^2}\left[A\left(\dfrac{1}{\lambda}\log s\right) + k^2\right]\right\} v_\varepsilon = \dfrac{1}{\lambda^2 s^2} f\left(\dfrac{1}{\lambda}\log s\right). \end{cases} \tag{5.12}$$

By a new change of variable and function, we are thus led to consider the following situation: u is a solution of

$$u'' + \beta u' + Au = f, \tag{5.13}$$

where

$$\begin{cases} \beta \geq 0, \ \beta \in \mathscr{E}_{M_k}, \\ (Av, v) \geq \alpha \|v\|^2 \quad \forall v \in V, \ t \in [0, T], \\ (A'v, v) \leq -\gamma \|v\|^2, \ \gamma > 0, \ v \in V, \ t \in [0T]. \end{cases} \tag{5.14}$$

and u_ε is a solution of

$$u_\varepsilon'' + (\beta_\varepsilon + \varepsilon\mathscr{A})\, u_\varepsilon' + (A_\varepsilon + \varepsilon k A)\, u_\varepsilon = f \tag{5.15}$$

where

$$\begin{cases} \beta_\varepsilon = \beta + \varepsilon\varphi, \ \varphi \in \mathscr{E}_{M_k}, \\ A_\varepsilon = A + \varepsilon a, \ a \in \mathscr{E}_{M_k}, \\ \mathscr{A} = \dfrac{1}{\lambda s} A \left(\dfrac{1}{\lambda} \log s\right) + k^2, \\ \left(\mathscr{A}(t)\, v, v\right) \geq \alpha_1 \|v\|^2, \ \alpha_1 > 0, \ v \in V, \ t \in [0, T], \\ \left(\mathscr{A}'(t)\, v, v\right) \leq -\gamma_1 \|v\|^2, \ \gamma_1 > 0, \ v \in V, \ t \in [0, T]. \end{cases} \tag{5.16}$$

2) We take the kth order t-derivative of (5.15) and we take the scalar product of both sides with $u_\varepsilon^{(k+1)}$; we take twice the real part of the result; next, we multiply by $T - t$ and integrate over $(0, T)$. Taking into account (5.16) and by a calculation analogous to the one in Section 2.3, it follows that, f being fixed, there exists $B \in \mathbf{R}_+$, B independent of ε, such that

(5.17) $\quad u_\varepsilon$ remains in a bounded set of $\mathscr{D}_{0,M_k}([0, T], B; V)$, as $\varepsilon \to 0$.

3) Setting

$$w_\varepsilon = u_\varepsilon - u, \tag{5.18}$$

it follows from (5.13) and (5.15) that

$$w_\varepsilon'' + \beta w_\varepsilon' + \mathscr{A} w_\varepsilon = \varepsilon g_\varepsilon, \tag{5.19}$$

where

$$g_\varepsilon = -(\varphi + \mathscr{A})\, u_\varepsilon' - (a + kA)\, u_\varepsilon. \tag{5.20}$$

According, to (5.17) we have:

(5.21) $\quad \begin{cases} \text{there exists } \mathscr{L}_1 > 0 \text{ such that } g_\varepsilon \text{ remains in a} \\ \text{bounded set of } \mathscr{D}_{0,M_k}([0, T], \mathscr{L}_1; V') \text{ as } \varepsilon \to 0. \end{cases}$

But then, according to Section 2.3, there exists $B_1 > 0$ such that

(5.22) $\quad \dfrac{1}{\varepsilon} w_\varepsilon$ remains in a bounded set of $\mathscr{D}_{0,M_k}([0, T], B_1; V)$ as $\varepsilon \to 0$,

therefore $w_\varepsilon \to 0$ in $\mathscr{D}_{0,M_k}([0, T], B_1; V)$, which proves (5.10). ☐

5.2 Approximation by Systems of Cauchy-Kowaleska Type (I)

In this section and the one following it, we consider problems *which are not of Cauchy-Kowaleska type* (see equations (5.29)—(5.30) and (5.46)—(5.47) below). We shall indicate how they can be *approximated* (in a sense to be specified below) by systems *which are* of Cauchy-Kowaleska type, with convergence of solutions to the solution of the initial problem *in spaces of class M_k*. (We also have convergence in Sobolev spaces, as introduced in Volumes 1 and 2, the data belonging to Sobolev spaces, but this is not discussed here.) ☐

In a bounded open set Ω in $\mathbf{R}^n$, consider the space

$$\Phi = \{\varphi \mid \varphi \in \mathscr{D}(\Omega)^n, \ \operatorname{div} \varphi = 0\}, \tag{5.23}$$

and next

$$H = \text{closure of } \Phi \text{ in } \left(L^2(\Omega)\right)^n, \tag{5.24}$$

$$V = \text{closure of } \Phi \text{ in } \left(H_0^1(\Omega)\right)^n, \tag{5.25}$$

the scalar product in H (resp. V) being

$$(u, v) = \sum_{j=1}^{n} \int_{\Omega} u_j \bar{v}_j \, \mathrm{d}x$$

$$\left(\text{resp.} \quad (u, v)_V = \sum_{j=1}^{n} (u_j, v_j)_{H_0^1(\Omega)}\right).$$

For $u, v \in V$, we set

$$a(u, v) = \sum_{i,j=1}^{n} \int_{\Omega} \frac{\partial u_j}{\partial x_i} \overline{\frac{\partial v_j}{\partial x_i}} \, \mathrm{d}x. \tag{5.26}$$

Since Ω is bounded and the elements of V vanish at the boundary of Ω, we have

$$a(v, v) \geq \alpha \, \|v\|_V^2, \quad \alpha > 0. \tag{5.27}$$

Theorem 1.2 is valid[1]; consequently, *for f given in $\mathscr{D}_{+,M_k}(\mathbf{R}; V')$, there exists a unique u belonging to $\mathscr{D}_{+,M_k}(\mathbf{R}; V)$, solution of*

$$\frac{\mathrm{d}}{\mathrm{d}t}(u, v) + a(u, v) = (f, v), \quad \forall v \in V. \quad ☐ \tag{5.28}$$

The interpretation of problem (5.28) is the following: Since (5.28) is satisfied for all test functions with *vanishing divergence*, the equality of functions (after integration by parts) holds "modulus a gradient";

[1] In a very simplified setting, since $a(u, v)$ *is independent of t*.

therefore

$$\frac{\partial u}{\partial t} - \Delta u + \operatorname{grad} p = f, \tag{5.29}$$

$$\operatorname{div} u = 0, \tag{5.30}$$

and according to (5.29), we have

$$\operatorname{grad} p \in \mathscr{D}_{+,M_k}\left(\mathbf{R}; (H^{-1}(\Omega))^n\right). \tag{5.31}$$

The system (5.29)—(5.30) *is not* of Cauchy-Kowaleska type: it does not contain a derivative $\partial p/\partial t$.

Remark 5.1. Other problems of this type are found among *transmission problems*; see Problem 17.11, Chapter 4—See also Problem 17.12, Chapter 4 and Section 5.3 below. □

Remark 5.2. The above system is none other than the linearized Navier-Stokes system. □

The preceding problem is *approximated* by problems of Cauchy-Kowaleska type in the following manner: we introduce

$$\begin{cases} \hat{H} = L^2(\Omega)^n \times L^2(\Omega), \\ \hat{V} = H_0^1(\Omega)^n \times L^2(\Omega), \end{cases} \tag{5.32}$$

and consider the problem

$$\begin{cases} \dfrac{\partial u_\varepsilon}{\partial t} - \Delta u_\varepsilon + \operatorname{grad} p_\varepsilon = f, \\ \\ \varepsilon \dfrac{\partial p_\varepsilon}{\partial t} + \operatorname{div} u_\varepsilon = 0, \quad \varepsilon > 0, \end{cases} \tag{5.33}$$

$$\{u_\varepsilon, p_\varepsilon\} \in \hat{V}^{((1))}. \tag{5.34}$$

If we set:

$$\vec{u}_\varepsilon = \{u_\varepsilon, p_\varepsilon\},$$

$$\Lambda_\varepsilon = \begin{pmatrix} \mathrm{d}/\mathrm{d}t & 0 \\ 0 & \varepsilon\, \mathrm{d}/\mathrm{d}t \end{pmatrix},$$

$$a(\vec{u}, \vec{v}) = a(u, v) + \int_\Omega \operatorname{grad} p . \bar{v}\, \mathrm{d}x + \int_\Omega \operatorname{div} u . \bar{q}\, \mathrm{d}x,$$

$$\text{for} \quad \vec{u} = \{u, p\}, \quad \vec{v} = \{v, q\} \in \hat{V}, \quad \vec{f} = \{f, 0\},$$

((1)) More precisely, function of t with values in $\hat{V}$.

we see that problem (5.33) may be written:

$$(5.35)\qquad \left(\Lambda_\varepsilon \vec{u}_\varepsilon, \vec{v}\right) + a\left(\vec{u}_\varepsilon, \vec{v}\right) = \left(\vec{f}, \vec{v}\right),$$

where

$$\left(\vec{f}, \vec{v}\right) = \int_\Omega \vec{f} \cdot \vec{v}\, dx.$$

We verify that

$$a\left(\vec{u}, \vec{v}\right) = a(u, v) + \int_\Omega p \cdot \operatorname{div} \bar{v}\, dx + \int_\Omega \operatorname{div} u \cdot \bar{q}\, dx,$$

so that $\vec{u}, \vec{v} \to a\left(\vec{u}, \vec{v}\right)$ is continuous on $\hat{V}$ and

$$a(v, v) \geq \alpha \|v\|^2_{(H_0^1(\Omega))^n}$$

and therefore

$$a\left(\vec{v}, \vec{v}\right) + c_1 \left\|\vec{v}\right\|^2_{\hat{H}} \geq \alpha \|v\|^2_{(H_0^1(\Omega))^n} + c_1 \|q\|^2_{L^2(\Omega)}.$$

The operator Λ_ε behaves like

$$\begin{pmatrix} d/dt & 0 \\ 0 & d/dt \end{pmatrix}^{((1))}.$$

Consequently Theorem 1.2 is valid and yields:

$$(5.36)\qquad \begin{cases} \text{if } f \in \mathscr{D}_{+,M_k}\left(\mathbf{R}; (H^{-1}(\Omega))^n\right), \\ \text{there exists a unique } \{u_\varepsilon, p_\varepsilon\}, \text{ solution of} \\ \text{(5.33), satisfying} \\ u_\varepsilon \in \mathscr{D}_{+,M_k}\left(\mathbf{R}; (H_0^1(\Omega))^n\right), \\ p_\varepsilon \in \mathscr{D}_{+,M_k}\left(\mathbf{R}; L^2(\Omega)\right). \end{cases}$$

We shall now prove

Theorem 5.2. *As* $\varepsilon \to 0$, *the solution* $\{u_\varepsilon, p_\varepsilon\}$ *of problem* (5.33) *given by* (5.36) *satisfies*

$$(5.37)\qquad u_\varepsilon \to u \text{ in } \mathscr{D}_{+,M_k}\left(\mathbf{R}; (H_0^1(\Omega))^n\right) \text{ weakly}^2,$$

$$(5.38)\qquad \operatorname{grad} p_\varepsilon \to \operatorname{grad} p \text{ in } \mathscr{D}_{+,M_k}\left(\mathbf{R}; (H^{-1}(\Omega))^n\right) \text{ weakly},$$

where $\{u, p\}$ *is the solution of* (5.29)–(5.30).

((1)) More precisely and more generally see Section 8 below.

[2] The sense of (5.37) will be specified in the course of the proof.

Proof. It follows immediately from (5.35) that, $\forall k$,

$$\left(\Lambda_\varepsilon \vec{u}_\varepsilon^{(k)}, \vec{v}\right) + a\left(\vec{u}_\varepsilon^{(k)}, \vec{v}\right) = \left(\vec{f}^{(k)}, \vec{v}\right), \tag{5.39}$$

from which, taking $\vec{v} = \vec{u}^{(k)}$:

$$\alpha \int_0^T \| u_\varepsilon^{(k)} \|^2_{(H_0^1(\Omega))^n} \, dt \le \int_0^T \| f^{(k)} \|_{(H^{-1}(\Omega))^n} \| u_\varepsilon^{(k)} \|_{(H_0^1(\Omega))^n} \, dt$$

and consequently, if we set

$$\| u_\varepsilon^{(k)} \|_{L^2(0,T;(H_0^1(\Omega))^n)} = \nu_1(u_\varepsilon^{(k)}),$$

$$\| f^{(k)} \|_{L^2(0,T;(H^{-1}(\Omega))^n)} = \nu_{-1}(f^{(k)}),$$

we have

$$\nu_1(u_\varepsilon^{(k)}) \le c\nu_{-1}(f^{(k)}). \tag{5.40}$$

Assume that

$$f \in \mathscr{D}_{0,M_k}\left([0, T], \mathscr{L}; (H^{-1}(\Omega))^n\right).$$

Then

$$\nu_{-1}(f^{(k)}) \le d\mathscr{L}^k M_k$$

and therefore there exists $\mathscr{L}_1$ such that

$$|||f|||^2_{\mathscr{L}_1} = \sum_{k=0}^{\infty} \frac{1}{(\mathscr{L}_1^k M_k)^2} \nu_{-1}(f^{(k)})^2 < \infty. \tag{5.41}$$

If we denote the space of functions f satisfying $|||f|||_{\mathscr{L}_1} < \infty$ by $\mathscr{D}^{L^2}_{0,M_k}([0, T], \mathscr{L}_1; (H^{-1}(\Omega))^n) = \Psi_{-1}$, Ψ_{-1} is a Hilbert space; if Ψ_1 denotes the space analogous to Ψ_{-1}, obtained by replacing $H^{-1}(\Omega)$ with $H_0^1(\Omega)$, we see that (5.40) entails that

(5.42) $\quad u_\varepsilon$ *remains in a bounded set of* Ψ_1 *as* $\varepsilon \to 0$.

Thus we can extract a subsequence, still denoted by u_ε, such that

(5.43) $\quad u_\varepsilon \to w$ in Ψ_1 weakly[1].

But it follows from (5.35) that

(5.44) $\quad \sqrt{\varepsilon}\, p_\varepsilon$ is bounded in $L^\infty(0, T; L^2(\Omega))$ weakly.

We may therefore assume that

(5.45) $\quad \begin{cases} \sqrt{\varepsilon}\, p_\varepsilon \text{ converges (and necessarily to 0)} \\ \text{in } L^\infty(0, T; L^2(\Omega)) \text{ weak star;} \end{cases}$

[1] Which makes (5.37) precise, after having shown that $w = u$.

Taking $v \in V$ in (5.35) (i.e. div $v = 0$), $\vec{v} = \{v, q\}$, we obtain:

$$(u_\varepsilon', v) + \varepsilon(p_\varepsilon', q) + a(u_\varepsilon, v) = (f, v)$$

and passing to the limit in, for example, the sense of distributions on $]0, T[$, it follows that

$$(w', v) + a(w, v) = (f, v), \ \forall v \in V.$$

Since $w(0) = 0$ we see that $w = u$ and (5.43) entails (5.37). Then (5.38) simply follows from the first equation in (5.33) which yields

$$\operatorname{grad} p_\varepsilon = f - \frac{\partial u_\varepsilon}{\partial t} + \Delta u_\varepsilon. \quad \square$$

5.3 Approximation by Systems of Cauchy-Kowaleska Type (II)

We now consider the setting of Problem 17.12, Chapter 4, for which we shall examine the properties of M_k-regularity in t. $\square$

We seek a function $u(x, t)$, $x \in \Omega$, $t \in \mathbf{R}$, satisfying

$$\Delta_x u(x, t) = 0, \ x \in \Omega, \ t \in \mathbf{R}, \tag{5.46}$$

$$\frac{\partial u}{\partial \nu} + \frac{\partial u}{\partial t} = g \text{ on } \Gamma, \ t \in \mathbf{R}. \tag{5.47}$$

Assume that (notation of Chapter 1 for the spaces $H^s(\Gamma)$):

$$g \in \mathscr{D}_{+,M_k}\big(\mathbf{R}; H^{-1/2}(\Gamma)\big). \tag{5.48}$$

Then problem (5.46), (5.47) *admits a unique solution which satisfies*

$$u \in \mathscr{D}_{+,M_k}\big(\mathbf{R}; H^1(\Omega)\big). \tag{5.49}$$

Indeed, let us introduce

$$w = \gamma_0 u = \text{trace of } u \text{ on } \Gamma. \tag{5.50}$$

As in Problem 17.12, Chapter 4, we introduce the operator

$$\mathscr{B} \in \mathscr{L}\big(H^{1/2}(\Gamma); H^{-1/2}(\Gamma)\big)^{((1))} \tag{5.51}$$

by

$$\mathscr{B}h = \frac{\partial \omega}{\partial \nu}, \ h \in H^{1/2}(\Gamma), \tag{5.52}$$

where

$$\Delta \omega = 0, \ \omega|_\Gamma = h. \tag{5.53}$$

((1)) We assume the boundary Γ of Ω to be sufficiently regular.

Then (5.46), (5.47) is equivalent to

$$\mathscr{B}w + \frac{\partial w}{\partial t} = g, \tag{5.54}$$

equation on Γ; since

$$(\mathscr{B}v, v) \geq \beta \|v\|^2_{H^{1/2}(\Gamma)}, \ \beta > 0, \ v \in H^{1/2}(\Gamma),$$

we can apply Theorem 1.2 (for a simplified case, since $\mathscr{B}$ *does not depend on* t) with:

$$V = H^{1/2}(\Gamma), \ A(t) = \mathscr{B}, \ H = L^2(\Gamma).$$

It follows that w exists and is unique, and that

$$w \in \mathscr{D}_{+,M_k}\left(\mathbf{R}; H^{1/2}(\Gamma)\right). \tag{5.55}$$

Now let us denote by G the operator

$$G \in \mathscr{L}\left(H^{1/2}(\Gamma); H^1(\Omega)\right)$$

defined by

$$\Delta(G\varphi) = 0 \text{ in } \Omega,$$

$$\gamma_0(G\varphi) = \varphi \text{ on } \Gamma.$$

Then

$$u(t) = Gw(t) \tag{5.56}$$

and (5.49) follows from (5.55) and (5.56). ▯

Problem (5.54) is an evolution equation *on the variety* Γ; in the form (5.46), (5.47), the problem *is not* of Cauchy-Kowaleska type. But it can be approximated by the following problems:

$$\varepsilon \frac{\partial u_\varepsilon}{\partial t} - \Delta u_\varepsilon = 0, \ x \in \Omega, \ t \in \mathbf{R}, \ \varepsilon > 0, \tag{5.57}$$

$$\frac{\partial u_\varepsilon}{\partial t} + \frac{\partial u_\varepsilon}{\partial \nu} = g, \ x \in \Gamma, \ t \in \mathbf{R}. \tag{5.58}$$

In variational form (see Chapter 3, Volume 1), problem (5.57), (5.58) is written

$$\varepsilon(u_\varepsilon', v) + (u_\varepsilon', v)_\Gamma + a(u_\varepsilon, v) = (g, v)_\Gamma, \ \forall v \in H^1(\Omega), \tag{5.59}$$

where

$$(\varphi, \psi)_\Gamma = \int_\Gamma \varphi\bar{\psi} \, d\Gamma,$$

$$a(\varphi, \psi) = \sum_{i=1}^{n} \int_\Omega \frac{\partial \varphi}{\partial x_i} \overline{\frac{\partial \psi}{\partial x_i}} \, dx.$$

It can be seen immediately (simple variant of Theorem 1.2[1]) that, *if* $g \in \mathscr{D}_{+,M_k}(\mathbf{R}; H^{1/2}(\Gamma))$, *problem* (5.59) *admits a unique solution* $u_\varepsilon \in \mathscr{D}_{+,M_k}(\mathbf{R}; H^1(\Omega))$. □

We have:

Theorem 5.3. *As* $\varepsilon \to 0$, *the solution* u_ε *of problem* (5.59) *satisfies*:

$$u_\varepsilon \to u \text{ in } \mathscr{D}_{+,M_k}\big(\mathbf{R}; H^1(\Omega)\big), \tag{5.60}$$

where u is the solution of (5.46), (5.47), (5.49).

Proof. It follows from (5.59) that

$$\varepsilon(u_\varepsilon^{(k+1)}, v) + (u_\varepsilon^{(k+1)}, v)_\Gamma + a(u_\varepsilon^{(k)}, v) = (g^{(k)}, v)_\Gamma, \tag{5.61}$$

from which, taking $v = u_\varepsilon^{(k)}$ and assuming that $g \in \mathscr{D}_{0,M_k}([0, T]; H^{-1/2}(\Gamma)$, we obtain:

$$\frac{\varepsilon}{2}\,|u_\varepsilon^{(k)}(T)|^2_{L^2(\Omega)} + \frac{1}{2}\,\|u_\varepsilon^{(k)}(T)\|^2_{L^2(\Gamma)} + \int_0^T a(u_\varepsilon^{(k)}, u_\varepsilon^{(k)})\,dt = \int_0^T (g^{(k)}, u_\varepsilon^{(k)})_\Gamma\,dt$$

from which we further obtain[2]

$$\alpha \int_0^T \|u_\varepsilon^{(k)}(t)\|^2_{H^1(\Omega)}\,dt \le \int_0^T \|g^{(k)}(t)\|_{H^{-1/2}(\Gamma)}\,\|u_\varepsilon^{(k)}(t)\|_{H^{1/2}(\Gamma)}\,dt$$

and since (Chapter 1, Section 8):

$$\|v\|_{H^{1/2}(\Gamma)} \le c_1\,\|v\|_{H^1(\Omega)},$$

it follows that

$$\int_0^T \|u^{(k)}(t)\|^2_{H^1(\Omega)}\,dt \le c_2 \int_0^T \|g^{(k)}(t)\|^2_{H^{-1/2}(\Gamma)}\,dt. \tag{5.62}$$

But on the other hand u satisfies

$$(u', v)_\Gamma + a(u, v) = (g, v)_\Gamma, \ \forall v \in H^1(\Omega),$$

therefore

$$(u^{(k+1)}, v)_\Gamma + a(u^{(k)}, v) = (g^{(k)}, v)_\Gamma, \ \forall v \in H^1(\Omega); \tag{5.63}$$

if we set

$$w_\varepsilon = u_\varepsilon - u,$$

then it follows from (5.61), (5.63) that

$$(w_\varepsilon^{(k+1)}, v)_\Gamma + a(w_\varepsilon^{(k)}, v) = -\varepsilon(u_\varepsilon^{(k+1)}, v),$$

[1] Moreover valid for $a(u, v)$ replaced by $a(t; u, v)$ depending on t, so that $A(t)$ is of class M_k.

[2] We always first reduce the problem to the case for which

$$a(v, v) \ge \alpha\,\|v\|^2_{H^1(\Omega)}, \quad \alpha > 0.$$

from which we deduce that

$$\alpha \int_0^T \| w^{(k)}(t) \|^2_{H^1(\Omega)} \, dt \leq \varepsilon \int_0^T \| u_\varepsilon^{(k+1)}(t) \|^2_{L^2(\Omega)} \, dt \leq$$

$$\leq (\text{by } (5.62)) \; c_3 \varepsilon \int_0^T \| g^{(k+1)}(t) \|^2_{H^{-1/2}(\Gamma)} \, dt \leq$$

$$\leq c_3 \varepsilon \mathscr{L}^{k+1} M_{k+1} \leq$$

$$\leq c_4 \varepsilon (\mathscr{L} H)^k \, M_k ,$$

whence the desired result. □

Remark 5.3. Analogous results hold for the problem (see Problem 17.12, Chapter 4):

$$\Delta u = 0 \text{ in } \Omega, \frac{\partial u}{\partial \nu} + \frac{\partial^2 u}{\partial t^2} = g \text{ on } \Gamma, \tag{5.64}$$

which can be approximated by

$$\begin{cases} \varepsilon \dfrac{\partial^2 u_\varepsilon}{\partial t^2} - \Delta u_\varepsilon = 0 \text{ in } \Omega, \\[2ex] \dfrac{\partial u_\varepsilon}{\partial \nu} + \dfrac{\partial^2 u_\varepsilon}{\partial t^2} = g \text{ on } \Gamma. \quad □ \end{cases} \tag{5.65}$$

Remark 5.4. One could also consider the "parabolic regularization" of problem (5.65):

$$\begin{cases} \varepsilon(u''_{\varepsilon,\varepsilon_1}, v) + (u''_{\varepsilon,\varepsilon_1}, v)_\Gamma + \varepsilon_1 a(u'_{\varepsilon,\varepsilon_1}, v) + a(u'_{\varepsilon,\varepsilon_1}, v) = (g, v)_\Gamma, \\ \forall v \in H^1(\Omega) \end{cases} \tag{5.66}$$

and let $\varepsilon, \varepsilon_1 \to 0$. □

Remark 5.5. Analogous considerations for the problem of "Schroedinger type on the variety Γ":

$$\begin{cases} \Delta u = 0 \text{ in } \Omega, \; t \in \mathbf{R}, \\[1ex] i \dfrac{\partial u}{\partial \nu} + \dfrac{\partial u}{\partial t} = g \text{ on } \Gamma, \; t \in \mathbf{R}. \quad □ \end{cases} \tag{5.67}$$

Remark 5.6. One obtains analogous convergence results for the *parabolic regularization* (see Chapter 5, Section 12.2) of problem (4.1) which is approximated by

$$(\varepsilon + i) \, A(t) \, u_\varepsilon + u'_\varepsilon = f. \quad □ \tag{5.68}$$

6. Transposition

6.1 Orientation

With the exception of the results of Section 3, all results obtained in the preceding sections can be "transposed" so as to yield solutions in the form of distributions or ultra-distributions of class M_k, when the right-hand side of the equation is a distribution or an ultra-distribution of class M_k.

We shall give some results[1] for ultra-distributions. □

6.2 The Parabolic Case

The notation is as in Section 1. We introduce the adjoint $A^*(t)$ of $A(t)$ defined by

$$(6.1) \qquad (A^*(t)\, u, v) = (u, A(t)\, v), \quad \forall u, v \in V.$$

Hypothesis (1.21) entails (and in fact is equivalent to):

$$(6.2) \qquad t \to A^*(t) \text{ belongs to the class } \mathscr{E}_{M_k}(\mathbf{R}; \mathscr{L}(V; V')).$$

Replacing the spaces $\mathscr{D}_{+,M_k}$ with $\mathscr{D}_{-,M_k}$ and "changing the sense of time" (thus $\mathrm{d}/\mathrm{d}t$ to $-\mathrm{d}/\mathrm{d}t$), Theorem 1.2 yields

$$(6.3) \qquad \begin{cases} \text{the operator } A^*(t) - \mathrm{d}/\mathrm{d}t \text{ is an isomorphism of} \\ \mathscr{D}_{-,M_k}(\mathbf{R}; V) \text{ onto } \mathscr{D}_{-,M_k}(\mathbf{R}; V'). \end{cases}$$

By transposition of (6.3), noting that the adjoint of $A^*(t) - \mathrm{d}/\mathrm{d}t$ is $A(t) + \mathrm{d}/\mathrm{d}t$, and recalling the definition of $\mathscr{D}'_{+,M_k}(\mathbf{R}; X) = (\mathscr{D}_{-,M_k}(\mathbf{R}; X'))'$ (see Chapter 7, Section 5), we obtain

Theorem 6.1. *Let $a(t; u, v)$ satisfy* (1.5), (1.21) *and assume that the sequence $\{M_k\}$ satisfies* (1.22), ..., (1.25); *then the mapping $u \to A(t)\, u + u'$ is an isomorphism of $\mathscr{D}'_{+,M_k}(\mathbf{R}; V)$ onto $\mathscr{D}'_{+,M_k}(\mathbf{R}; V')$.* □

Example 6.1. Let us give an explicit application of Theorem 6.1 in the setting of Section 1.5 with $m = 1$.

Let

$$(6.4) \qquad \begin{cases} A(t)\, u = -\sum \dfrac{\partial}{\partial x_i}\left(a_{ji}(x, t)\, \dfrac{\partial u}{\partial x_j}\right), \\ t \to a_{ij}(\cdot, t) \in \mathscr{E}_{M_k}(\mathbf{R}_t; L^\infty(\Omega)), \\ \displaystyle\sum_{i,j=1}^{n} a_{ij}(x, t)\, \xi_i \bar{\xi}_j \geq \alpha\, |\xi|^2, \quad \alpha > 0,\ x \in \Omega,\ t \in \mathbf{R}. \end{cases}$$

[1] The—in fact very simple!—method is general.

Let f and g be given with

$$(6.5)\qquad \begin{cases} f \in \mathscr{D}'_{+,M_k}\big(\mathbf{R};\, L^2(\Omega)\big), \text{ or more generally,} \\ f \in \mathscr{D}'_{+,M_k}\big(\mathbf{R};\, \Xi^{-1}(\Omega)\big) \text{ (see Chapter 3, Section 4.7.3,} \\ \text{with } \Gamma_0 = \phi \text{ for } \Xi^{-1}(\Omega)\big), \end{cases}$$

$$(6.6)\qquad g \in \mathscr{D}'_{+,M_k}\big(\mathbf{R};\, H^{-1/2}(\Gamma)\big).$$

For $v \in V = H^1(\Omega)$, set

$$(6.7)\qquad L(v) = (f, v) + (g, \gamma_0 v)_\Gamma,$$

where $(f, v) \in \mathscr{D}'_{+,M_k}(\mathbf{R})$ is defined by

$$\langle (f, v), \varphi\rangle = \langle f, \varphi \otimes \bar{v}\rangle,\ \varphi \in \mathscr{D}_{-,M_k}(\mathbf{R})$$

and similarly

$$\langle (g, \gamma_0 v)_\Gamma, \varphi\rangle = \langle g, \varphi \otimes \gamma_0 \bar{v}\rangle;$$

formula (6.7) defines $L \in \mathscr{D}'_{+,M_k}(\mathbf{R};\, V')$ and Theorem 6.1 shows the *existence and uniqueness of* $u \in \mathscr{D}'_{+,M_k}(\mathbf{R};\, V)$ *satisfying*

$$(6.8)\qquad (u', v) + a(t;\, u, v) = L(v),\ \forall v \in V,$$

$$\left(\text{where } a(t;\, \varphi, \psi) = \sum_{i,j=1}^{n} \int_\Omega a_{ij}(x, t)\, \frac{\partial \varphi}{\partial x_j}\, \frac{\partial \bar{\psi}}{\partial x_i}\, \mathrm{d}x\right).$$

It follows from (6.8) that, in the sense of $\mathscr{D}'_{+,M_k}(\mathbf{R};\, \mathscr{D}'(\Omega))$ in particular, we have

$$(6.9)\qquad u' + A(t)\, u = f.$$

It then follows from (6.9) that $A(t)\, u = f - u' \in \mathscr{D}'_{+,M_k}(\mathbf{R};\, \Xi^{-1}(\Omega))$; thus u has the properties

$$(6.10)\qquad u \in \mathscr{D}'_{+,M_k}\big(\mathbf{R};\, H^1(\Omega)\big),\ A(t)\, u \in \mathscr{D}'_{+,M_k}\big(\mathbf{R};\, \Xi^{-1}(\Omega)\big).$$

It can be deduced from (6.10) that $\dfrac{\partial u}{\partial \nu_{A(t)}}$ (extension by continuity of $\varphi \to \dfrac{\partial \varphi}{\partial \nu_{A(t)}} = \sum_{i,j=1}^{n} a_{ij}(x, t)\, \dfrac{\partial \varphi}{\partial x_j} \cos(\nu, x_i)$ for regular functions φ in $\bar{\Omega} \times \mathbf{R}_t$) is well-defined and that

$$(6.11)\qquad \frac{\partial u}{\partial \nu_{A(t)}} \in \mathscr{D}'_{+,M_k}\big(\mathbf{R};\, H^{-1/2}(\Gamma)\big).$$

The proof of (6.11) is rather lengthy, except for the case where $A(t) = A$ is independent of t, to which we shall limit our presentation.

We first recall (see Chapter 7, Remarks 5.3 and 5.5) that when X is, in particular, a Hilbert space,

$$\mathscr{D}'_{+,M_k}(\mathbf{R}; X) = \mathscr{L}\big(\mathscr{D}_{-,M_k}(\mathbf{R}); X\big).$$

Thus if u satisfies (6.10), then, $\forall \psi \in \mathscr{D}_{-,M_k}(\mathbf{R})$, $u(\psi)$ satisfies

$$u(\psi) \in H^1(\Omega),$$

$$A\big(u(\psi)\big) = (Au)\,(\psi) \in \Xi^{-1}(\Omega)\ ^{((1))}$$

and the linear mapping

$$\psi \to u(\psi)$$

is *continuous* from $\mathscr{D}_{-,M_k}(\mathbf{R}) \to Y$, where

$$Y = \{w \mid w \in H^1(\Omega),\ Aw \in \Xi^{-1}(\Omega)\},$$

provided with the (Hilbert) norm of the "graph":

$$(\|w\|^2_{H^1(\Omega)} + \|Aw\|^2_{\Xi^{-1}(\Omega)})^{1/2}$$

and this is *equivalent* to (6.10).

But, according to the results of Chapter 2, $w \to \partial w/\partial \nu_A$ is a continuous linear mapping of $Y \to H^{-1/2}(\Gamma)$ and consequently

$$\psi \to \frac{\partial}{\partial \nu_A} u(\psi)$$

is a continuous linear mapping of $\mathscr{D}_{-,M_k}(\mathbf{R}) \to H^{-1/2}(\Gamma)$ and thus defines the ultra-distribution $\partial u/\partial \nu_A$, element of $\mathscr{D}'_{+,M_k}(\mathbf{R}; H^{-1/2}(\Gamma))$.

We can now complete the interpretation of (6.8). Taking ψ in $\mathscr{D}_{-,M_k}(\mathbf{R})$, it follows from (6.8) (still for the case "$A(t)$ independent of t") that

$$\big(u'(\psi), v\big) + a\big(u(\psi), v\big) = \big(f(\psi), v\big) + \big(g(\psi), \gamma_0 v\big)_\Gamma;$$

on the other hand, taking the scalar product of both sides of equation (6.9) with $\bar{\psi} \otimes \bar{v}$, we obtain

$$\big(u'(\psi), v\big) + \big(Au(\psi), v\big) = \big(f(\psi), v\big),$$

from which it follows that

$$\big(Au(\psi), v\big) = a\big(u(\psi), v\big) - \big(g(\psi), \gamma_0 v\big)_\Gamma$$

and consequently

$$\frac{\partial u}{\partial \nu_A}(\psi) = g(\psi),\ \forall \psi \in \mathscr{D}_{-,M_k}(\mathbf{R}).$$

((1)) This is where the hypothesis "$A(t)$ independent of t" intervenes.

So that we have obtained *the existence and uniqueness of u in $\mathscr{D}'_{+,M_k}(\mathbf{R}; H^1(\Omega))$ satisfying*

$$u' + Au = f,$$

$$\frac{\partial u}{\partial \nu_A} = g. \quad \square$$

Remark 6.1. We have transferred the initial data to the second member, a classical procedure for the solution of the Cauchy problem. □

Example 6.2. Let $\Gamma_0 \subset \Gamma$, as in Chapter 3, Section 4.7.3. According to G. Geymonat, we then take (in the notation of this Section and of Chapter 1):

$$f \in \mathscr{D}'_{+,M_k}\left(\Xi^{-1}(\Omega)\right),$$

$$g \in \mathscr{D}'_{+,M_k}\left(H_{00}^{-1/2}(\Gamma_0)\right);$$

then we see that there exists a unique u in $\mathscr{D}'_{+,M_k}(\mathbf{R}; H^1(\Omega))$, solution of (6.9), with, formally:

$$\begin{cases} \dfrac{\partial u}{\partial \nu_{A(t)}} = g \text{ on } \Gamma_1 \times \mathbf{R}_t \ (\Gamma_1 = \Gamma - \Gamma_0), \\ u = 0 \text{ on } \Gamma_0 \times \mathbf{R}_t. \end{cases}$$

If A is independent of t, a precise interpretation, such as for Example 6.1, can be given. □

6.3 The Second Order in t Case and the Schroedinger Case

We now consider the setting of Theorem 2.2. Then

$$A^*(t) = A(t)$$

and we can invert the sense of time:

$$u \to A(t)\, u + u'' \text{ is an isomorphism of} \tag{6.12}$$

$$\mathscr{D}_{-,M_k}(V) \to \mathscr{D}_{-,M_k}(V').$$

By transposition of (6.12), we obtain

Theorem 6.2. *Let hypotheses* (1.5), (2.1) *and* (1.21) *hold, and assume that the sequence $\{M_k\}$ satisfies* (1.22), ..., (1.25). *Then the mapping $u \to A(t)\, u + u''$ is an isomorphism of $\mathscr{D}'_{+,M_k}(V)$ onto $\mathscr{D}'_{+,M_k}(V')$.*

By transposition of Theorem 4.2 (and changing i to $-i$, which is permissible) we obtain

Theorem 6.3. *Under the hypotheses of Theorem* 6.2, *the mapping $u \to iA(t)\, u + u'$ is an isomorphism of $\mathscr{D}'_{+,M_k}(V)$ onto $\mathscr{D}'_{+,M_k}(V')$.* □

Remark 6.2. Examples of applications of Theorems 6.2 and 6.3 analogous to Examples 6.1 and 6.2 can be given. □

7. Semi-Groups

7.1 Orientation

We shall now consider evolution equations of the form

$$\frac{du}{dt} + Au = 0, \; t > 0, \tag{7.1}$$

$$u(0) = u_0, \tag{7.2}$$

where $-A$ is the infinitesimal generator of a semi-group in a *Banach space* E.

7.2 The Space of Vectors of Class M_k

Let E be a Banach space on $\mathbf{C}$, with norm denoted by $\| \|$. In order to simplify the presentation, we assume that E is *reflexive*.

In the space E, define a continuous semi-group $G(t)$, i.e.:

$$\begin{cases} G(t) \in \mathscr{L}(E; E), \; \forall t \geq 0, \\ \forall e \in E, \text{ the function } t \to G(t)e \text{ is a continuous} \\ \text{mapping of } t \geq 0 \to E, \\ G(0) = I, \\ G(t)\, G(s) = G(t+s), \; \forall t, s \geq 0. \end{cases} \tag{7.3}$$

If we are willing to change $G(t)$ to $\exp(-\omega t)\, G(t)$, suitable ω, we may always assume that

$$\|G(t)\| < \mu < \infty \; \left(\|G(t)\| = \text{norm of } G(t) \text{ in } \mathscr{L}(E; E)\right). \tag{7.4}$$

$-A$ denotes the infinitesimal generator of $G(t)$. Then the solution u of (7.1), (7.2) may be written

$$u(t) = G(t) u_0. \tag{7.5}$$

Remark 7.1. Instead of (7.1), we could, more generally, consider the equation

$$\frac{du}{dt} + Au = f; \tag{7.1a}$$

then, at least formally, the solution of (7.1a), (7.2) is expressed by

$$u(t) = G(t)\, u_0 + \int_0^t G(t - \sigma)\, f(\sigma)\, d\sigma. \tag{7.5a}$$

If, for example, f is assumed to be continuous from $t \geq 0 \to E$, this formula can be justified in the following manner.

Introduce

$$\mathscr{G} = \begin{cases} G(t) & \text{for } t \geq 0 \\ 0 & \text{for } t < 0. \end{cases} \tag{7.6}$$

Then (see Chapter 4, Section 3.1):

$$\begin{cases} \dfrac{\mathrm{d}}{\mathrm{d}t}\mathscr{G} + A\mathscr{G} = \delta(t) \otimes I_E \quad (I_X = \text{identity from } X \to X), \\ \mathscr{G}\left(\dfrac{\mathrm{d}}{\mathrm{d}t} + A\right) = \delta(t) \otimes I_{\mathrm{D}(A)}, \end{cases} \tag{7.7}$$

where

$$\begin{cases} \mathrm{D}(A) = \text{domain of } A = \{e \mid h^{-1}(G(h)\,e - e) \text{ converges} \\ \text{in } E \text{ as } h \to 0\}, \end{cases} \tag{7.8}$$

and then

$$A\,e = \lim_{h \to 0} h^{-1}(G(h)e - e)\,^{((1))}. \tag{7.9}$$

Now, if

$$\tilde{u}(t) = \begin{cases} u(t), & t \geq 0, \\ 0, & t < 0, \end{cases} \qquad \tilde{f}(t) = \begin{cases} f(t), & t \geq 0, \\ 0, & t < 0, \end{cases} \tag{7.10}$$

problem (7.1), (7.2) may be written

$$\frac{\mathrm{d}\tilde{u}}{\mathrm{d}t} + A\tilde{u} = \tilde{f} + \delta(t)\,u_0, \tag{7.11}$$

of which *the* solution in $\mathscr{D}'_+(\mathrm{D}(A))$ is

$$\tilde{u} = \mathscr{G} * (\tilde{f} + \delta(t)\,u_0), \tag{7.12}$$

which is a justification of (7.1a). ☐

We aim to study the properties of *regularity* and *M_k-regularity* of the function $t \to u(t)$ and, more particularly, of the function $t \to G(t)\,u_0$. Thus, we shall assume that $f = 0$. ☐

The function $t \to G(t)\,u_0$ is once continuously differentiable from $t \geq 0 \to E$ if and only if $u_0 \in \mathrm{D}(A)$ and then

$$\frac{\mathrm{d}}{\mathrm{d}t} G(t)\,u_0 = -G(t)\,Au_0. \tag{7.13}$$

We iterate on this remark; we set

$$\mathrm{D}(A^k) = \{e \mid e \in \mathrm{D}(A),\ \ Ae \in \mathrm{D}(A), \ldots, A^{k-1}e \in \mathrm{D}(A)\}, \tag{7.14}$$

((1)) The operator A is closed (see Hille-Phillips [1], K. Yosida [1]); $\mathrm{D}(A)$ is provided with the norm of the graph $||e|| + ||Ae||$, which makes it a Banach space.

which is a Banach space with the norm

$$(7.15) \qquad \sum_{j=0}^{k} \|A^j e\|;$$

then

$$(7.16) \qquad \begin{cases} \mathrm{D}(A^k) \text{ is the space of vectors } e \text{ such that } t \to G(t)e \\ \text{is } k\text{-times continuously differentiable from } t \geq 0 \to E. \end{cases}$$ ▯

Next we set

$$(7.17) \qquad \mathrm{D}(A^\infty) = \bigcap_{k=0}^{\infty} \mathrm{D}(A^k),$$

which is a Fréchet space with the sequence of norms (7.15); we have:

$$(7.18) \qquad \begin{cases} \mathrm{D}(A^\infty) \text{ is the space of vectors } e \text{ such that } t \to G(t)e \\ \text{is infinitely differentiable from } t \geq 0 \to E. \end{cases}$$

The following is a classical result.

Proposition 7.1. *The space* $\mathrm{D}(A^\infty)$ *is dense in* E.

Proof. Let $\varrho_n \in \mathscr{D}(]0, +\infty[)$, $\varrho_n(t) \geq 0$, ϱ_n with support in $[\alpha_n, \beta_n]$, $0 < \alpha_n < \beta_n$, $\beta_n \to 0$, $\int_{\alpha_n}^{\beta_n} \varrho_n(t)\,\mathrm{d}t = 1$ (ϱ_n is a regularizing sequence).

In general, if φ is a continuous scalar function on $t \geq 0$, with compact support (or "sufficiently small" at infinity), one sets:

$$(7.19) \qquad G(\varphi)e = \int_0^\infty G(t)\, e\varphi(t)\,\mathrm{d}t,$$

an integral with values in E.

If $\varphi \in \mathscr{D}(]0, +\infty[)$, we can easily verify that, $\forall e \in E$, $G(\varphi)\, e \in \mathrm{D}(A)$ and

$$(7.20) \qquad AG(\varphi)\, e = G(\varphi')\, e.$$

We can iterate: $G(\varphi)e \in \mathrm{D}(A^\infty)$ and, $\forall e \in E$, we have:

$$(7.21) \qquad A^k G(\varphi)e = G(\varphi^k)e, \ \forall k.$$

Thus, for arbitrary e in E, $G(\varrho_n)e \in \mathrm{D}(A^\infty)$. Now

$$G(\varrho)\, e - e = \int_0^\infty \left(G(t)e - G(0)e\right) \varrho_n(t)\,\mathrm{d}t,$$

therefore

$$\|G(\varrho_n)e - e\| \leq \sup_{t \in [0,\beta_n]} \|G(t)e - e\| \to 0 \quad \text{as} \quad n \to \infty,$$

whence the desired result. ▯

We now introduce

Definition 7.1. We call *vector of class* M_k, $\{M_k\}$ being a logarithmically convex sequence, every element $e \in E$ for which there exist constants c and L such that

$$\|A^k e\| \leq c L^k M_k, \quad \forall k. \tag{7.22}$$

Of course, it is not obvious (and in fact not true!) that such *non-zero* vectors always exist. Nevertheless, we shall set (for the time being an *algebraic* definition):

$$\begin{cases} \mathrm{D}(A^\infty; M_k) = \text{space (maybe reduced to } \{0\}) \text{ of} \\ \text{"vectors of class } M_k\text{" in the sense of Definition 7.1.} \end{cases} \tag{7.23}$$

□

We easily verify that

$$\begin{cases} \mathrm{D}(A^\infty; M_k) \text{ coincides with the space of } e\text{'s} \in E \\ \text{such that the function } t \to G(t)e \text{ is of class } M_k \\ \text{from } t \geq 0 \to E. \end{cases} \tag{7.24}$$

□

We have the following: *condition of non-triviality* of $\mathrm{D}(A^\infty; M_k)$:

Theorem 7.1. *If the sequence* $\{M_k\}$ *satisfies* (1.22) *and* (1.23), *the space* $\mathrm{D}(A^\infty; M_k)$ *is dense in* E.

Proof. Let $\mathrm{D}_{M_k}(]0, \infty[) = \{\varphi \mid \varphi \in \mathscr{D}(]0, \infty[),\ \varphi$ of class M_k on $t \geq 0\}$. Since M_k is not quasi-analytic, this space is different from $\{0\}$, and there exists (see S. Mandelbrojt [1], C. Roumien [1]) a sequence $\varrho_n \in \mathscr{D}_{M_k}(]0, \infty[)$ which satisfies the conditions given in the proof of Proposition 7.1.

If $e \in E$, we have: $G(\varrho_n)\, e \to e$ in E, and therefore we shall have the desired result if we can show that

$$G(\varphi)e \in \mathrm{D}(A^\infty; M_k),\ \forall e \in E,\ \forall \varphi \in \mathscr{D}_{M_k}(]0, \infty[).$$

But we have (7.21), which shows that

$$\frac{\mathrm{d}^k}{\mathrm{d}t^k} G(t)\, G(\varphi)e = G(\varphi^{(k)})e,$$

so that

$$\left\| \frac{\mathrm{d}^k}{\mathrm{d}t^k} G(t)\ G(\varphi)e \right\| \leq \mu \|e\| \int_0^\infty |\varphi^{(k)}(t)|\, \mathrm{d}t \leq$$

$$\leq C L^k M_k \left(\text{since } \varphi \in \mathscr{D}_{M_k}(]0, \infty[)\right),$$

therefore $t \to G(t)\, G(\varphi)e$ is of class M_k. □

We note that the preceding proof also shows that $\mathrm{D}(A^\infty, M_k)$ is dense in $\mathrm{D}(A^\infty)$.

Remark 7.2. The hypothesis of non-quasi-analyticity of the sequence $\{M_k\}$ is not always necessary for the non-triviality of $D(A^\infty; M_k)$.

Indeed, let $M_k = 1$, $\forall k$.

Assume that A admits a complete system of eigenfunctions:

$$Aw_j = \lambda_j w_j, \quad w_j \in D(A). \tag{7.25}$$

Then

$$w_j \in D(A^\infty; M_k), \; \forall j,$$

thus $D(A^\infty; M_k)$ contains the space generated by the w_j's and is therefore dense in E. ☐

Remark 7.3. Here is an example for which

$$D(A^\infty; M_k) = \{0\}, \; \{M_k\} \text{ quasi-analytic}. \tag{7.26}$$

Let $E = L^p(0, \infty)$, $1 \leq p < \infty$, and let $G(t)$ be the semi-group of translations defined by

$$G(t)\, f(x) = \begin{cases} 0 \;\text{ for }\; x < t, \\ f(x-t) \;\text{ for }\; x > t, \end{cases} \qquad \text{a.e. in } x,\; f \in L^p(0, \infty).$$

Then

$$D(A) = \{f \mid f, f' \in L^p(0, \infty),\; f(0) = 0\},\; Af = f'.$$

If we let $M_k = k!$, the elements of $D(A^\infty; M_k)$ must be analytic functions f on $x \geq 0$ with $f^{(k)}(0) = 0$, $\forall k$, therefore $f = 0$. ☐

Remark 7.4. Topology on $D(A^\infty; M_k)$

Taking into account (7.24), we introduce

$$D^L(A^\infty; M_k) = \left\{e \mid \sup_k \frac{\|A^k e\|}{L^k M_k} < \infty\right\}, \tag{7.27}$$

which is a Banach space with the norm

$$\|e\| = \sup_k \frac{\|A^k e\|}{L^k M_k}.$$

We then provide $D(A^\infty; M_k)$ with the topology

$$D(A^\infty; M_k) = \operatorname*{ind\,lim}_{L \to \infty} D^L(A^\infty; M_k). \tag{7.28}$$

The mapping

$$e \to \text{“}t \to G(t)e\text{''} \tag{7.29}$$

(by definition!) maps $D(A^\infty; M_k) \to \mathcal{D}_{M_k}([0, T]; E)$.

We easily verify, using Chapter 7, Section 4, that the topology defined in (7.28) coincides with the topology such that the mapping (7.29) is an isomorphism of $D(A^\infty; M_k)$ onto its image, provided with the topology induced by $\mathcal{D}_{M_k}([0, T]; E)$. ☐

Remark 7.5. The preceding considerations extend to the case in which E is not a Banach space (for semi-groups in locally convex spaces, see K. Yosida [1]). ☐

We shall now prove

Theorem 7.2. *If the injection of* $\mathrm{D}(A) \to E$ *is compact, so is the injection of* $\mathrm{D}^L(A^\infty; M_k) \to \mathrm{D}^{L'}(A^\infty; M_k)$, $L < L'$ (and then, see Chapter 7, Section 1.2, the space $\mathrm{D}(A^\infty; M_k)$ is an inductive limit of a regular sequence of Banach spaces).

Proof. For the purposes of this proof, let us write D^L instead of $\mathrm{D}^L(A^\infty; M_k)$, etc. For $u, v \in \mathrm{D}^L$, we have $u, v \in D^{L'}$ and

$$\|u - v\|_{D^{L'}} = \sup_k \frac{\|A^k(u - v)\|}{L'^k M_k} \leq$$

$$\leq \sup_{0 \leq k \leq N} \frac{\|A^k(u - v)\|}{L'^k M_k} + \sup_{k > N} \frac{\|A^k(u - v)\|}{L^k M_k} \left(\frac{L}{L'}\right)^N$$

and therfore

$$\|u - v\|_{\mathrm{D}^{L'}} \leq \sup_{0 \leq k \leq N} \frac{\|A^k(u - v)\|}{L'^k M_k} + \left(\frac{L}{L'}\right)^N \|u - v\|_{\mathrm{D}^L}. \tag{7.30}$$

Now let u_n be a bounded sequence in D^L, thus

$$\|A^k u_n\| \leq c_1 L^k M_k, \ \forall k, \forall n.$$

In particular u_n is bounded in $\mathrm{D}(A^q)$, $\forall q$. Since the injection "$\mathrm{D}(A) \to E$" is compact, so is the injection $\mathrm{D}(A^{q+1}) \to \mathrm{D}(A^q)$ and consequently we can extract a subsequence u_ν from u_n such that

$$u_\nu \to u \text{ in } \mathrm{D}(A^q) \text{ strongly, } \forall q$$

and

$$\|A^k u\| \leq c_1 L^k M_k, \ \forall k, \text{ therefore } u \in \mathrm{D}^L.$$

Let us show that $u_\nu \to u$ in $\mathrm{D}^{L'}$ *strongly*. To this end, let $\varepsilon > 0$ be given. Choose N such that

$$2c_1 \left(\frac{L}{L'}\right)^N \leq \varepsilon/2.$$

Then

$$\|u_\nu - u\|_{\mathrm{D}^{L'}} \leq \sup_{0 \leq k \leq N} \frac{\|A^k(u_\nu - u)\|}{L'^k M_k} + \frac{\varepsilon}{2}$$

and since $u_\nu \to u$ in $\mathrm{D}(A^q)$ strongly, we have

$$\sup_{0 \leq k \leq N} \frac{\|A^k(u_\nu - u)\|}{L^k M_k} \leq \frac{\varepsilon}{2} \quad \text{for } \nu \geq \nu(\varepsilon),$$

whence $\|u_\nu - u\|_{\mathrm{D}^{L'}} \leq \varepsilon$ for $\nu \geq \nu(\varepsilon)$. ☐

Remark 7.6. The principle of this proof is analogous to the one given for the spaces $\mathscr{D}_{M_k}(\Omega)$; see Chapter 7, Section 1.2. ☐

7.3 The Semi-Group G in the Spaces $D(A^\infty; M_k)$. Applications

7.3.1 General Results

Theorem 7.3. *Let the sequence* $\{M_k\}$ *satisfy* (1.22), (1.23), (1.24). *Then, for every* $t \geq 0$:

(7.31) $$G(t) \in \mathscr{L}(\mathrm{D}(A^\infty; M_k); \mathrm{D}(A^\infty; M_k)).$$

$t \to G(t)$ *is an infinitely differentiable function of* $t \geq 0 \to \mathscr{L}(\mathrm{D}(A^\infty; M_k); \mathrm{D}(A^\infty; M_k))$; *finally*

(7.32) $$G'(0) = -A.$$

Proof. Let $e \in \mathrm{D}^L(A^\infty; M_k)$. Then

(7.33) $$A^k G(t)\, e = G(t)\, A^k e,$$

therefore

$$\| A^k G(t) e \| \leq c\mu L^k M^k,$$

which shows that

(7.34) $$G(t) \in \mathscr{L}(\mathrm{D}^L(A^\infty; M_k); \mathrm{D}^L(A^\infty; M_k)),$$

whence (7.31).

Let us now that show

(7.35) $$t \to G(t) \text{ is a continuous mapping of}$$
$$t \geq 0 \to \mathscr{L}(\mathrm{D}^L(A^\infty; M_k); \mathrm{D}^{LH}(A^\infty; M_k))^{((1))}.$$

Indeed, if we set

$$\|G(t+h) - G(t)\| = \text{norm of } G(t+h) - G(t) \text{ in}$$
$$\mathscr{L}(\mathrm{D}^L(A^\infty; M_k); \mathrm{D}^{LH}(A^\infty; M_k)),$$

we obtain:

$$\|G(t+h) - G(t)\| = \sup_e \sup_k \left\| \frac{A^k(G(t+h) - G(t))\, e}{L^k H^k M_k} \right\| \|e\|^{-1}_{\mathrm{D}^L(A^\infty, M_k)} \leq$$

$$\leq \sup_e \sup_k \frac{\left\| \int_t^{t+h} G(\sigma)\, (A^{k+1} e)\, d\sigma \right\|}{L^k H^k M_k} \|e\|^{-1}_{\mathrm{D}^L(A^\infty, M_k)} \leq$$

$$\leq \mu h \sup_e \sup_k \frac{\|A^{k+1} e\|}{L^k H^k M_k} \|e\|^{-1}_{\mathrm{D}^L(A^\infty, M_k)} \leq$$

$$\leq L\mu h$$

((1)) Provided with the uniform norm of the operators.

(since $\|A^{k+1}e\| \leq \|e\|_{D^L(A^\infty, M_k)} L^{k+1} M_{k+1}$ and $M_{k+1} \leq H^k M_k$). Therefore we have (7.35) and then

$$t \to G(t) \text{ is a continuous mapping of } t \geq 0 \to \mathscr{L}(D(A^\infty; M_k); D(A^\infty; M_k)). \tag{7.36}$$

But

$$\frac{d^k}{dt^k} G(t) = (-1)^k G(t) A^k, \tag{7.37}$$

and since $A^k \in \mathscr{L}(D(A^\infty; M_k); D(A^\infty; M_k))$ (thanks to (1.24)), it follows from (7.37) and (7.36) that $t \to G(t)$ is an infinitely differentiable mapping of $t \geq 0 \to \mathscr{L}(D(A^\infty; M_k); D(A^\infty; M_k))$.

Finally (7.32) follows from (7.37). ☐

Corollary 7.1. *If* $u_0 \in D(A^\infty; M_k)$ *and if the sequence* $\{M_k\}$ *satisfies* (1.22), (1.23), (1.24), *then the solution* u *of the problem*

$$\frac{du}{dt} + Au = 0, \; u(0) = u_0, \tag{7.38}$$

is a C^∞*-function of* $t \geq 0 \to D(A^\infty; M_k)$. ☐

Remark 7.7. The solution $u(t)$ of (7.38) is also (by definition of $D(A^\infty; M_k)$) of class M_k with values in E. ☐

Remark 7.8. With stronger hypotheses on the sequence M_k, we can strengthen Theorem 7.2. For example, *let there exist a constant* d *such that*

$$M_{k+j} \leq d^{k+j} M_k M_j \quad \forall k, j. \tag{7.39}$$

Then the semi-group $t \to G(t)$ *is of class* M_k *from* $t \geq 0 \to D(A^\infty; M_k)$.

Indeed, let e be given in $D(A^\infty; M_k)$, thus in a space $D^L(A^\infty; M_k)$. We shall verify (and this is sufficient to prove our assertion) that $t \to G(t)e$ belongs to $\mathscr{D}_{M_k}([0, 1]; D^{dL}(A^\infty; M_k))$.

Indeed

$$\sup_k \frac{1}{(dL)^k M_k} \|G(t)^{(k)} e\|_{D^{dL}(A^\infty, M_k)} = \sup_k \frac{1}{(dL)^k M_k} \sup_j \frac{\|G(t)^{(k)} A^j e\|}{(dL)^j M_j} \leq$$

$$\leq \mu \sup_{j,k} \left(\frac{\|A^{j+k} e\|}{L^{j+k} M_{j+k}} \cdot \frac{M_{j+k}}{d^{j+k} M_j M_k} \right) \leq$$

$$\leq \mu \sup_q \frac{\|A^q e\|}{L^q M_q} = \mu \|e\|_{D^L(A^\infty, M_k)}. \quad ☐$$

Note that if $\{M_k\}$ is a Gevrey sequence:

$$M_k = ((pk)!)^\beta, \text{ integer } p, \; \beta \geq 1,$$

then (7.39) holds; see Chapter 8, Section 1.2.

7.3.2 Application to Parabolic Operators

Let

$$(7.40) \qquad \begin{cases} A = A(x, \partial/\partial x) = \text{elliptic operator of order } 2m, \\ B_j,\ 0 \leq j \leq m-1, \text{ be a system of boundary} \\ \text{conditions covering } A. \end{cases}$$

Assume that $\{A, B_j\}$ is a variational system (see Chapter 2, Section 9). Thus we can find:

$$\begin{cases} \text{a space } V = \text{closed subspace of } H^m(\Omega) \text{ of } v\text{'s} \\ \text{such that } B_j v = 0, \text{ for } j \in \text{a (possibly empty)} \\ \text{set of indices of } [0, 1, \ldots, m-1], \end{cases}$$

a continuous sesquilinear form $u, v \to a(u, v)$ on V, such that

$$(7.41) \qquad \begin{cases} \mathrm{D}(A) = \{v \mid v \in H^{2m}(\Omega),\ B_j v = 0,\ 0 \leq j \leq m-1\} \\ \text{is equal to } \{v \mid v \in H^{2m}(\Omega) \cap V,\ a(v, w) = (Av, w),\ \forall w \in V\}. \end{cases}$$

We shall assume that

$$\operatorname{Re} a(v, v) + \lambda_0 |v|^2 \geq \alpha \|v\|^2, \quad \alpha > 0,$$

$\|v\| =$ norm of v in V, $|v| =$ norm of v in $H = L^2(\Omega)$ [(1)]. Then

Proposition 7.2. *The operator* $-A$, *with domain* $\mathrm{D}(A)$ *defined in* (7.41), *is the infinitesimal generator of a semi-group of contractions in* H.

Proof. Indeed, the equation

$$Au + \lambda u = f, \quad f \in H,\ u \in \mathrm{D}(A),$$

admits a unique solution for $\operatorname{Re} \lambda \geq \lambda_0$ and which satisfies

$$\|u\| \leq \frac{1}{\operatorname{Re} \lambda + c} |f|, \ \operatorname{Re} \lambda \geq \lambda_0.$$

The desired result then follows from the Hille-Yosida theorem (see Yosida [1]). ▯

Now assume that

$$(7.42) \qquad \begin{cases} \text{the boundary } \Gamma \text{ of } \Omega \text{ is of Gevrey class of} \\ \text{order } s,\ s \geq 1, \text{ and the coefficients of } A \text{ and } B_j \\ \text{are of Gevrey class of order } s \text{ in } \bar{\Omega} \text{ or on } \Gamma. \end{cases}$$

[(1)] For the choice of other "pivot spaces", see Chapter 2, Section 9.

Let

$$M_k = \big((2km)!\big)^s. \tag{7.43}$$

According to Theorem 1.2, Chapter 8, we have:

$$\mathrm{D}(A^\infty; M_k) = \{v \mid v \in \mathscr{D}_s(\bar{\Omega}),\ B_j A^k v = 0,\ \forall k,\ 0 \le j \le m-1\}, \tag{7.44}$$

where (see Chapter 8)

$\mathscr{D}_s(\bar{\Omega})$ = space of Gevrey functions of order s in $\bar{\Omega}$;

(7.44) can also be stated in the equivalent form:

$$\mathrm{D}(A^\infty; M_k) = \mathrm{D}(A^\infty) \cap \mathscr{D}_s(\bar{\Omega}). \tag{7.44a}$$

Remark 7.8 then yields the following result:

Theorem 7.4. *Let u_0 belong to $\mathrm{D}(A^\infty; M_k)$, defined by (7.44). Then the solution $u(x, t) = u$ of*

$$\begin{cases} \dfrac{\partial u}{\partial t} + A\left(x, \dfrac{\partial}{\partial x}\right) u = 0, \ x \in \Omega,\ t > 0, \\ B_j u = 0,\ x \in \Gamma,\ t > 0,\ 0 \le j \le m-1, \\ u(x, 0) = u_0(x),\ x \in \Omega, \end{cases} \tag{7.45}$$

is a Gevrey function in t of order $2sm$ for $t \ge 0$ with values in the space of Gevrey functions of order s in $\bar{\Omega}$. ☐

7.3.3 Application to Schroedinger Operators

Analogous results can be obtained for the equation

$$\frac{\partial u}{\partial t} + iAu = 0, \tag{7.46}$$

if the system $\{A, B_j\}$ is self-adjoint, the other hypotheses remaining unchanged. ☐

7.3.4 Application to Operators of the Second Order in t

Let V be given as in Section 7.3.2.; thus

$$\begin{cases} H_0^m(\Omega) \subset V \subset H^m(\Omega), \\ V \text{ being defined by differential boundary conditions} \\ \text{with Gevrey coefficients of order } s \ge 1, \end{cases} \tag{7.47}$$

and let A, $\mathrm{D}(A)$, $a(u, v)$ also be defined as in Section 7.3.2.

Assume that:

$$\left\{\begin{array}{l}\text{the coefficients of } A \text{ and } B_j \text{ are Gevrey functions of}\\ \text{order } s \text{ and } \Gamma \text{ is of Gevrey class of order } s,\end{array}\right. \tag{7.48}$$

$$\text{for suitable } \lambda,\ a(v, v) + \lambda \|v\|^2_{L^2(\Omega)} \geq \alpha \|v\|^2_V,\ \forall v \in V, \alpha > 0, \tag{7.49}$$

$$a(u, v) = \overline{a(v, u)},\ \forall u, v \in V. \tag{7.50}$$

We would like to study the regularity properties of the problem:

$$u'' + Au = 0,\ x \in \Omega,\ t > 0, \tag{7.51}$$

$$B_j u = 0,\ x \in \Gamma,\ t > 0,\ 0 \leq j \leq m - 1, \tag{7.52}$$

$$u(x, 0) = u_0(x),\ u'(x, 0) = u_1(x),\ x \in \Omega. \quad \square \tag{7.53}$$

We introduce:

$$\left\{\begin{array}{l} E = V \times H,\ \big(H = L^2(\Omega)\big),\\ \mathscr{A} = \begin{pmatrix} 0 & -I \\ A & 0 \end{pmatrix},\\ \mathrm{D}(\mathscr{A}) = \mathrm{D}(A) \times V. \end{array}\right. \tag{7.54}$$

Proposition 7.3. *The operator $-\mathscr{A}$ is the infinitesimal generator of a semi-group (in fact a group) in E.*

Proof. Changing u to $\exp(kt)\, u$, equation (7.51) becomes

$$u'' + 2ku' + (A + k^2)\, u = 0;$$

k is chosen so that

$$\big((A + k^2)\, v, v\big) \geq \alpha \|v\|^2_V.$$

We are thus led back to

$$u'' + \beta u' + Au = 0, \tag{7.55}$$

with

$$\beta \geq 0,\ (Av, v) \geq \alpha \|v\|^2_V. \tag{7.56}$$

(7.55) is written as a first order system in t; setting

$$u = u^1,\ \frac{\mathrm{d}u^1}{\mathrm{d}t} = u^2,\ \vec{u} = \{u^1, u^2\},$$

(7.55) is equivalent to

$$\frac{\mathrm{d}\vec{u}}{\mathrm{d}t} + \mathscr{A}_\beta \vec{u} = 0, \tag{7.57}$$

where

$$\mathscr{A}_\beta = \begin{pmatrix} 0 & -I \\ A & \beta \end{pmatrix}. \tag{7.58}$$

E is provided with the norm

$$\|\vec{u}\|_E = (a(u^1, u^1) + \| u^2 |^2)^{1/2} \quad (\text{where } |f| = \|f\|_{L^2(\Omega)}). \tag{7.59}$$

We shall show that $-\mathscr{A}_\beta$ is the infinitesimal generator of a group of *contractions* in E (provided with the norm (7.49)).

To this end, let us examine the equation

$$\mathscr{A}_\beta \vec{u} + \lambda \vec{u} = \vec{f}, \ \vec{f} = \{f^1, f^2\} \text{ given in } E, \tag{7.60}$$

that is

$$-u^2 + \lambda u^1 = f^1, \quad A u^1 + (\lambda + \beta)\, u^2 = f^2, \tag{7.61}$$

whence

$$A u^1 + \lambda(\lambda + \beta)\, u^1 = f^2 - (\lambda + \beta)\, f^1, \tag{7.62}$$

equation which admits a unique solution if $\operatorname{Re} \lambda > 0$ or if $\operatorname{Re} \lambda < -\beta$.

Taking the scalar product of the first (resp. second) equation (7.61) with Au^1(resp. u^2) and adding, we obtain

$$-(u^2, Au^1) + \lambda a(u^1, u^1) + (Au^1, u^2) + (\lambda + \beta)\, |u^2|^2 = (f^1, Au^1) + (f^2, u^2),$$

whence

$$(\operatorname{Re} \lambda) \|\vec{u}\|_E^2 + \beta\, |u^2|^2 = \operatorname{Re} \left(\vec{u}, \vec{f}\right)_E.$$

It follows that

$$\|\vec{u}\|_E \leq \frac{1}{\operatorname{Re} \lambda} \|\vec{f}\|_E \quad \text{if } \operatorname{Re} \lambda > 0,$$

$$\|\vec{u}\|_E \leq \frac{1}{(|\operatorname{Re} \lambda| - \beta)} \|\vec{f}\|_E \quad \text{if } \operatorname{Re} \lambda < -\beta,$$

which proves our assertion. ▯

Remark 7.9. The preceding calculation is of course very similar to those of Chapter 3, Section 8.3; see also the Comments to Chapter 3. ▯

If we now take

$$M_k = ((km)!)^s, \tag{7.63}$$

with

$$\begin{cases} s \geq 1 & \text{if } m \geq 2, \\ s > 1 & \text{if } m = 1, \end{cases} \tag{7.64}$$

then we can apply Remark 7.8. Thus:

$$(7.65)\qquad \begin{cases} \textit{if } \vec{u}^0 = \{u_0, u_1\} \in \mathrm{D}(\mathscr{A}^\infty; M_k), \textit{ the solution} \\ t \to \vec{u}(t) \textit{ of } (7.57), \textit{ with } \vec{u}(0) = \vec{u}^0, \textit{ is of class} \\ M_k \textit{ with values in } \mathrm{D}(\mathscr{A}^\infty; M_k). \quad \square \end{cases}$$

There remains to interpret the fact of belonging to $\mathrm{D}(\mathscr{A}^\infty; M_k)$. We shall need

Lemma 7.1. *Let $\{M_k\}$ be a logarithmically convex sequence satisfying* (1.24). *Let $\mathscr{A}$ be the infinitesimal generator of a bounded semi-group $G(t)$ in a Banach space E. Then*

$$(7.66)\qquad \mathrm{D}(\mathscr{A}^\infty; M_k) = \{e \mid e \in \mathrm{D}(\mathscr{A}^\infty), \|\mathscr{A}^{2k} e\| \le c L^{2k} M_{2k}, \forall k\}.$$

(In other words, instead of considering *all* powers $\mathscr{A}^k$, it is sufficient to consider the *even* powers $\mathscr{A}^{2k}$ ((1)).

Proof. If we are willing to change $G(t)$ to $\exp(-\omega t)\, G(t)$, which does not affect the space $\mathrm{D}(\mathscr{A}^\infty; M_k)$ of vectors of class M_k, we may assume that

$$(7.67)\qquad \|G(t)\|_{\mathscr{L}(E;E)} \le \mu e^{-\omega t} \quad (\omega > 0).$$

Let us for the moment admit that

$$(7.68)\qquad \begin{cases} \text{there exists a constant } c_1 \text{ such that, } \forall e \in \mathrm{D}(\mathscr{A}^2), \\ \|\mathscr{A} e\| \le c_1 \|\mathscr{A}^2 e\|^{1/2} \|e\|^{1/2}. \end{cases}$$

Let e satisfy

$$(7.69)\qquad \|\mathscr{A}^{2k} e\| \le c L^{2k} M_{2k}, \ \forall k.$$

We have

$$\|\mathscr{A}^{2k-1} e\| = \|\mathscr{A}(\mathscr{A}^{2k-2} e)\| \le \big(\text{by } (7.68)\big)\, c_1 \|\mathscr{A}^2(\mathscr{A}^{2k-2} e)\|^{1/2} \times$$
$$\times \|\mathscr{A}^{2k-2} e\|^{1/2} \le \big(\text{by } (7.69)\big)\, c_1 (c L^{2k} M_{2k})^{1/2} \times$$
$$\times (c L^{2k-2} M_{2k-2})^{1/2}.$$

But $M_{2k} \le H^{2k} M_{2k-1}$, $M_{2k-2} \le \dfrac{M_0}{M_1} M_{2k-1}$ (since, according to the logarithmic convexity, $\dfrac{M_{k+1}}{M_k} \ge \dfrac{M_k}{M_{k-1}} \ge \cdots \ge \dfrac{M_1}{M_0}$), so that

$$\|\mathscr{A}^{2k-1} e\| \le c c_1 \left(\frac{M_0}{M_1}\right)^{1/2} L^{2k-1} M_k M_{2k-1},$$

((1)) More generally, $\mathrm{D}(\mathscr{A}^\infty; M_k) = \{e \mid e \in \mathrm{D}(\mathscr{A}^\infty), \|\mathscr{A}^{kq} e\| \le c L^{kq} M_{kq}, \forall k\}$, q being any fixed positive integer.

therefore $e \in \mathrm{D}(\mathscr{A}^\infty; M_k)$. Thus we have the desired result under the assumption (7.68).

Proof of (7.68). If $e \in \mathrm{D}(\mathscr{A}^2)$, the function

$$t \to G(t) e = u(t)$$

satisfies

$$u, \frac{\mathrm{d}u}{\mathrm{d}t}, \frac{\mathrm{d}^2 u}{\mathrm{d}t^2} \in L^2(0, \infty; E),$$

thanks to (7.67).

Applying inequality (3.18) of Chapter 1 (with $X = Y = E$, $m = 2$, $j = 1$), we obtain:

$$\| u'(0) \| \leq c_2 \, \| u \|_{L^2(0,\infty;E)}^{1/2} \, \| u'' \|_{L^2(0,\infty;E)}^{1/2}.$$

Now $u'(0) = -\mathscr{A} e$, $\| u \|_{L^2(0,\infty;E)} \leq c_3 \| e \|$, $\| u'' \|_{L^2(0,\infty;E)} \leq c_3 \| \mathscr{A}^2 e \|$, from which the result follows. ☐

We now apply Lemma 7.1 to the operator

$$\mathscr{A} = \begin{pmatrix} 0 & -I \\ A & 0 \end{pmatrix}$$

(i.e. in the form (7.54), without a preliminary reduction of the problem by changing u to $\exp(kt)\, u$). We have

$$\mathscr{A}^{2k} = (-1)^k \begin{pmatrix} A^k & 0 \\ 0 & A^k \end{pmatrix}$$

and consequently, according to Lemma 7.1, we see that

$$\begin{aligned} \mathrm{D}(\mathscr{A}^\infty; M_k) = \{ \vec{u} \mid \vec{u} = \{u^1, u^2\}, \; & u^i \in \mathrm{D}(A^\infty), \\ \| A^k u^1 \|_V & \leq c L^{2k} M_{2k} \;\; \forall k, \\ \| A^k u^2 \|_{L^2(\Omega)} & \leq c L^{2k} M_{2k} \;\; \forall k \}. \end{aligned} \tag{7.70}$$

Theorem 7.5. *Assume that* (7.47), ..., (7.50) *hold and that the sequence* $\{M_k\}$ *is given by* (7.63) *with* (7.64). *Then, in the notation of* (7.54), *if*

$$\vec{u} = \{u^1, u^2\} \in \mathrm{D}(\mathscr{A}^\infty; M_k),$$

we have:

$$u^i \in \mathrm{D}(A^\infty) \cap \mathscr{D}_s(\bar{\Omega}). \tag{7.71}$$

Proof. For u^2, this follows from (7.70) and Theorem 1.2, Chapter 8.

For u^1, we note that, *in particular*,

(7.72) $$\|A^k u^1\|_V \geq c\, \|A^k u^1\|_{L^2(\Omega)},$$

so that the result is again a consequence of Theorem 1.2, Chapter 8.

Note that *nothing has been lost* by using the (crude) inequality (7.72), since according to Chapter 2, we have in particular

$$\|A^k u^1\|_V \leq c\, \|A^{k+1} u^1\|_{L^2(\Omega)},$$

so that it does not matter whether we take the inequalities in $L^2(\Omega)$ or in V. □

According to Theorem 7.4, we may now state (7.65) in the form:

Theorem 7.6. *Let the hypotheses of Theorem* 7.5 *be satisfied. Consider the problem* (7.51), (7.52), *where* $u_0, u_1 \in \mathrm{D}(A^\infty)$, i.e.

$$u_i \in \mathscr{D}(\overline{\Omega}),\ B_j A^k u_0 = 0,\ B_j A^k u_1 = 0,\ \forall k,\ 0 \leq j \leq m-1.$$

Assume that u_0 *and* u_1 *are Gevrey functions of order* s *in* $\overline{\Omega}$. *Then* $t \to u(.,t)$, *solution of the problem, is a Gevrey function of order* sm *for* $t \geq 0$ *with values in the space of Gevrey functions of order* s *in* $\overline{\Omega}$.

In this statement, $s \geq 1$ *if* $m \geq 2$ *and* $s > 1$ *if* $m = 1$. □

7.4 The Transposed Settings. Applications

7.4.1 The Space $D(A^{*\infty}; M_k)'$

Let $G^*(t)$ be the adjoint semi-group of $G(t)$ in $\mathscr{L}(E'; E')$. If $-A^*$ is the infinitesimal generator of $G^*(t)$, we see that A^* is the *adjoint*, in the sense of unbounded operators, of the operator A in E, with domain $\mathrm{D}(A)$.

We thus introduce $\mathrm{D}(A^{*\infty}; M_k)$ as above, replacing A with A^*.

We assume that the sequence $\{M_k\}$ satisfies (1.22), (1.23), (1.24), so that, according to Theorem 7.1, $\mathrm{D}(A^{*\infty}; M_k)$ *is dense* in $\mathrm{D}(A^{*\infty})$, $\mathrm{D}(A^{*q})$ and E'. Therefore, by duality (or anti-duality), we can consider the dual (or anti-dual) space $\mathrm{D}(A^{*\infty}; M_k)'$. We obtain:

(7.73)

$$\mathrm{D}(A^\infty; M_k) \subset \mathrm{D}(A^\infty) \subset \mathrm{D}(A^q) \subset E \subset \mathrm{D}(A^{*q})' \subset \mathrm{D}(A^{*\infty})' \subset \mathrm{D}(A^{*\infty}; M_k)'.$$

The spaces $\mathrm{D}(A^{*q})' \ldots$ are provided with their *strong* dual (or anti-dual) structure. □

By applying the Hahn-Banach theorem and some standard properties of topological vector spaces, we obtain the following *structural result*:

$$(7.74)\quad \begin{cases} \text{every element } f \text{ of } \mathrm{D}(A^{*\infty}; M_k)' \text{ may be represented,} \\ \text{non-uniquely, by} \\ f = \sum_{k=0}^{\infty} A^k e_k, \ A^0 = \text{identity}, \\ \text{where} \\ e_k \in E, \ \sum_{k=0}^{\infty} L^k M_k \|e_k\| < \infty, \ \forall L. \end{cases}$$

In (7.74), the equality "$f = \sum_{k=0}^{\infty} A^k e_k$" is understood in the sense: if $e \in \mathrm{D}(A^{*\infty}; M_k)$, then

$$(7.75)\qquad \langle f, e\rangle \ (= \text{value of } f \text{ at } e) = \sum_{k=0}^{\infty} \langle e_k, A^{*k} e\rangle .$$

Conversely, if the e_k's are given as in (7.74), f given by $f = \sum_{k=0}^{\infty} A^k e_k$ (in the sense of (7.75)) belongs to $\mathrm{D}(A^{*\infty}; M_k)'$. □

7.4.2 The Semi-Group $G(t)$ in $D(A^{*\infty}; M_k)'$

In order to avoid difficulties of a "topological vector space" nature, we shall assume that

$$(7.76)\qquad \textit{the injection } \mathrm{D}(A^*) \to E' \textit{ is compact.}$$

Remark 7.10. This hypothesis is realized in all the examples to be considered, as long as Ω is *bounded* and *sufficiently regular.*

Also note that the results to be obtained below are valid *without* hypothesis (7.76) by replacing (which in fact may be unnecessary!) the differentiability by *scalar* differentiability, etc. □

Theorem 7.7. *Let the sequence* $\{M_k\}$ *satisfy* (1.22), (1.23), (1.24) *and assume that* (7.76) *holds. Then* $G(t)$ *is a semi-groups in* $\mathrm{D}(A^{*\infty}; M_k)'$. *For every* $f \in \mathrm{D}(A^{*\infty}; M_k)'$, $t \to G(t) f$ *is an infinitely differentiable function of* $t \geq 0 \to \mathrm{D}(A^{*\infty}; M_k)'$.

By transposition of the property "$A^* \in \mathscr{L}(\mathrm{D}(A^{*\infty}; M_k); \mathrm{D}(A^{*\infty}; M_k))$*", it can be seen that* A *extends by continuity to a continuous linear operator, still denoted by* A, *of* $\mathrm{D}(A^{*\infty}; M_k)'$ *into itself. The infinitesimal generator of* $G(t)$ *in* $\mathrm{D}(A^{*\infty}; M_k)'$ *is* $-A$.

If the sequence $\{M_k\}$ *satisfies* (7.39), *the function*

$$(7.77)\qquad t \to \langle G(t) f, v\rangle, \ f \in \mathrm{D}(A^{*\infty}; M_k)', \ v \in \mathrm{D}(A^{*\infty}; M_k),$$

is of class M_k.

Proof. Under hypothesis (7.76), the space $D(A^{*\infty}; M_k)'$ is (see Remark 7.5) a Fréchet space; it is therefore, in particular, quasi-complete and, according to a lemma of Grothendieck [3] (see also L. Schwartz [2], page 146), in order to show that the function

$$t \to G(t)\, f,\ f \in D(A^{*\infty}; M_k)',$$

of $\{t \geq 0\} \to D(A^{*\infty}; M_k)'$ is of class C^∞, it is sufficient to show that the function

$$t \to \langle G(t)\, f, v\rangle,\ v \in D(A^{*\infty}; M_k)$$

is of class C^∞. Now

$$\langle G(t)\, f, v\rangle = \langle f, G^*(t)\, v\rangle$$

and the result follows from Theorem 7.2 (with G^*, A^*, ... instead of $G, A, \ldots$).

If (7.39) holds, then the function (7.77) is of class M_k, according to Remark 7.8. ☐

The following is another property of the semi-group $G(t)$ in the space $D(A^{*\infty}; M_k)'$:

Theorem 7.8. *Let the hypotheses of Theorem* 7.6 *be satisfied and assume that* (7.39) *holds. Then, for every function* $\varphi \in \mathscr{D}_{M_k}(]0, \infty[)$ (see the proof of Theorem 7.1), *we have*

$$G(\varphi) \in \mathscr{L}\big(D(A^{*\infty}; M_k)';\ D(A^{\infty}; M_k)\big). \tag{7.78}$$

Proof. Since, according to Remark 7.5, the space $D(A^{*\infty}; M_k)'$ is a Fréchet space, the closed graph theorem holds from $D(A^{*\infty}; M_k)' \to D(A^\infty; M_k)$. Now $G(\varphi) \in \mathscr{L}(D(A^{*\infty}; M_k)';\ D(A^{*\infty}; M_k)')$, according to Theorem 7.7, and therefore it suffices to show that

(7.79) for f given in $D(A^{*\infty}; M_k)'$, $G(\varphi)\, f$ belongs to $D(A^\infty; M_k)$.

But let us apply (7.74). Then

$$G(\varphi)\, f = \sum_{k=0}^{\infty} G(\varphi^{(k)})\, e_k \tag{7.80}$$

and we shall show that (7.80) converges in a space $D^L(A^\infty; M_k)$ for a suitable L.

By hypothesis

$$|\varphi^{(k)}(t)| \leq c L^k M_k,\ \ \forall k, \forall t.$$

We shall verify that $G(\varphi)\, f \in D^{dL}(A^\infty; M_k)$. Indeed, subject to convergence, we have:

$$A^j G(\varphi)\, f = \sum_{k=0}^{\infty} G(\varphi^{(k+j)})\, e_k,$$

from which we obtain (the c's denoting various constants):

$$\|A^j G(\varphi) f\| \leq c \sum_{k=0}^{\infty} L^{k+j} M_{k+j} \|e_k\| \leq$$

$$\leq (\text{by } (7.39)) \; c\, d^j L^j \left(M_j \sum_{k=0}^{\infty} L^k M_k \|e_k\|\right) \leq$$

$$\leq (\text{by } (7.74)) \; c\, d^j L^j M_j, \text{ whence the desired result.} \quad \square$$

Remark 7.11. It is possible to go a little further if the semi-group G is *analytic*, i.e. if $\forall e \in E$, the function $t \to G(t)e$ is analytic from $t > 0 \to E$ i.e.

$$\|G^{(k)}(t)\| \leq c L^k k!, \; \forall k, t \in [t_0, t_1], \; 0 < t_0 \leq t_1 < \infty. \tag{7.81}$$

(For analytic semi-groups, see K. Yosida [1].) Then

Theorem 7.9. *Let the semi-group $G(t)$ be analytic; under the hypotheses of Theorem 7.7 and assuming the existence of a constant d_1 such that*

$$k! \leq d_1^k M_k, \; \forall k, \tag{7.82}$$

we have:

$$G(t) \in \mathscr{L}\big(\mathrm{D}(A^{*\infty}; M_k)'; \; \mathrm{D}(A^\infty; k!)\big), \; t > 0^{((1))}. \tag{7.83}$$

Furthermore, for $f \in \mathrm{D}(A^{*\infty}; M_k)'$, the function $t \to G(t) f$ is analytic from $\{t > 0\} \to \mathrm{D}(A^\infty; k!)$.

Proof. Indeed, in order to show (7.83), it is sufficient, as in the proof of Theorem 7.8, to show that for f given in $\mathrm{D}(A^{*\infty}; M_k)'$, $G(t) f$ belongs to $\mathrm{D}(A^\infty; k!)$.

Now, according to (7.74),

$$G(t) f = \sum_{k=0}^{\infty} (-1)^k G^{(k)}(t) e_k$$

and, subject to convergence:

$$A^j G(t)^{(r)} f = \sum_{k=0}^{\infty} (-1)^{k+j} G^{(k+j+r)}(t) e_k.$$

Therefore, for $t \in [t_0, t_1]$, $0 < t_0 < t_1 < \infty$, we have:

$$\|A^j G(t)^{(r)} f\| \leq \sum_{k=0}^{\infty} c L^{k+j+r} (k+j+r)! \|e_k\| \leq$$

$$\leq c \sum_{k=0}^{\infty} L^{k+j+r} 2^{k+j+r} k!\, j!\, r!\, \|e_k\| \leq$$

$$\leq (\text{by } (7.82)) \; c(2L)^{j+r} j!\, r! \sum_{k=0}^{\infty} (2 d_1 L)^k M_k \|e_k\| \leq$$

$$\leq \tilde{c}(2L)^{j+r} j!\, r!,$$

which proves the theorem. $\square$

((1)) If the semi-group is analytic, $\mathrm{D}(A^\infty; k!) \supset G(t_0) \cdot E$, $\forall t_0 > 0$, therefore does not reduce to $\{0\}$.

7.4.3 Applications

All results obtained in Section 7.4.2 apply in the settings of Theorems 7.4 and 7.6.

If, for example, in the setting of Theorem 7.4, u_0 is given in $D(A^*)'$, resp. $D(A^{*\infty})'$, resp. $D(A^{*\infty}; M_k)'$, then there exists a unique solution to problem (7.45), which is continuous, resp. C^∞, resp. of class M_k, with values in $D(A^*)'$, resp. $D(A^{*\infty})'$, resp. $D(A^{*\infty}; M_k)'$. ▯

Therefore, if, for u_0 given in $D(A^{*\infty}; M_k)'$ (for example), the solution $t \to u(t)$ is a *very regular* function—of class M_k—with values in the "very coarse" space $D(A^{*\infty}; M_k)'$, it is also a "*very general distribution in t*" with values in a "*very regular space*[1]; more precisely: for $\varphi \in \mathscr{D}_{M_k}(]0, \infty[)$,

$$\int_0^\infty u(t)\, \varphi(t)\, dt = G(\varphi) \cdot u_0$$

belongs to $D(A^\infty; M_k)$.

Thus, for fixed t, $u(t)$ does not satisfy the boundary conditions of the problem, but $\int_0^\infty u(t)\, \varphi(t)\, dt$ does. ▯

It must also be carefully noted that, since $\mathscr{D}(\Omega)$ *is not dense in* $D(A^*)$, $D(A^{*\infty})$, $D(A^{*\infty}; M_k)$, the spaces $D(A^*)'$, ... *are not* spaces of distributions on Ω, but are spaces which contain "transversal layers to Γ", respectively of finite order, arbitrary finite order and infinite order.

Here is a *very simple example*. We take

$$\Omega =]0, \infty[,$$

an unbounded open set (see Remark 7.10), and

$$A = -\frac{d^2}{dx^2}, \quad D(A) = H^2(\Omega) \cap H_0^1(\Omega). \tag{7.84}$$

Then $A = A^*$, and we take $u_0 \in D(A^*)'$ defined by

$$\langle u_0, v \rangle = -v'(0), \; v \in H^2(\Omega) \cap H_0^1(\Omega). \tag{7.85}$$

The space $(D(A^*))'$ can be identified with $\mathscr{D}'(\mathbf{R})/[\delta]$, quotient of $\mathscr{D}'(\mathbf{R})$ by the line $[\delta]$ generated by the Dirac mass δ at the origin.

Then the solution of problem (7.45) is

$$u(x, t) = -\frac{1}{2\sqrt{\pi t}} \frac{x}{t} \exp\left(-\frac{x^2}{4t}\right), \; t > 0. \tag{7.86}$$

As $t \to 0$, $u(., t)$ does not converge in the sense of $\mathscr{D}'(]0, +\infty[)$, but in the sense of $D(A^*)'$; if we extend $u(., t)$ to $\tilde{u}(., t)$ by 0 for $t < 0$, we obtain: $\tilde{u}(., t) \to u_0$ in $\mathscr{D}'(\mathbf{R})/[\delta]$ (dual of the space of functions in $\mathscr{D}(\mathbf{R})$ *vanishing at* 0).

---

[1] This is one aspect of the "uncertainty principle", in the form encountered by L. Schwartz in the theory of kernels.

Remarks of the same type hold if we take u_0 in $\mathrm{D}(A^{*\infty})'$ (resp. $\mathrm{D}(A^{*\infty}; M_k)'$) by

$$\langle u_0, v\rangle = \sum_{\text{finite}} c_k v^{(2k+1)}(0)\,, \tag{7.87}$$

(resp. by

$$\langle u_0, v\rangle = \sum_{k=0}^{\infty} c_k v^{(2k+1)}(0), \quad \sum_{k=0}^{\infty} L^k M_k \, |c_k| < \infty, \forall L). \quad \square \tag{7.88}$$

Let us at this point insist on the importance of the space $\mathrm{D}(A^\infty; k!)$ which appears in the case of analytic semi-groups (see Theorem 7.9), and of its dual. We shall again encounter these spaces on several occasions in the case of parabolic equations (see Section 10 and Chapter 10). It would therefore be of great interest to characterize the space $\mathrm{D}(A^\infty; k!)$ when A and $\mathrm{D}(A)$ are given as in Section 7.3.2; see Problem 6.2, Chapter 8; in the case of space dimension 1, $\mathrm{D}(A^\infty; k!)$ is a space of certain entire functions; see Lions-Magenes [2], Section 7.

7.5 Another M_k-Regularity Result

Let us now consider the problem

$$\frac{\mathrm{d}u}{\mathrm{d}t} + A(t)\, u = f(t), \quad t \in \,]0, T[, \tag{7.89}$$

$$u(0) = 0\,, \tag{7.90}$$

where, *for each t, $-A(t)$ is the infinitesimal generator of a semi-group in E.*

More precisely, we make the following hypotheses:

$$\begin{cases} \text{for each } t \in [0, T],\ A(t) \text{ is closed, with domain} \\ \mathrm{D}\big(A(t)\big) \text{ dense in } E,\ \lambda - A(t) \text{ being invertible} \\ \text{for } \lambda \in \Sigma, \text{ where } \Sigma = \{\lambda \mid \theta \le \arg\lambda \le 2\pi - \theta\}, \\ \theta \in \,]0, \pi/2[; \end{cases} \tag{7.91}$$

$$\begin{cases} t \to A(t)^{-1} \ \big(\text{which exists according to (7.91)}\big) \text{ is an} \\ \text{infinitely differentiable mapping of } [0, T] \to \mathscr{L}(E; E); \end{cases} \tag{7.92}$$

$$\begin{cases} \text{there exist } c \text{ and } L \text{ such that} \\ \left\| \dfrac{\mathrm{d}^k}{\mathrm{d}t^k} \big(\lambda - A(t)\big)^{-1} \right\| \le \dfrac{c}{|\lambda|} L^k M_k, \ \forall k. \end{cases} \tag{7.93}$$

Then the following result, which is due to Tanabe [1], holds:

Theorem 7.10. *Assume that hypotheses* (7.91), (7.92), (7.93) *hold and let the sequence $\{M_k\}$ satisfy* (1.22), (1.24), (1.25), (7.39) *and $M_k \le M_{k+1}$, $\forall k$. Let f be of class M_k in $[0, T]$ with values in E. Then there exist constants* *c and L such that*

(7.94) $$\left\| \frac{\mathrm{d}^k}{\mathrm{d}t^k} u(t) \right\| \leq c L^k M_k t^{1-k}. \quad \square$$

We refer the reader to Tanabe [1] for the proof of this theorem; he uses sharp estimates on the "kernel" $U(t, s)$ defined by

$$\begin{cases} \dfrac{\partial}{\partial t} U(t, s) + A(t)\, U(t, s) = 0, \; 0 \leq s < t \leq T, \\ U(s, s) = I, \;\; 0 \leq s \leq T. \quad \square \end{cases}$$

Remark 7.12. Here the sequence $\{M_k\}$ may be quasi-analytic. In particular, we can take $M_k = k!$. We then obtain the fact that the solution $u(t)$ is analytic in $t > 0$ with values in E. $\square$

Remark 7.13. The estimate (7.94) brings into evidence a singularity at $t = 0$. This singularity really exists and is not due to the method of proof. Indeed, assuming for the sake of simplicity that $A(t) = A$, it *formally* follows from (7.89) and (7.90) that

(7.95) $$u^k(0) = \sum_{j=0}^{k} (-1)^{k-j} A^{k-j} f^{(j)}(0),$$

so that there is a singularity at $t = 0$, unless we assume that

$$f^{(j)}(0) \in \mathrm{D}(A^\infty), \;\; \forall j.$$

In fact this condition is not sufficient if we want that

$$\left\| \frac{\mathrm{d}^k}{\mathrm{d}t^k} u(t) \right\| \leq c L^k M_k, \; \forall k \text{ and } \forall t \in [0, T].$$

For the cases studied in the preceding sections, we have assumed that $f \in \mathscr{D}_{+,M_k}(V')$; assuming that 0 is the left endpoint of the support, we then had

$$f^{(j)}(0) = 0, \; \forall j. \quad \square$$

Remark 7.14. Consider problem (7.89) with

(7.96) $$u(0) = u_0, \;\; u_0 \in E.$$

Then there exist constants c_1, L_1 such that

(7.97) $$\left\| \frac{\mathrm{d}^k}{\mathrm{d}t^k} u(t) \right\| \leq c L^k M_k t^{1-k} + c_1 L_1^k M_k t^{-k}, \; \forall k.$$

Here again (as in Remark 7.13) the singularity at $t = 0$ really exists, except if u_0 and $f(0), f'(0), \ldots$ satisfy *compatibility relations* which we have not given (see Problems 14.8 and 14.9). $\square$

Remark 7.15. Here is a very partial result concerning the preceding remarks. In the setting of Section 1, consider the problem (7.89), (7.96); assume that $A(t)$ is given in $\mathscr{L}(V; V')$ by $a(t; u, v)$ and that the function $t \to a(t; u, v)$ is of class C^1 in $[0, T]$, $\forall u, v \in V$. Further, assume that

(7.98) $$u_0 \in V, \ A(0)\, u_0 \in V,$$

(7.99) $$f, \frac{\mathrm{d}f}{\mathrm{d}t} \in L^2(0, T; V'), \ f(0) \in V.$$

Then

(7.100) $$u' \in L^2(0, T; V), \ u'' \in L^2(0, T; V').$$

Indeed, we can find a function Φ having the following properties:

(7.101) $$\begin{cases} \Phi \text{ is a twice continuously differentiable function} \\ \text{of } t \leq 0 \to V, \\ \Phi(t) = 0 \text{ if } t < -1 \text{ (for example)}, \\ \Phi(0) = u_0, \\ \Phi'(0) = f(0) - A(0)\, u_0. \end{cases}$$

Next, we introduce

(7.102) $$w(t) = \begin{cases} \Phi(t), \text{ if } t < 0, \\ u(t), \text{ if } 0 < t < T, \end{cases} \qquad F = \begin{cases} A(t)\, \Phi(t) + \Phi'(t), \ t < 0^{((1))}, \\ f(t), \text{ if } 0 < t < T. \end{cases}$$

Since, if u is the solution, according to (7.101), we have:

$$\Phi(0) = u(0), \ A(0)\, \Phi(0) + \Phi'(0) = f(0),$$

we see that

$$w' = \begin{cases} \Phi', \text{ if } t < 0 \\ u', \text{ if } 0 < t < T, \end{cases} \qquad F' = \begin{cases} A'\Phi + A\Phi' + \Phi'', \text{ if } t < 0 \\ f', \text{ if } 0 < t < T \end{cases}$$

and therefore

$$F, \frac{\mathrm{d}F}{\mathrm{d}t} \in L^2(-\infty, T; V')$$

and

$$w' + A(t)\, w = F \ \text{ in } \ -\infty < t < T.$$

Furthermore, the supports of w and F are bounded on the left. Thus we may apply the method of differential quotients (see Section 1), from which we deduce (7.100). □

((1)) We assume, which is permissible, that $t \to a(t; u, v)$ is extended to a function of class C^1 on $-\infty < t \leq T$.

8. M_k-Classes and Laplace Transformation

8.1 Orientation-Hypotheses

We now consider operators A which are *more general* than the (negatives of the) infinitesimal generators of semi-groups studied in the preceding section.

More precisely, we consider the setting of Chapter 4, Section 3.

Let H be a reflexive Banach space, A an unbounded, closed operator in H with domain $\mathrm{D}(A)$ (provided with the norm of the graph).

Assume that

$$(8.1)\qquad \begin{cases} A + p \text{ is an isomorphism of } \mathrm{D}(A) \to H \text{ for } \operatorname{Re} p > \xi_0, \\ \text{such that} \\ \|(A+p)^{-1}\|_{\mathscr{L}(H;H)} \leq \text{polynomial } (|p|). \end{cases}$$

We shall see that under these conditions, the "basic results" pertaining to M_k-regularity (and the transposed results) remain valid.

Consequently, the M_k-regularity results are, in particular, valid for the *general parabolic systems* with coefficients independent of t, studied in Chapter 4, Section 4. We do not specify the corresponding results, since they are immediate. □

Remark 8.1. In no way have we tried to put the theory into the framework of maximum generality. For example, one could consider the framework of Lions [5], Chapter 11, Section 5. □

8.2 M_k-Regularity Results

Theorem 8.1. *Assume that* (8.1) *holds and that the sequence* $\{M_k\}$ *satisfies* (1.22), (1.23) *and* (1.24). *Then* $A + \mathrm{d}/\mathrm{d}t$ *is an isomorphism of* $\mathscr{D}_{+,M_k}(\mathbf{R}; \mathrm{D}(A))$ *onto* $\mathscr{D}_{+,M_k}(\mathbf{R}; H)$.

Proof. Reasoning as in the proof of Theorem 1.2 for the "topological" aspects, it all comes down to verifying: if $f \in \mathscr{D}_{+,M_k}(\mathbf{R}; H)$, then the solution u (which exists and is unique in $\mathscr{D}'_{+}(\mathbf{R}; \mathrm{D}(A))$, according to Theorem 3.1 of Chapter 4) of

$$(8.2)\qquad Au + u' = f$$

satisfies

$$(8.3)\qquad u \in \mathscr{D}_{+,M_k}(\mathbf{R}; \mathrm{D}(A)).$$

Since if $u \in \mathscr{D}_{+,M_k}(\mathbf{R}; H)$, so does u' (thanks to (1.24)), and thus $Au = f - u' \in \mathscr{D}_{+,M_k}(\mathbf{R}; H)$, we shall have (8.3) if we can show that

$$(8.4)\qquad u \in \mathscr{D}_{+,M_k}(\mathbf{R}; H).$$

There is no loss of generality by assuming that $u = 0$ for $t < 0$.

According to equations (3.4), Chapter 4, the solution u of (8.2) is

$$u = \mathscr{G} * f, \tag{8.5}$$

where $\mathscr{G}$ (inverse Laplace transform of $(A + p)^{-1}$) satisfies

$$\begin{cases} \mathscr{G} \in \mathscr{L}\big(\mathscr{D}_-(\mathbf{R}); \mathscr{L}(H; \mathrm{D}(A))\big), \\ \mathscr{G} = 0 \text{ for } t < 0. \end{cases} \tag{8.6}$$

But then, according to the theory of L. Schwartz of distiibutions with values *in Banach spaces, $\mathscr{G}$ is of finite order on every compact set*; thus, for arbitrarily fixed finite T, we can find $\Phi(t)$ such that

(8.7) $t \to \Phi(t)$ is a continuous mapping of $[0, T] \to \mathscr{L}\big(H; \mathrm{D}(A)\big)$,

and such that, for suitable l,

$$\begin{cases} \mathscr{G} = \dfrac{\mathrm{d}^l}{\mathrm{d}t^l} \tilde{\Phi} \text{ on }]-\infty, T[, \\ \tilde{\Phi} = \begin{cases} \Phi \text{ for } t \geq 0 \\ 0 \text{ for } t < 0. \end{cases} \end{cases} \tag{8.8}$$

Then (8.5) yields

$$u = \Phi * \frac{\mathrm{d}^l}{\mathrm{d}t^l} f \text{ for } t < T. \tag{8.9}$$

But

$$\frac{\mathrm{d}^l f}{\mathrm{d}t^l} = g \in \mathscr{D}_{+,M_k}(\mathbf{R}; H), \tag{8.10}$$

and therefore

$$u = \Phi * g \text{ for } t < T. \tag{8.11}$$

Thus there remains to be shown that if

$$\| g^{(k)}(t) \|_H \leq c L^k M_k, \ \forall k, \ t \leq T,$$

we have analogous inequalities for u given by (8.11), which is immediate, since

$$u^{(k)}(t) = \Phi * g^{(k)}(t) = \int_0^t \Phi(t - \sigma) \, g^{(k)}(\sigma) \, \mathrm{d}\sigma,$$

therefore

$$\| u^{(k)}(t) \|_H \leq \sup_{t \in [0,T]} \| \Phi(t) \|_{\mathscr{L}(H;H)} \sup_{\sigma \in [0,t]} \| g^{(k)}(\sigma) \|_H \leq c_1 c L^k M_k \ \forall k. \quad \square$$

8.3 Transposition

We note that $A^* + p$ is an isomorphism of $\mathrm{D}(A^*)$ onto H' for $\operatorname{Re} p > \xi_0$, with $\|(A^* + p)^{-1}\|_{\mathscr{L}(H;H')} \leq$ polynomial $(|p|)$, thus (immediate variant of Theorem 8.1):

$$\text{(8.12)} \qquad \begin{cases} A^* - \mathrm{d}/\mathrm{d}t \text{ is an isomorphism of } \mathscr{D}_{-,M_k}(\mathbf{R}; \mathrm{D}(A^*)) \\ \text{onto } \mathscr{D}_{-,M_k}(\mathbf{R}; H'). \end{cases}$$

Similarly $A^* + p$ is an isomorphism of H' onto $\mathrm{D}(A)'$ for $\operatorname{Re} p > \xi_0$ and

$$\|(A^* + p)^{-1}\|_{\mathscr{L}(\mathrm{D}(A)';\mathrm{D}(A'))} \leq \text{polynomial } (|p|),$$

so that

$$\text{(8.13)} \qquad \begin{cases} A^* - \mathrm{d}/\mathrm{d}t \text{ is an isomorphism of } \mathscr{D}_{-,M_k}(\mathbf{R}; H') \\ \text{onto } \mathscr{D}_{-,M_k}(\mathbf{R}; \mathrm{D}(A)'). \end{cases}$$

By transposition, we obtain

Theorem 8.2. *Under the hypotheses of Theorem* 8.1, *the operator* $A + \mathrm{d}/\mathrm{d}t$ *is an isomorphism of* $\mathscr{D}'_{+,M_k}(\mathbf{R}; \mathrm{D}(A))$ *onto* $\mathscr{D}'_{+,M_k}(\mathbf{R}; H)$. ☐

Remark 8.2. It is easy to show that, under the same hypotheses, $A + \mathrm{d}/\mathrm{d}t$ is an isomorphism of

$$\mathscr{D}_{+,M_k}(\mathbf{R}; \mathrm{D}(A^q)) \to \mathscr{D}_{+,M_k}(\mathbf{R}; \mathrm{D}(A^{q-1})), \quad q \geq 1,$$

$$\mathscr{D}_{+,M_k}(\mathbf{R}; \mathrm{D}(A^{*q})') \to \mathscr{D}_{+,M_k}(\mathbf{R}; (\mathrm{D}(A^{*q+1}))'), \quad q \geq 1$$

and analogous results hold if we replace $\mathscr{D}_{+,M_k}$ with $\mathscr{D}'_{+,M_k}$. ☐

9. General Operator Equations

9.1 General Results

We consider the setting of Chapter 3, Section 1. Thus we consider the Hilbert spaces

$$\text{(9.1)} \qquad \mathscr{V} \subset \mathscr{H} \subset \mathscr{V}',$$

(f, v) denoting the scalar product between $f \in \mathscr{V}'$ and $v \in \mathscr{V}$ and $\| \|_{\mathscr{V}}$, $\| \|_{\mathscr{H}}, \ldots$ denoting the norm in $\mathscr{V}, \mathscr{H}, \ldots$

Consider:

$$\text{(9.2)} \qquad \begin{cases} \text{a bounded, continuous semi-group } G(s) \\ \text{in } \mathscr{V}', \mathscr{H}, \mathscr{V}, \text{ of contractions in } \mathscr{H}; \end{cases}$$

$$\text{(9.3)} \qquad \begin{cases} \text{an operator } \mathscr{A} \in \mathscr{L}(\mathscr{V}; \mathscr{V}'), \text{ with} \\ (\mathscr{A}v, v) \geq \alpha \|v\|_{\mathscr{V}}^2, \ \alpha > 0, \ \forall v \in \mathscr{V}^{((1))}. \end{cases}$$

In Chapter 3, Section 1, we have introduced the spaces

$$\mathrm{D}(\Lambda, \mathscr{V}), \mathrm{D}(\Lambda, \mathscr{H}), \ldots,$$

domain of Λ in $\mathscr{V}, \mathscr{H}, \ldots,$ where $-\Lambda$ is the infinitesimal generator of the semi-group G.

We know (Theorem 1.1, Chapter 3) that

$$\text{(9.4)} \qquad \begin{cases} \textit{under hypotheses (9.2), (9.3), the operator} \\ \Lambda + \mathscr{A} \textit{ is an isomorphism of } \mathscr{V} \cap \mathrm{D}(\Lambda; \mathscr{V}') \textit{ onto } \mathscr{V}'. \quad \square \end{cases}$$

We propose to give certain *regularity results* pertaining to the solution u of

$$\text{(9.5)} \qquad \begin{cases} \Lambda u + \mathscr{A} u = f, \\ f \in \mathscr{V}', \ u \in \mathscr{V} \cap \mathrm{D}(\Lambda; \mathscr{V}'). \quad \square \end{cases}$$

We shall use the following notation (already used in Chapter 3, Lemma 1.2 and in Section 7). Let $\varphi \in \mathscr{D}(]0, \infty[)$; set

$$\text{(9.6)} \qquad G(\varphi) = \int_0^\infty G(t)\, \varphi(t)\, \mathrm{d}t,$$

integral with values in $\mathscr{L}(\mathscr{V}; \mathscr{V})$, $\mathscr{L}(\mathscr{H}; \mathscr{H})$, $\mathscr{L}(\mathscr{V}'; \mathscr{V}')$.

We shall use a regularizing sequence $\varrho_n \in \mathscr{D}_{0,M_k}$ (see the proof of Theorem 7.1[2]), the sequence $\{M_k\}$ being assumed *non-quasi-analytic*.

Thus:

$$\text{(9.7)} \qquad \varrho_n \in \mathscr{D}_{0,M_k}, \ \varrho_n \to \delta \text{ in the sense of measures on } [0, +\infty[.$$

We shall also make use of the usual bracket notation: if $\mathscr{B}_1$ and $\mathscr{B}_2$ are two operators, we set:

$$\text{(9.8)} \qquad [\mathscr{B}_1, \mathscr{B}_2] = \mathscr{B}_1 \mathscr{B}_2 - \mathscr{B}_2 \mathscr{B}_1. \quad \square$$

((1)) $\mathscr{A}$ plays the role of M in Chapter 3; we have changed the notation in order to avoid any confusion with the sequence $\{M_k\}$.

[2] For, for example Theorem 9.1, it is not necessary to take ϱ_n in $\mathscr{D}_{0,M_k}$; it is sufficient to take ϱ_n in $\mathscr{D}(]0, \infty[)$; but this choice permits us to pick ϱ_n *once and for all* without loss in generality.

The first regularity theorem is

Theorem 9.1. *Assume that* (9.2), (9.3) *hold and that*

$$\begin{cases} \| [\mathscr{A}, G(\varrho_n')]\, v \|_{\mathscr{V}'} \leq c \, \| v \|_{\mathscr{V}}, \; \forall v \in \mathscr{V}, \\ \mathscr{A} \in \mathscr{L}(\mathrm{D}(\Lambda; \mathscr{V}); \mathrm{D}(\Lambda; \mathscr{V}')), \end{cases} \tag{9.9}$$

($c =$ constant independent of n). *Then, if*

$$f \in \mathrm{D}(\Lambda; \mathscr{V}'), \tag{9.10}$$

we have

$$u \in \mathrm{D}(\Lambda; \mathscr{V}) \cap \mathrm{D}(\Lambda^2; \mathscr{V}'). \tag{9.11}$$

Proof. Applying the operator $G(\varrho_n')$ to both sides of the equation and noting that $[\Lambda, G(\psi)] = 0$, it follows from (9.5) that

$$\Lambda G(\varrho_n')\, u + \mathscr{A} G(\varrho_n')\, u = G(\varrho_n')\, f + [\mathscr{A}, G(\varrho_n')]\, u. \tag{9.12}$$

But we also deduce from (9.5) (see Chapter 3, Section 1) that

$$\| u \|_{\mathscr{V}} \leq \frac{1}{\alpha} \| f \|_{\mathscr{V}'}; \tag{9.13}$$

using (9.9), the same inequality applied to (9.12) yields:

$$\| G(\varrho_n')\, u \|_{\mathscr{V}} \leq \frac{1}{\alpha} \left(\| G(\varrho_n')\, f \|_{\mathscr{V}'} + c \, \| u \|_{\mathscr{V}} \right). \tag{9.14}$$

But

$$G(\varrho_n')\, f \to G(\delta')\, f = \Lambda f \text{ in } \mathscr{V}',$$

and consequently (9.14) implies:

$$\| G(\varrho_n')\, u \|_{\mathscr{V}} \leq \text{constant}. \tag{9.15}$$

It follows (see for example Chapter 3, Section 1) that

$$u \in \mathrm{D}(\Lambda; \mathscr{V})$$

and we deduce from (9.12) that

$$\Lambda^2 u + \mathscr{A} \Lambda u = \Lambda f + [\mathscr{A}, \Lambda]\, u. \tag{9.16}$$

This implies that $u \in \mathrm{D}(\Lambda^2; \mathscr{V}')$. □

Remark 9.1. If we take (in the notation of Section 1):

$$\begin{cases} \mathscr{V} = L^2(0, T; V), \; \mathscr{H} = L^2(0, T; H), \\ \mathscr{A} u = \text{“} t \to A(t)\, u(t) \text{”}, \\ \Lambda = \mathrm{d}/\mathrm{d}t, \; \mathrm{D}(\Lambda; \mathscr{H}) = \{ f \mid f, f' \in L^2(0, T; H), f(0) = 0 \}, \end{cases} \tag{9.17}$$

then hypothesis (9.9) means that

$$\|A'(t)\|_{\mathscr{L}(V;V')} \leq \text{constant}, \quad t \in [0, T]. \tag{9.18}$$

Then the above proof is a variant of the one given in 2), proof of Theorem 1.1; indeed, if we take for ϱ_n the function (evidently not in $\mathscr{D}_{0,M_k}$!!, but as already pointed out, this hypothesis is in no way indispensable for the moment):

$$\varrho_n(t) = \begin{cases} n \text{ in } [0, 1/n] \\ 0 \text{ for } t > 1/n, \end{cases}$$

then

$$G(\varrho_n') f = \frac{f(t - 1/n) - f(t)}{1/n}. \quad \square$$

Remark 9.2. Hypothesis (9.9) may also be formulated, in a somewhat less rigorous fashion, as

$$\|[\mathscr{A}, \Lambda]\, v\|_{\mathscr{V}'} \leq c\, \|v\|_{\mathscr{V}}. \quad \square \tag{9.19}$$

Remark 9.3. Of course the preceding result may be iterated; in this way we obtain:

Theorem 9.2. *Assume that* (9.2), (9.3) *are satisfied and that*

$$\begin{cases} \|[\mathscr{A}, G(\varrho_n^{(k)})]\, v\|_{\mathscr{V}'} \leq c_k(\|v\|_{\mathscr{V}} + \|\Lambda v\|_{\mathscr{V}} + \cdots + \|\Lambda^{k-1} v\|_{\mathscr{V}}), \\ \forall v \in \mathrm{D}(\Lambda^{k-1}; \mathscr{V});\ \mathscr{A} \in \mathscr{L}(\mathrm{D}(\Lambda^j, \mathscr{V}); \mathrm{D}(\Lambda^j; \mathscr{V}')), \\ 0 \leq j \leq k-1. \end{cases} \tag{9.20}$$

Then, if

$$f \in \mathrm{D}(\Lambda^k; \mathscr{V}'), \tag{9.21}$$

we have

$$u \in \mathrm{D}(\Lambda^k; \mathscr{V}) \cap \mathrm{D}(\Lambda^{k+1}; \mathscr{V}'). \tag{9.22}$$

Corollary 9.1. *Assume that* (9.2), (9.3) *are satisfied and that* (9.20) *holds* $\forall k$ (without uniformity in k; the c_k's depend on k). *Then, if*

$$f \in \mathrm{D}(\Lambda^\infty; \mathscr{V}'), \tag{9.23}$$

we have

$$u \in \mathrm{D}(\Lambda^\infty; \mathscr{V}). \quad \square \tag{9.24}$$

Remark 9.3. In the setting of Remark 9.1, the conditions of Corollary 9.1 are satisfied if $t \to A(t)$ is an infinitely differentiable function of

$$[0, T] \to \mathscr{L}(V; V'). \quad \square$$

Let us now give a general M_k-regularity result.

Theorem 9.3. *Let* (9.2), (9.3) *be satisfied. Let* $\{M_k\}$ *satisfy* (1.22), ..., (1.25). *Assume that there exist c and L such that*

$$(9.25) \quad \begin{cases} \| [\mathscr{A}, G(\varrho_n^{(k)})]\, u \|_{\mathscr{V}'} \leq c \sum_{j=0}^{k-1} \binom{k}{j} L^{k-j} M_{k-j} \|\Lambda^j u\|_{\mathscr{V}}, \\ \forall u \in D(\Lambda^{k-1}; \mathscr{V}), \text{ and } \mathscr{A} \in \mathscr{L}\big(\mathrm{D}(\Lambda^j; \mathscr{V}); \mathrm{D}(\Lambda^j; \mathscr{V})\big), \\ 0 \leq j \leq k-1, \end{cases}$$

inequality (9.25) *being satisfied for all k. Then, if*

$$(9.26) \quad f \in \mathrm{D}(\Lambda^\infty, M_k; \mathscr{V}'),$$

we have

$$(9.27) \quad u \in \mathrm{D}(\Lambda^\infty, M_k; \mathscr{V}).$$

Proof. The proof is in a certain sense an "axiomatisation" of the proof of Theorem 1.2.

By hypothesis, there exist d, $\mathscr{L}$ such that

$$(9.28) \quad \|\Lambda^k f\|_{\mathscr{V}'} \leq \mathrm{d} \mathscr{L}^k M_k, \ \forall k.$$

By induction on k, we shall verify that

$$(9.29) \quad \|\Lambda^k u\|_{\mathscr{V}} \leq \delta B^k M_k, \ \forall k,$$

where

$$(9.30) \quad \delta = 2\mathrm{d}/\alpha, \ B = \max\left(\left(1 + \frac{2cc_1}{\alpha}\right), L \ \mathscr{L}\right).$$

Indeed we have:

$$(9.31) \quad \Lambda(\Lambda^k u) + \mathscr{A} \Lambda^k u = [\mathscr{A}, \Lambda^k]\, u + \Lambda^k f$$

and for $u \in \mathrm{D}(\Lambda_k; \mathscr{V})$, it follows from (9.25) that

$$(9.32) \quad \| [\mathscr{A}, \Lambda^k]\, u \|_{\mathscr{V}'} \leq c \sum_{j=0}^{k-1} \binom{k}{j} L^{k-j} M_{k-j} \|\Lambda^j u\|_{\mathscr{V}}.$$

Then (9.13) applied to (9.31) yields:

$$(9.33) \quad \|\Lambda^k u\|_{\mathscr{V}} \leq \frac{1}{\alpha} \Big[\|\Lambda^k f\|_{\mathscr{V}'} + c \sum_{j=0}^{k-1} \binom{k}{j} L^{k-j} M_{k-j} \|\Lambda^j u\|_{\mathscr{V}}\Big].$$

Relation (9.29) holds for $k = 0$. Admitting it up to $(k-1)$, (9.33) yields

$$\|\Lambda^k u\|_{\mathscr{V}} \le \frac{1}{\alpha}\left(d\mathscr{L}^k M_k + c\sum_{j=0}^{k-1}\binom{k}{j} L^{k-j} M_{k-j}\,\delta B^j M_j\right) \le$$

$$\le \frac{\delta}{2}\mathscr{L}^k M_k + \frac{cc_1\delta}{\alpha} M_k \sum_{j=0}^{k-1} L^{k-j} B^j \le$$

$$\le \frac{\delta}{2} B^k M_k + \frac{cc_1\delta}{\alpha} B^k M_k \frac{1}{\dfrac{B}{L}-1} \le \delta B^k M_k. \quad \square$$

9.2 Application. Periodic Problems

We consider the setting of Chapter 3, Section 6. Thus:

$$\mathscr{V} = L^2(0, T; V) \quad \mathscr{H} = L^2(0, T; H),$$

$$(9.34) \qquad \mathscr{A} = A(t) \ \big(\text{i.e. } \mathscr{A}u = \text{“}t \to A(t)\, u(t)\text{”}\big), \text{ with}$$

$$\big(A(t)\, v, v\big) \ge \alpha\, \|v\|^2, \ v \in V, \ t \in [0, T], \ \alpha > 0;$$

next we let

$$(9.35) \qquad \Lambda = \mathrm{d}/\mathrm{d}t,$$

$$\mathrm{D}(\Lambda; \mathscr{H}) = \{f \mid f, f' \in L^2(0, T; H), f(0) = f(T)\}.$$

Hypotheses (9.2) and (9.3) are satisfied. We are within the conditions of applicability of (9.4). Therefore, for f given in $L^2(0, T; V')$, there exists a unique u in $L^2(0, T; V)$ with $\mathrm{d}u/\mathrm{d}t \in L^2(0, T; V')$, satisfying

$$(9.36) \qquad \frac{\mathrm{d}u}{\mathrm{d}t} + A(t)\, u = f, \ u(0) = u(T). \quad \square$$

Now, via an applications of Theorem 9.3, we shall prove

Theorem 9.4. *Assume that* (9.34) *holds and that*

$$(9.37) \qquad \begin{cases} t \to A(t) \text{ is a function of class } M_k \text{ of } [0, T] \to \mathscr{L}(V; V'), \\ \text{with } A^{(j)}(0) = A^{(j)}(T), \ \forall j. \end{cases}$$

Further assume that the sequence $\{M_k\}$ *satisfies* (1.22), ..., (1.25). *Let* f *be given and satisfy*:

$$(9.38) \qquad \begin{cases} \text{there exist } d \text{ and } \mathscr{L} \text{ such that} \\ \left(\int_0^T \|f^{(k)}(t)\|^2_{\mathscr{V}'}\, \mathrm{d}t\right)^{1/2} \le d\mathscr{L}^k M_k, \ \forall k, \\ f^{(k)}(0) = f^{(k)}(T), \ \forall k. \end{cases}$$

Then the solution u of (9.36) *is a function of class M_k of $[0, T] \to V$ and satisfies $u^{(k)}(0) = u^{(k)}(T)$, $\forall k$.*

Proof. Thanks to the hypothesis "$A^{(j)}(0) = A^{(j)}(T)$" $\forall j$, we have:

$$\mathscr{A} \in \mathscr{L}\big(\mathrm{D}(\Lambda^j; \mathscr{V}); \mathrm{D}(\Lambda^j; \mathscr{V}')\big) \ \forall j$$

and the first part of (9.24) follows from the fact that $t \to A(t)$ is of class M_k from $[0, T] \to \mathscr{L}(V; V')$. Then the theorem is a consequence of Theorem 9.3. ☐

Example 9.1. Consider the setting of Example 6.1, with the coefficients $a_{ij}(x, t)$ satisfying

$$\frac{\partial^k}{\partial t^k} a_{ij}(x, 0) = \frac{\partial^k}{\partial t^k} a_{ij}(x, T) \ \forall k.$$

Let us, for example, take $V = H^1(\Omega)$, $H = L^2(\Omega)$ and $a(\varphi, \psi)$ as in Example 6.1.

Let f_1 and g be given with:

$$\text{(9.39)} \quad \begin{cases} \left(\int_0^T |f_1^{(k)}(t)|^2 \, \mathrm{d}t\right)^{1/2} \le d_1 \mathscr{L}_1^k M_k \ \forall k, \\ f_1^{(k)}(0) = f_1^{(k)}(T) \ \forall k, \end{cases}$$

$$\text{(9.40)} \quad \begin{cases} \left(\int_0^T \|g^{(k)}(t)\|^2_{H^{-1/2}(\Gamma)} \, \mathrm{d}t\right)^{1/2} \le d_2 \mathscr{L}_2^k M_k \ \forall k, \\ g^{(k)}(0) = g^{(k)}(T) \ \forall k. \end{cases}$$

We apply Theorem 9.4, taking f given by

$$\text{(9.41)} \qquad \big(f(t), v\big) = \int_\Omega f_1(t) \, \bar{v} \, \mathrm{d}x + \int_\Gamma g(t) \, \gamma_0 \bar{v} \, \mathrm{d}\Gamma, \ v \in H^1(\Omega).$$

Then the solution u of (9.36) satisfies:

$$\text{(9.42)} \qquad \big(u'(t), v\big) + a\big(t; u(t), v\big) = \big(f(t), v\big), \ \forall v \in V,$$

where $(f(t), v)$ is given by (9.41).

It follows from (9.42) that

$$\text{(9.43)} \qquad \frac{\partial u}{\partial t} + A\left(x, t, \frac{\partial}{\partial x}\right) u = f_1.$$

u is a function of class M_k of $[0, T] \to V$, therefore, according to (9.43), $t \to A(x, t\, \partial/\partial x)\, u(t)$ is of class M_k from $[0, T] \to L^2(\Omega)$ ((1)).

((1)) $L^2(\Omega)$ could be replaced with $\Xi^{-1}(\Omega)$, in the hypotheses on f.

Then, according to the results of Chapter 2, we can define

$$\frac{\partial u}{\partial \nu_{A(t)}} \in H^{-1/2}(\Gamma) \tag{9.44}$$

and, from (9.42), (9.43), it is easy to show that

$$\frac{\partial u}{\partial \nu_{A(t)}} = g. \tag{9.45}$$

In short, for f_1 and g given with (9.39), (9.40), there exists a unique u, of class M_k, mapping $[0, T] \to V = H^1(\Omega)$ and satisfying (9.43), (9.45) and

$$u^{(k)}(x, 0) = u^{(k)}(x, T), \ \forall k. \quad \square$$

9.3 Transposition

Under the hypotheses of Theorem 9.3, $\Lambda + \mathscr{A}$ is an isomorphism of $\mathrm{D}(\Lambda^\infty, M_k; \mathscr{V})$ onto $\mathrm{D}(\Lambda^\infty, M_k; \mathscr{V}')$. If we make *"adjoint hypotheses"*, then $\Lambda^* + \mathscr{A}^*$ is an isomorphism of $\mathrm{D}(\Lambda^{*\infty}, M_k; \mathscr{V})$ onto $\mathrm{D}(\Lambda^{*\infty}, M_k; \mathscr{V}')$ and therefore, by transposition:

$$\begin{cases} \Lambda + \mathscr{A} \textit{ is an isomorphism of } \big(\mathrm{D}(\Lambda^{*\infty}, M_k; \mathscr{V}')\big)' \\ \textit{onto } \big(\mathrm{D}(\Lambda^{*\infty}, M_k; \mathscr{V})\big)'. \quad \square \end{cases} \tag{9.46}$$

Example 9.2. Let us reconsider Example 9.1 formally; we obtain the existence and uniqueness of u satisfying:

$$\begin{cases} \int_0^T [-(u, \varphi') + a(t; u, \varphi)] \, \mathrm{d}t = \int_0^T (f_1, \varphi) \, \mathrm{d}t + \int_0^T (g, \gamma_0 \varphi)_\Gamma \, \mathrm{d}t - \\ - \big(h_0, \varphi(0)\big), \ h_0 \in \Xi^{-1}(\Omega), \end{cases} \tag{9.47}$$

for $\varphi \in \mathrm{D}(\Lambda^{*\infty}, M_k; \mathscr{V})$, i.e.

$$\begin{cases} \left(\int_0^T \|\varphi^{(k)}(t)\|_V^2 \, \mathrm{d}t\right)^{1/2} \le c L^k M_k \ \forall k, \\ \varphi^{(k)}(0) = \varphi^{(k)}(T) \ \forall k. \end{cases}$$

In this fashion, we obtain u, solution—in a sense which can be made precise [1]—of (9.43) with (9.45) and

$$u(T) - u(0) = h_0. \quad \square \tag{9.48}$$

---

[1] By procedures analogous to those of Section 10 below.

10. The Case of a Finite Interval $]0, T[$

10.1 Orientation. General Problems

Let us consider a family of operators $A(t) \in \mathscr{L}(V; V')$, as in Section 1 but with $t \in [0, T]$.

We make the hypotheses:

$$\operatorname{Re}\left(A(t)\, v, v\right) + \lambda\, |v|^2 \geq \alpha\, \|v\|^2,\ \alpha > 0,\ v \in V, \tag{10.1}$$

$$t \to A(t) \text{ is of class } M_k \text{ from } [0, T] \to \mathscr{L}(V; V'), \tag{10.2}$$

$$\text{the sequence } \{M_k\} \text{ satisfies (1.22), ..., (1.25)}. \tag{10.3}$$

We consider the problem:

$$\begin{cases} A^*(t)\, v - v' = \varphi,\ t \in]0, T[, \\ v(T) = 0, \end{cases} \tag{10.4}$$

where

$$\varphi \in \mathscr{D}_{M_k}(]0, T[; V'). \tag{10.5}$$

We *denote by* X the space described by the solution v of (10.4) as φ describes $\mathscr{D}_{M_k}(]0, T[; V')$.

According to Theorem 1.2, we know that

$$X \subset \mathscr{D}_{M_k}([0, T]; V) \text{ (algebraic inclusion)}. \tag{10.6}$$

Providing X with the topology translated from $\mathscr{D}_{M_k}(]0, T[; V')$ by the mapping $\varphi \to v$, we have (done what was necessary to have):

$$\begin{cases} \text{the operator } A^* - \mathrm{d}/\mathrm{d}t \text{ is an isomorphism of } X \\ \text{onto } \mathscr{D}_{M_k}(]0, T[; V'). \end{cases} \tag{10.7}$$

By transposition of (10.7), we obtain:

Theorem 10.1. *Let hypotheses* (10.1), (10.2) *and* (10.3) *be satisfied. Let* $v \to L(v)$ *be a continuous antilinear form on* X. *Then there exists a unique element* $u \in \mathscr{D}'_{M_k}(]0, T[; V)$ *such that*

$$\langle u, A^*(t)\, v - v' \rangle = L(v),\ \forall v \in X, \tag{10.8}$$

(where in (10.8) the brackets denote the scalar product in the antiduality between $\mathscr{D}'_{M_k}(]0, T[; V)$ and $\mathscr{D}_{M_k}(]0, T[; V')$). ☐

The problems are now completely analogous, in principle, to those already encountered in Volumes 1 and 2 and in Chapter 8 and Section 9.3:

(i) choice of L in (10.8);

(ii) interpretation of (10.8).

Formally we shall take

$$L(v) = \int_0^T \langle f, v \rangle \, dt + (u_0, v(0)), \tag{10.9}$$

where f is given as a "suitable" functional with values in V', u_0 is given in the dual to the space described by $v(0)$.

Next, still formally, we have

$$\begin{cases} A(t)\, u + u' = f, \\ u(0) = u_0. \end{cases} \tag{10.10}$$

Thus, in principle, this procedure yields "the most general" right-hand members f and u_0 for which problem (10.10) is well-defined and admits a unique solution. ☐

Remark 10.1. Of course, in the examples of differential operators where the coefficients are "regular in x", we can go further, as we shall see in Chapters 10 and 11. ☐

Remark 10.2. Remarks analogous to the preceding ones can be made, starting from the results of Sections 2 and 4 (second order equation and Schroedinger equation).

We shall not develop these possibilities. ☐

Orientation

We shall successively examine points (i) and (ii), moreover subject to additional hypotheses. ☐

10.2 Space Described by $v(0)$ as v Describes X

We have completely resolved the problem of the characterization of $v(0)$ as v describes X only under two very restrictive types of hypotheses. See Problem 14.12.

10.2.1 Case of an Analytic Semi-Group

We make the hypothesis:

(10.11) $t \to a(t; u, v)$ *is constant in the neighborhood of* $t = 0$;

we set

(10.12) $a(t; u, v) = a(u, v)$, $A(t) = A$, in the neighborhood of $t = 0$.

Note that thanks to (10.1):

(10.13) $\begin{cases} -A \textit{ is the infinitesimal generator of an analytic} \\ \textit{semi-group in } H \textit{ (or in } V), \end{cases}$

and this, in fact, is *the only hypothesis* which will intervene.

Theorem 10.2. *Under the hypotheses* (10.1), (10.2), (10.3), (10.11) (*or more generally* $A(t) = A$ *in the neighborhood of* $t = 0$, A *satisfying* (10.13)), $v(0)$ *describes the space* $\mathrm{D}(A^{*\infty}; k!)$ *as* v *describes* X.

Proof. Let $v \in X$, therefore solution of (10.4). Since φ has compact support in $[0, T[$, we have:

(10.14) $$A^*(t)\, v - v' = 0 \text{ in the neighborhood of } t = 0,$$

and, according to (10.11), we thus have

(10.15) $$A^* v = v' \text{ in the neighborhood of } t = 0,$$

and therefore

(10.16) $$A^{*k}\, v(0) = v^{(k)}(0), \ \forall k.$$

But, according to (10.13) (and the same property holding for A^*), we see that $t \to v(t)$ is analytic in the neighborhood of $t = 0$, with values in, for example, H and therefore there exist c and L such that

(10.17) $$|v^{(k)}(0)| \leq c L^k k!, \ \forall k.$$

From (10.16) and (10.17), it follows that $v(0) \in \mathrm{D}(A^{*\infty}; k!)$.

Conversely, let v_0 be given in $\mathrm{D}(A^{*\infty}; k!)$, Then the function

(10.18) $$w(t) = \sum_{k=0}^{\infty} \frac{A^{*k} v_0}{k!} t^k,$$

is analytic in $[0, t_0]$, t_0 sufficiently small, with values in V, since $|A^{*k} v_0| \leq$ $\leq c L^k k!$ and therefore $\|A^{*k} v_0\| \leq c L^k k!$. If θ is a function of class M_k in $[0, T]$, $\theta(t) = 1$ in $[0, t_0/3]$, $\theta(t) = 0$ for $t > 2t_0/3$, and if we define

$$v(t) = \begin{cases} \theta(t)\, w(t) \text{ in } [0, t_0], \\ 0 \text{ in } [t_0, T], \end{cases}$$

we obtain

$$A^* v - v' = \psi, \ \psi \in \mathscr{D}_{M_k}(]0, T[; V'), \ v(T) = 0,$$

therefore $v \in X$ and since $v(0) = v_0$, we have the desired result. ☐

10.2.2 Case of a Group

Let A be independent of t. Thus we take

(10.19) $$a(t; u, v) = a(u, v), t \in [0, T].$$

We assume that

(10.20) $$-A \text{ is the infinitesimal generator of a group in } V.$$

Then we have:

Theorem 10.3. *Under hypothesis* (10.20), $v(0)$ *describes the space* $D(A^{*\infty}; M_k)$.

Proof. Indeed $A^*v - v' = 0$ in the neighborhood of $t = 0$; therefore we have (10.16) again and since v is of class M_k with values in V, we have:

$$\|v^{(k)}(0)\| \leq cL^k M_k, \quad \forall k, \tag{10.21}$$

whence $v(0) \in D(A^{*\infty}; M_k)$.

Conversely, if v_0 is given in $D(A^{*\infty}; M_k)$, we have:

$$\begin{cases} A^*w - w' = 0, \\ w(0) = v_0, \end{cases} \tag{10.22}$$

a *well-posed* problem, thanks to (10.20); since $v_0 \in D(A^{*\infty}; M_k)$, $t \to w(t)$ is a function of class M_k of $[0, T] \to V$; then we take $v = \theta w$, θ as in the proof of Theorem 10.2, and we proceed as in this proof. □

10.3 The Space Ξ_{M_k}

Orientation

Our aim is to construct a space which contains X and *"the smallest possible"*, such that $\mathcal{D}_{M_k}(]0, T[; V)$ *is dense in this space.*

We shall only use the fact that we have (10.6).

More generally, we introduce a Banach space F and *construct a space* $\Xi_{M_k}(]0, T[; F)$ with the following properties:

$$\mathcal{D}_{M_k}([0, T]; F) \subset \Xi_{M_k}(]0, T[; F), \tag{10.23}$$

$$\mathcal{D}_{M_k}(]0, T[; F) \text{ is dense in } \Xi_{M_k}(]0, T[; F). \tag{10.24}$$ □

First we define

$$\Xi^{L,a}_{M_k} = \left\{ v \mid v \text{ is } C^\infty \text{ from } [0, T] \to F, v = 0 \text{ in } [T - a, T], \right. \\ \left. \frac{1}{L^k M_k} \|\varrho^k v^{(k)}\|_{L^2(0,T;F)} \leq c, \forall k \right\}, \tag{10.25}$$

where ϱ is given by

$$\varrho(t) = \begin{cases} t/\beta, \ 0 \leq t \leq \beta < T, \\ 1, \ \beta \leq t \leq T. \end{cases} \tag{10.26}$$

Provided with the norm

$$\sup_k \frac{1}{L^k M_k} \|\varrho^k v^{(k)}\|_{L^2(0,T;F)} \tag{10.27}$$

it is a Banach space.

Next we take

$$(10.28)\qquad \Xi_{M_k}(]0, T[; F) = \Xi_{M_k} = \operatorname*{ind\,lim}_{L_n \uparrow +\infty} \left(\operatorname*{ind\,lim}_{a_n \downarrow 0} \Xi_{M_k}^{L_n, a_n} \right).$$

We obviously have (10.23).

Theorem 10.4. *Assume that* (10.3) *is satisfied. Then property* (10.24) *holds.*

Proof. The proof is analogous to the one in Section 4.2, Chapter 8.

Let v be given in Ξ_{M_k}, thus

$$(10.29)\qquad v \in \Xi_{M_k}^{L,a}, \text{ for suitable } L \text{ and } a.$$

Let

$$(10.30)\qquad L_1 = \max(L, 1).$$

Later on we shall show that:

$$(10.31)\qquad \begin{cases} \text{there exists a sequence } \theta_n(t) \text{ satisfying} \\ 0 \leq \theta_n(t) \leq 1,\ \theta_n(t) = 1 \text{ for } t \geq 2/n,\ \theta_n(t) = 0 \text{ for } t \leq 1/n, \\ |t^k \theta_n^{(k)}(t)| \leq C_2 L_1^k M_k,\ \forall k. \end{cases}$$

Then $\theta_n v \in \mathscr{D}_{M_k}(]0, T[; F)$ and we shall have the theorem if we can show that $\theta_n v \to v$ in Ξ_{M_k}. More precisely, we shall show that

$$(10.32)\qquad \theta_n v \to v \text{ in } \Xi_{M_k}^{L_1+\eta, a}, \text{ arbitrarily fixed } \eta > 0.$$

Indeed, let

$$y_{n,k} = \frac{1}{(L_1 + \eta)^k M_k} \| \varrho^k((\theta_n v)^{(k)} - v^{(k)}) \|_{L^2(0,T;F)}.$$

We have:

$$y_{n,k} \leq z_{n,k} + \zeta_{n,k},$$

where

$$z_{n,k} = \frac{1}{(L_1 + \eta)^k M_k} \| (\theta_n - 1)\, \varrho^k v^{(k)} \|_{L^2(0,T;F)},$$

$$\zeta_{n,k} = \frac{1}{(L_1 + \eta)^k M_k} \sum_{j=1}^{k} \binom{k}{j} \| \theta_n^{(j)} \varrho^k v^{(k-j)} \|_{L^2(0,T;F)}.$$

Next, we set

$$y_n = \sup_{k \geq 0} y_{n,k},\ z_n = \sup_{k \geq 0} z_{n,k},\ \zeta_n = \sup_{k \geq 0} \zeta_{n,k}.$$

To show (10.32) is equivalent to showing that

$$(10.33)\qquad y_n \to 0 \text{ as } n \to \infty.$$

Now $y_n \leq z_n + \zeta_n$ and we shall have (10.33) if we can show that

$$z_n \to 0 \quad \text{as} \quad n \to \infty, \tag{10.34}$$

$$\zeta_n \to 0 \quad \text{as} \quad n \to \infty. \tag{10.35}$$

But, according to (10.29),

$$z_{n,k} \leq \frac{1}{(L_1 + \eta)^k M_k} cL^k M_k,$$

therefore, the c's denoting different constants:

$$z_n \leq \sup_{0 \leq k \leq N} z_{n,k} + c \left(\frac{L}{L_1 + \eta} \right)^N$$

and this proves (10.34) if we can show that

$$z_{n,k} \to 0, \; n \to \infty, \; k \textit{ fixed},$$

which is obvious.

Let us now prove (10.35). We have

$$\zeta_{n,k} = \frac{1}{(L_1 + \eta)^k M_k} \sum_{j=1}^{k} \binom{k}{j} \| \theta_n^{(j)} \varrho^j \varrho^{k-j} v^{(k-j)} \|_{L^2(0,T;F)} \leq$$

$$\leq \frac{c}{(L_1 + \eta)^k M_k} \sum_{j=1}^{k} \binom{k}{j} L_1^j M_j L^{k-j} M_{k-j} \leq$$

$$\leq c \frac{1}{(L_1 + \eta)^k} \sum_{j=1}^{k} L_1^j L^{k-j} \leq ck \frac{L_1^k}{(L_1 + \eta)^k}.$$

Therefore, for sufficiently large N:

$$\zeta_n \leq \sup_{0 \leq k \leq N} \zeta_{n,k} + cN \left(\frac{L_1}{L_1 + \eta} \right)^N$$

and the result follows if we can show that:

$$\zeta_{n,k} \to 0, \; n \to \infty, \; k \textit{ fixed}.$$

But, for fixed k,

$$\zeta_{n,k} \leq (\text{constant}) \sum_{j=1}^{k} \left(\int_0^{2/n} t^{2(k-j)} \| v^{(k-j)}(t) \|_F^2 \, dt \right)^{1/2} \to 0.$$

Therefore the theorem is proved, *subject to proving* (10.31).

At the risk of having to translate and change n to $2n$ later, it is sufficient to show the existence of χ_n with

$$\begin{cases} \chi_n(t) = 1 \text{ for } t \geq 2/n, \\ 0 \leq \chi_n(t) \leq 1, \\ \chi_n^{(j)}(0) = 0, \; \forall j, \end{cases} \tag{10.36}$$

$$|t^k \chi_n^{(k)}(t)| \leq c_2 L_1^k M_k, \; \forall k. \tag{10.37}$$

To this end, we start with a function $\varrho \in \mathscr{D}_{M_k}(\mathbf{R})$, with support in $[0, 1]$,

$$\varrho \geq 0, \int_0^1 \varrho \, dt = 1, |\varrho^{(k)}(t)| \leq c_3 \left(\frac{L_1}{2}\right)^k M_k, \ \forall k.$$

Next, we set

$$\varrho_n(t) = n\varrho(nt),$$

$$\psi_n(t) = \begin{cases} 0 \text{ if } t \leq 0, \\ nt \text{ if } 0 \leq t \leq 1/n, \\ 1 \text{ if } t \geq 1/n, \end{cases}$$

and *define*

$$\chi_n = \varrho_n * \psi_n.$$

Conditions (10.36) hold. Let us verify (10.37). We have:

$$\chi_n^{(k)}(t) = \int_0^t \psi_n(t - \sigma) \, \varrho_n^{(k)}(\sigma) \, d\sigma,$$

whence

$$|\chi_n^{(k)}(t)| \leq n^k \int_0^{1/n} |\varrho^{(k)}(n\sigma)| \, n \, d\sigma = n^k \int_0^1 |\varrho^{(k)}(\sigma)| \, d\sigma \leq cn^k \left(\frac{L_1}{2}\right)^k M_k,$$

therefore if $0 \leq t \leq 2/n$,

$$|t^k \chi_n^{(k)}(t)| \leq c \left(\frac{2}{n}\right)^k n^k \left(\frac{L_1}{2}\right)^k M_k, \text{ whence (10.37).} \quad \square$$

Corollary 10.1. *If* (10.3) *holds, the space* $(\Xi_{M_k})'$ *can be identified to a subspace of* $\mathscr{D}'_{M_k}(]0, T[; F')$. $\square$

10.4 Choice of L

We now take in (10.8) the choice (10.9) of $L(v)$, where:

(10.38) $$f \in \Xi'_{M_k}(0, T; V') = (\Xi_{M_k}(0, T; V))',$$

(10.39) $$\begin{cases} u_0 \in (D(A^{*\infty}; k!))' \text{ in the case of Theorem 10.2,} \\ u_0 \in (D(A^{*\infty}; M_k))' \text{ in the case of Theorem 10.3.} \end{cases}$$

There remains to interpret the problem; in order to fix our ideas, we shall consider the case where $A(t) = A$ does not depend on t.

10.5 The Space Y and Trace Theorems

10.5.1 In the Setting of Theorem 10.2

Let (10.8) hold, with the choice (10.9), (10.38), (10.39). It follows that

$$Au + u' = f \text{ in } \mathscr{D}'_{M_k}(]0, T[; V'). \tag{10.40}$$

Consequently u belongs to the space Y defined by

$$Y = \{v \mid v \in \mathscr{D}'_{M_k}(]0, T[; V), Av + v' \in \Xi'_{M_k}(]0, T[; V')\}. \tag{10.41}$$

We provide Y with the coarsest locally convex topology which makes the mappings $v \to 0$ and $v \to Av + v'$, of

$$Y \to \mathscr{D}'_{M_k}(]0, T[; V) \text{ and } Y \to \Xi'_{M_k}(]0, T[; V'),$$

continuous, each of these spaces being provided with the strong dual topology. ☐

In order to state the density result which we are aiming at, let us make the hypothesis in the setting of Theorem 10.2:

$$\begin{cases} \textit{if in } [0, a], \psi \textit{ satisfies} \\ A^*\psi - \psi' = 0 \textit{ with } \psi(0) = 0 \\ \textit{and if } \psi \textit{ is of class } M_k \textit{ in } [0, a] \textit{ with values in } \mathrm{D}(A^*), \\ \textit{then } \psi(t) = 0 \textit{ for } t \in [0, a]. \end{cases} \tag{10.42}$$

In this case A^* is said to have the *retrograde uniqueness* property.

This hypothesis is satisfied for *parabolic operators* (see Lions-Malgrange [1]).

Theorem 10.5. *Assume that* (10.3) *and* (10.42) *are satisfied. Then the space* $\mathscr{D}_{M_k}([0, T]) \otimes V$ (where $\mathscr{D}_{M_k}([0, T])$ is the space of scalar functions of class M_k in $[0, T]$) *is dense in* Y.

Proof. Let M be a continuous linear form on Y, vanishing on $\mathscr{D}_{M_k}([0, T]) \otimes V$.

We need to show that $M = 0$. Now M may be represented in the form

$$\begin{cases} M(u) = \langle \varphi, u\rangle + \langle \psi, Au + u'\rangle, \\ \varphi \in \mathscr{D}_{M_k}(]0, T[; V'), \quad \psi \in \Xi_{M_k}(]0, T[; V) = \Xi_{M_k}, \end{cases} \tag{10.43}$$

where the first (resp. second) bracket denotes the duality between $\mathscr{D}_{M_k}(]0, T[; V')$ (resp. Ξ_{M_k}) and $\mathscr{D}'_{M_k}(]0, T[; V)$ (resp. Ξ'_{M_k}).

Let $\tilde{\varphi}$ and $\tilde{\psi}$ be the extensions of φ and ψ to $\mathbf{R}_t$ by 0 outside $[0, T]$. If $w \in \mathscr{D}_{M_k}(\mathbf{R}_t) \otimes V$, its restriction u to $[0, T]$ belongs to $\mathscr{D}_{M_k}([0, T]) \otimes V$, therefore $M(u) = 0$ and therefore

$$\langle \tilde{\varphi}, w\rangle + \langle \tilde{\psi}, Aw + w'\rangle = 0,$$

therefore

(10.44) $$\tilde{\varphi} + A^*\psi - \frac{d\tilde{\psi}}{dt} = 0 \text{ on } \mathbf{R}_t.$$

But for example, $\tilde{\psi} \in L^2(-1, T; V)$, and by (10.44), $\frac{d\tilde{\psi}}{dt} \in L^2(-1, T; V')$ and therefore

(10.45) $$\psi(0) = 0.$$

On the other hand, the support of $\tilde{\varphi}$ is bounded on the right (by $T - a$, suitable $a > 0$), so that (10.44) yields (since $\tilde{\varphi} \in \mathscr{D}_{M_k}(\mathbf{R}; V')$):

(10.46) $$\tilde{\psi} \in \mathscr{D}_{+,M_k}(\mathbf{R}; V).$$

But, *in the neighbourhood* of, 0 $\varphi = 0$, therefore

$$A^*\psi - \frac{d\psi}{dt} = 0,$$

which, together with (10.45), (10.46) and hypothesis (10.42), yields

$$\varphi = 0 \textit{ in the neighbourhood of } 0.$$

Therefore

$$\psi \in \mathscr{D}_{M_k}(]0, T[; V)$$

and consequently, for $u \in \mathscr{D}'_{M_k}(]0, T[; V)$, we have:

$$\langle \psi, Au + u' \rangle = \langle A^*\psi - \psi', u \rangle.$$

But then (10.43) yields

$$M(u) = \langle \varphi + A^*\psi - \psi', u \rangle = 0 \text{ (by (10.44))}. \quad \square$$

We are now in a position to prove the following *trace theorem.*

Theorem 10.6. *Assume that* (10.3) *holds, A being given as in Theorem* 10.2[1]. *The mapping $u \to u(0)$ of $\mathscr{D}_{M_k}([0, T]) \otimes V \to V$ extends by continuity to a continuous linear mapping, still denoted by $u \to u(0)$, of Y (defined by* (10.41)) *into* $(D(A^{*\infty}; k!))'$.

Proof. Let $u \in Y$ be fixed.

By the proof of Theorem 10.2, we can find a continuous linear mapping $v_0 \to \mathscr{C}v_0$ of $D^L(A^{*\infty}; k!) \to X$ such that

(10.47) $$(\mathscr{C}v_0)(0) = v_0.$$

Set

(10.48) $$Z(v_0) = \langle A^*\mathscr{C}v_0 - (\mathscr{C}v_0)', u \rangle - \langle \mathscr{C}v_0, Au + u' \rangle,$$

[1] Then (10.42) holds.

the first (resp. second) bracket denoting the scalar product between $\mathscr{D}_{M_k}(]0, T[; V')$ T(resp. Ξ_{M_k}) and $\mathscr{D}'_{M_k}(]0, T[; V]$ (resp. Ξ'_{M_k}).

The number $Z(v_0)$ depends only on v_0, i.e. if we replace $\mathscr{C}v_0$ by $w \in X$ with $w(0) = v_0$, then

$$Z(v_0) = \langle A^*w - w', u\rangle - \langle w, Au + u'\rangle.$$

Indeed if $\psi = \mathscr{C}v_0 - w$, this amounts to showing that

$$\langle A^*\psi - \psi', u\rangle - \langle \psi, Au + u'\rangle = 0. \tag{10.49}$$

But we have:

$$A^*\psi - \psi = 0 \text{ in the neighbourhood of } 0,$$

$$\varphi(0) = 0,$$

therefore, by the retrograde uniqueness property, $\psi = 0$ in the neighbourhood of 0, whence (10.49).

The form

$$v_0 \to Z(v_0)$$

is continuous on $\mathrm{D}^L(A^{*\infty}; k!)$ and since it is independent of L, we see that $v_0 \to Z(v_0)$ is a continuous linear form on $\mathrm{D}(A^{*\infty}; k!)$. Therefore

$$Z(v_0) = (\tau u, v_0), \tau u \in \big(\mathrm{D}(A^{*\infty}; k!)\big)'. \tag{10.50}$$

But if $u \in \mathscr{D}_{M_k}([0, T]) \otimes V$, we immediately see that

$$\tau u = u(0).$$

Finally, if $u \to 0$ in Y, then $Z(v_0) \to 0$ uniformly for v_0 in a bounded set of $\mathrm{D}(A^{*\infty}; k!)$—indeed v_0 then remains in a bounded set of $\mathrm{D}^L(A^{*\infty}; k!)$, suitably fixed L, and we can choose $\mathscr{C}v_0$ in a bounded set of X, from which the desired result follows. $\square$

10.5.2 In the Setting of Theorem 10.3

We again have the definition of Y given by (10.41); hypothesis (10.42) is evidently satisfied; the analogue of Theorem 10.5 holds and, in the same way as above we arrive at

Theorem 10.6a. *Under the hypotheses of Theorem* 10.3, *the linear mapping* $u \to u(0)$ *of* $\mathscr{D}_{M_k}([0, T] \otimes V \to V$ *extends by continuity to a continuous linear mapping, still denoted by* $u \to u(0)$, *of* $Y \to \mathrm{D}(A^{*\infty}; M_k)'$.

10.6 Non-Homogeneous Problems

We can now state the principal results:

Theorem 10.7. *Assume that* (10.3) *holds, A being given as in Theorem* 10.2. *Let f be given in* $\Xi'_{M_k}(]0, T[; V')$, u_0 *be given in* $\mathrm{D}(A^{*\infty}; k!)'$.

There exists a unique u in Y satisfying

(10.51) $\quad Au + u' = f$ *in the sense of* $\mathscr{D}'_{Mk}(]0, T[; V')$,

(10.52) $\quad u(0) = u_0$ *in the sense of Theorem* 10.6. □

Similarly, from Theorem 10.6a, we obtain:

Theorem 10.7a. *Under the hypotheses of Theorem* 10.3, *let f be given in $\Xi'_{Mk}(]0, T[; V')$, u_0 be given in $\mathrm{D}(A^{*\infty}; M_k)'$. There exists a unique u in Y, solution of $Au + u' = f$ in the sense of $\mathscr{D}'_{Mk}(]0, T[; V')$ and of $u(0) = u_0$ in the sense of Theorem* 10.6a.

Remark 10.3. Thus we see that in the *parabolic case*, we may take u_0 in $\mathrm{D}(A^{*\infty}; k!)$, whereas in the second order in t case (which is reduced to the first order as in Section 7.3.4) or in the Schroedinger case (and in particular in the hyperbolic case) we may take the initial data in $\mathrm{D}(A^{*\infty}; M_k)'$. □

Let us give some applications of Theorems 10.7 and 10.7a.

Example 10.1. (Non-homogeneous Neumann problem; parabolic case).

Let us consider the setting of Section 7.3.2, with A a second order elliptic operator and $V = H^1(\Omega)$; for $u, v \in H^1(\Omega)$, we set

$$a(u, v) = \sum_{i,j=1}^{n} \int_{\Omega} a_{ij}(x) \frac{\partial u}{\partial x_j} \frac{\partial \bar{v}}{\partial x_i} \, \mathrm{d}x. \tag{10.53}$$

We apply Theorem 10.7, with f given by

$$(f, v) = (f_0, v) + (g, \gamma_0 v)_\Gamma, \tag{10.54}$$

where

$$f_0 \in \Xi'_{Mk}(]0, T[; L^2(\Omega))^{((1))}, \tag{10.55}$$

$$g \in \Xi'_{Mk}(]0, T[; H^{-1/2}(\Gamma)). \tag{10.56}$$

Then the interpretation of problem (10.51), (10.52) is the following (without specifying the adequate trace theorems). u satisfies

$$\begin{cases} \dfrac{\partial u}{\partial t} - \displaystyle\sum_{i,j=1}^{n} \frac{\partial}{\partial x_i}\left(a_{ij} \frac{\partial u}{\partial x_j}\right) = f_0 \text{ in } Q, \\ \dfrac{\partial u}{\partial \nu_A} = g \text{ on } \Sigma, \text{ where } \dfrac{\partial u}{\partial \nu_A} = \displaystyle\sum_{i,j=1}^{n} a_{ij} \frac{\partial u}{\partial x_j} \cos(n, x_i), \\ u(x, 0) = u_0(x) \text{ on } \Omega. \end{cases} \tag{10.57}$$

□

((1)) More generally, we could take $f_0 \in \Xi'_{Mk}(]0, T[; \Xi^{-1}(\Omega))$, with $\Xi^{-1}(\Omega)$ defined as in Chapter 2, Section 6.3.

Example 10.2. (Non-homogeneous Dirichlet problem; parabolic case).

If we take A as in the preceding example, but with $V = H_0^1(\Omega)$, then Theorem 10.7 yields the existence and uniqueness of

$$u \in \mathscr{D}'_{M_k}(]0, T[; H_0^1(\Omega))$$

such that

$$\frac{\partial u}{\partial t} - \sum_{i,j=1}^{n} \frac{\partial}{\partial x_i}\left(a_{ij}(x)\frac{\partial u}{\partial x_j}\right) = f_0 \text{ in } Q, \tag{10.58}$$

where

$$f_0 \in \Xi'_{M_k}(]0, T[; H^{-1}(\Omega)),$$

with

$$u(0) = 0 \text{ on } \Sigma \tag{10.59}$$

and

$$u(x, 0) = u_0(x) \text{ on } \Omega.$$

But we can consider *non-homogeneous data on* Σ (as opposed to (10.59)) in the following way.

Generally speaking, let A—*independent of* t—be in $\mathscr{L}(V; V')$, such that (10.1) holds. Then, as we have seen in Section 7, A satisfies *for example*

$$A \in \mathscr{L}(H; \mathrm{D}(A^*)') \tag{10.60}$$

and $-A$ is the infinitesimal generator of a *semi-group in* V'

We may apply Theorem 10.7 by replacing V with H and V' with $\mathrm{D}(A^*)'$.

The space $\mathrm{D}(A^{*\infty}; k!)$ becomes the space of v's belonging to V' such that $A^*v \in V', \ldots, A^{*k}v \in V', \ldots$ and such that there exists an L with

$$\|A^{*k}v\|_{V'} \leq cL^k k!.$$

But thanks to the "coerciveness inequalities", this *does not change* the space $\mathrm{D}(A^{*\infty}; k!)$ Therefore the analogue to Theorem 10.7 is:

$$\left\{\begin{array}{l}\text{let } f \text{ be given in } \Xi'_{M_k}(]0, T[; \mathrm{D}(A^*)') \text{ and } u_0 \\ \text{in } \mathrm{D}(A^{*\infty}; k!)'. \text{ There exists a unique } u \text{ in} \\ \mathscr{D}'_{M_k}(]0, T[; H), \text{ such that } Au + u' \in \Xi'_{M_k}(]0, T[; \mathrm{D}(A^*)'), \\ \text{with } Au + u' = f,\ u(0) = u_0. \end{array}\right. \tag{10.61}$$

(Note that even more generally, we may consider A as belonging to $\mathscr{L}(\mathrm{D}(A^{*m})'; \mathrm{D}(A^{*m+1})')$ and take f in $\Xi'_{M_k}(]0, T[; \mathrm{D}(A^{*m+1})')$, and then find u in $\mathscr{D}'_{M_k}(]0, T[; \mathrm{D}(A^{*m})')$).

If we come back to the concrete situation, we may take f by

$$(f, v) = (f_0, v) - \left(g, \frac{\partial v}{\partial \nu_{A^*}}\right)_\Gamma, \quad v \in \mathrm{D}(A^*) = H^2(\Omega) \cap H_0^1(\Omega), \tag{10.62}$$

where

(10.63) $$f_0 \in \Xi'_{M_k}(]0, T[; H^{-1}(\Omega)), \; g \in \Xi'_{M_k}(]0, T[; H^{-1/2}(\Gamma)).$$

The interpretation of the equation $Au + u' = f$ is:

$$(u, A^*v) + (u', v) = (f, v), \; \forall v \in D(A^*)$$

and consequently u satisfies (10.58) and the conditions

(10.64) $$\begin{cases} u = g \text{ on } \Sigma, \\ u(x, 0) = u_0(x) \text{ on } \Omega. \quad \square \end{cases}$$

Remark 10.4. If we take f in $\Xi'_{M_k}(]0, T[; D(A^{*m+1})')$, then we may take $g \in \Xi'_{M_k}(]0, T[; H^{-2m-1/2}(\Gamma))$ in (10.64). $\square$

Remark 10.5. Of course the preceding examples can be extended to elliptic operators A of *arbitrary order*. $\square$

Example 10.3. If we new consider the operator given by

$$u \to \frac{\partial^2 u}{\partial t^2} - \sum_{i,j=1}^{n} \frac{\partial}{\partial x_i}\left(a_{ij}(x)\frac{\partial u}{\partial x_j}\right),$$

where (in addition to the ellipticity condition):

(10.65) $$a_{ij} = \overline{a_{ij}}, \; \forall i, j,$$

then we can apply Theorem 10.7a, as long as we write the equation

(10.66) $$\frac{\partial^2 u}{\partial t^2} - \sum_{i,j=1}^{n} \frac{\partial}{\partial x_i}\left(a_{ij}(x)\frac{\partial u}{\partial x_j}\right) = f$$

as a first order system in t.

If we first take $V = H^1(\Omega)$ (as in Example 10.1), then we are solving (without specifing the adequate trace theorems) equation (10.66) with the boundary conditions:

(10.67) $$\begin{cases} \dfrac{\partial u}{\partial \nu_A} = g \text{ on } \Sigma, \; g \text{ given in } \Xi'_{M_k}(]0, T[; H^{-1/2}(\Gamma)), \\ u(x, 0) = u_0(x) \text{ on } \Omega, \; u_0 \in D(A^{*\infty}; M_k)'^{\,((1))}, \\ \dfrac{\partial u}{\partial t}(x, 0) = u_1(x) \text{ on } \Omega, \; u_1 \in D(A^{*\infty}; M_k)'. \end{cases}$$

If we take $V = H_0^1(\Omega)$, then we find—as in Example 10.2—$u = 0$ on Σ as boundary condition.

((1)) We write $u_0(x)$ by an abuse of language. $u_0(x)$ is evidently an "entity" of a very general nature ...

But, by the same principle as in Example 10.2, we can solve equation (10.66) with $u \in \mathscr{D}'_{M_k}(]0, T[; L^2(\Omega))$ and

$$\begin{cases} u = g \text{ on } \Sigma, \ g \text{ given in } \Xi'_{M_k}(]0, T[; H^{-1/2}(\Gamma)) \\ \text{and } u(x, 0) = u_0(x), \ \dfrac{\partial u}{\partial t}(x, 0) = u_1(x), \ u_0, u_1 \\ \text{given in } \mathrm{D}(A^{*\infty}; M_k)'. \quad \square \end{cases} \tag{10.68}$$

Remark 10.6. Following the same principle, we can solve the analogous problems for the Schroedinger equation. $\square$

11. Distribution and Ultra-Distribution Semi-Groups

11.1 Distribution Semi-Groups

Let E be a Banach space and A a *closed* unbounded operator in E, with domain $\mathrm{D}(A)$ dense in E. We provide $\mathrm{D}(A)$ with the norm of the graph, which makes it a Banach space.

We shall make use of the space

$$\mathscr{L}(\mathscr{D}_-(\mathbf{R}); E) \tag{11.1}$$

of distributions according to Schwartz [3] on $\mathbf{R}$ with values in E. If E is *reflexive*, then, as already noted in Chapter 7, Remark 5.1, $\mathscr{L}(\mathscr{D}_-(\mathbf{R}); E) = (\mathscr{D}_-(\mathbf{R}; E'))' = \mathscr{D}'_+(\mathbf{R}; E)$. However in the sequel the reflexivity of E plays no role. Thus, in order to avoid any confusion, we shall use the notation (11.1).

The *problem* is the following: for f given in $\mathscr{L}(\mathscr{D}_-(\mathbf{R}); E)$, find the necessary and sufficient condition for the existence of a unique distribution u having the properties:

$$u \in \mathscr{L}(\mathscr{D}_-(\mathbf{R}); \mathrm{D}(A)), \tag{11.2}$$

$$Au + u' = f, \quad (u' = \mathrm{d}u/\mathrm{d}t), \tag{11.3}$$

$$\inf_t \{\text{support of } u \text{ in } t\} \le \inf_t \{\text{support of } f \text{ in } t\}, \tag{11.4}$$

$$\begin{cases} f \to u \text{ is a continuous mapping of} \\ \mathscr{L}(\mathscr{D}_-(\mathbf{R}); E) \to \mathscr{L}(\mathscr{D}_-(\mathbf{R}); \mathrm{D}(A)). \quad \square \end{cases} \tag{11.5}$$

Theorem 11.1. (Chazarain [1]) *In order that conditions* (11.2), ..., (11.5) *be satisfied, it is necessary and sufficient that A has the following properties: there exists a set $\mathscr{R} \subset \mathbf{C}$, of the form*

$$\mathscr{R} = \{\lambda \mid \lambda = \xi + i\eta, \ \xi \ge \alpha \log|\eta| + \beta, \ \xi \ge \xi_0, \ \alpha, \beta, \xi_0 \in \mathbf{R}\}, \tag{11.6}$$

such that $A + \lambda$ is an isomorphism of $\mathrm{D}(A)$ *onto* E *for* $\lambda \in \mathscr{R}$ *and that*

$$\|(A+\lambda)^{-1}\|_{\mathscr{L}(E;E)} \leq \text{polynomial in } |\lambda|, \lambda \in \mathscr{R}. \tag{11.7}$$

Proof of the necessity of (11.6), (11.7). We divide the proof into several points.

1) It is easily seen that it is equivalent to assume (11.2), ..., (11.5) or the existence of a distribution $\mathscr{G}$ ("elementary solution") with the following properties (see Chapter 4, Section 3):

$$\begin{cases} \mathscr{G} \in \mathscr{L}(\mathscr{D}_-(\mathbf{R}); \mathscr{L}(E; \mathrm{D}(A))), \\ \mathscr{G} = 0 \quad \text{for} \quad t < 0, \\ \left(\dfrac{\mathrm{d}}{\mathrm{d}t} + A\right) * \mathscr{G} = \delta \otimes I_E \text{ (where } I_X = \text{identity from } X \to X), \\ \mathscr{G} * \left(\dfrac{\mathrm{d}}{\mathrm{d}t} + A\right) = \delta \otimes I_{\mathrm{D}(A)}. \end{cases} \tag{11.8}$$

We shall show that (11.8) implies (11.6), (11.7).

2) Formally $(\lambda + A)^{-1}$ is given by $\mathscr{G}(\mathrm{e}^{-\lambda t})$, but (and this is the main difficulty) $\mathscr{G}(\mathrm{e}^{-\lambda t})$ may have no meaning for any $\lambda \in \mathbf{C}$.

The essential idea behind the proof is to "approach" $\mathscr{G}(\mathrm{e}^{-\lambda t})$ with $\mathscr{G}(\theta_\lambda)$, where θ_λ is a *truncation of* $\mathrm{e}^{-\lambda t}$ such that $\theta_\lambda \in \mathscr{D}_-$.

More precisely, we introduce once and for all a function $b \in \mathscr{D}(\mathbf{R})$ with the properties:

$$\begin{cases} b \in \mathscr{D}(\mathbf{R}), \text{ support of } b \text{ in } [t_0, t_1], \ 0 < t_0 < t_1, \\ b \geq 0, \ \int b \,\mathrm{d}t = 1. \end{cases} \tag{11.9}$$

Next, we introduce:

$$b_\eta = \mathrm{e}^{-i\eta t} b \text{ (so that } b_0 = b), \tag{11.10}$$

$$B_\eta(\lambda) = \int \mathrm{e}^{\lambda t} b_\eta(t) \,\mathrm{d}t = \int \mathrm{e}^{\xi t} b(t) \,\mathrm{d}t = B(\xi). \tag{11.11}$$

Let θ_λ be the solution *with support bounded on the right* (by t_1) of

$$\frac{\mathrm{d}\theta_\lambda}{\mathrm{d}t} + \lambda\theta_\lambda = -\frac{1}{B(\xi)} b_\eta(t). \tag{11.12}$$

If Y is the Heaviside function ($Y(t) = 1$ for $t > 0$, $Y(t) = 0$ for $t < 0$), then we have:

$$\left(\frac{\mathrm{d}}{\mathrm{d}t} + \lambda\right)((1 - Y)\,\mathrm{e}^{-\lambda t}) = -\delta,$$

so that (11.12) yields

$$\theta_\lambda = \frac{1}{B(\xi)} \left((1 - Y)\, e^{-\lambda t}\right) * b_\eta, \tag{11.13}$$

or, expanding:

$$\theta_\lambda = e^{-\lambda t} - \frac{e^{-i\eta t}}{B(\xi)} \left((e^{-\xi t} Y) * b\right). \tag{11.14}$$

Thus we see that

$$\theta_\lambda \text{ has support in } [t_0, t_1],\ \theta_\lambda \in \mathscr{D}(\mathbf{R}), \tag{11.15}$$

and $\theta_\lambda(t) = e^{-\lambda t}$ for $t < t_0$.

We now form $\mathscr{G}(\theta_\lambda)$, and we shall show that it is an "approximated resolvent".

Noting that, according to (11.8), $A\mathscr{G}(\varphi) = \mathscr{G}(\varphi') + \varphi(0)\, I_E$, $\varphi \in \mathscr{D}(\mathbf{R})$, we obtain:

$$(\lambda + A)\, \mathscr{G}(\theta_\lambda) = \mathscr{G}(\theta_\lambda' + \lambda\theta_\lambda) + I_E,$$

from which, together with (11.12), we have:

$$(\lambda + A)\, \mathscr{G}(\theta_\lambda) = I_E - \frac{1}{B(\xi)} \mathscr{G}(b_\eta), \tag{11.16}$$

and, in the same way,

$$\mathscr{G}(\theta_\lambda)\, (\lambda + A) = I_{D(A)} - \frac{1}{B(\xi)} \mathscr{G}(b_\eta). \tag{11.17}$$

Now the plan of the proof is as follows: we shall successively show:

$$\begin{cases} \text{there exists a region } \mathscr{R} \text{ given by (11.6) such that} \\ \left\| \dfrac{1}{B(\xi)} \mathscr{G}(b_\eta) \right\|_{\mathscr{L}(E;E)} \le \dfrac{1}{2}, \quad \left\| \dfrac{1}{B(\xi)} \mathscr{G}(b_n) \right\|_{\mathscr{L}(D(A);D(A))} \le \dfrac{1}{2}, \\ \text{for } \lambda = \xi + i\eta \in \mathscr{R}; \end{cases} \tag{11.18}$$

then, by (11.6), (11.17), (11.18), $(\lambda + A)^{-1}$ exists for $\lambda \in \mathscr{R}$, and we have

$$\|(\lambda + A)^{-1}\|_{\mathscr{L}(E;E)} \le \|\mathscr{G}(\theta_\lambda)\|;$$

then we shall verify that we have the estimate:

$$\|\mathscr{G}(\theta_\lambda)\|_{\mathscr{L}(E;E)} \le \text{polynomial in } |\lambda|,\ \lambda \in \mathscr{R}. \tag{11.19}$$

3) *Proof of* (11.18). Since $\mathscr{G}$ vanishes for $t < 0$, there exists (see L. Schwartz [5], Chazarain [1]) a constant c and an integer m such that

$$\|\mathscr{G}(\varphi)\|_{\mathscr{L}(E;E)} \le c|||\varphi|||_m, \tag{11.20}$$

where

$$(11.21)\qquad \begin{cases} |||\varphi|||_m = \sum_{k=0}^{m} \sup_{t\in[0,t_1]} |\varphi^{(k)}(t)|, \\ \varphi \in \mathscr{D}(\mathbf{R}),\ \varphi = 0 \text{ for } t \geq t_1. \end{cases}$$

Then

$$(11.22)\qquad \frac{1}{|B(\xi)|} \|\mathscr{G}(b_\eta)\|_{\mathscr{L}(E;E)} \leq \frac{c}{|B(\xi)|} |||b_\eta|||_m.$$

But, for $\xi \geq 0$, $B(\xi) \geq e^{\xi t_1} \int b\, dt = e^{\xi t_1}$, and (10.22) yields:

$$\frac{1}{B(\xi)} \|\mathscr{G}(b_\eta)\|_{\mathscr{L}(E;E)} \leq c e^{-\xi t_1}(1 + |\eta|)^m.$$

Thus we shall have obtained the first inequality in (10.18) (and the second inequality can be verified in the same way) if we choose ξ, η to satisfy:

$$(11.23)\qquad \xi \geq 0,\ C e^{-\xi t_1}(1 + |\eta|)^m \leq \frac{1}{2}.$$

Then we choose $\mathscr{R}$ in (11.6) so that (11.23) holds.

4) *Proof of* (11.19). It follows from (11.20) and (11.14) that:

$$\|\mathscr{G}(\theta_\lambda)\|_{\mathscr{L}(E;E)} \leq c\, |||e^{-\lambda t}|||_m + \frac{c}{b(\xi)} |||e^{-i\eta t}((e^{-\xi t} Y) * b)|||_m \leq$$

$$\leq \text{(the } c\text{'s denoting different constants)} \leq$$

$$\leq c(1 + |\lambda|)^m + \frac{c}{B(\xi)} e^{-\xi t_1} (1 + |\eta|)^m.$$

But, for $\lambda \in \mathscr{R}$, we have (11.23), and since $B(\xi) \geq e^{\xi t_1}$, we obtain

$$\|\mathscr{G}(\theta_\lambda)\|_{\mathscr{L}(E;E)} \leq c(1 + |\lambda|)^m + c,$$

whence (11.19). □

Proof of the sufficiency of (11.6), (11.7). For $\lambda \in \mathscr{R}$, we set:

$$(11.24)\qquad R(\lambda) = (\lambda + A)^{-1},$$

and, for $\varphi \in \mathscr{D}(\mathbf{R})$,

$$(11.25)\qquad \Phi(\lambda) = \int e^{\lambda t} \varphi(t)\, dt.$$

We introduce a contour γ as indicated in fig. 1; γ is contained in $\mathcal{R}$ and has a "vertical" part $\xi = \xi_0$ and two arcs

$$e^{-\xi} |\eta|^{\alpha} = \text{constant}.$$

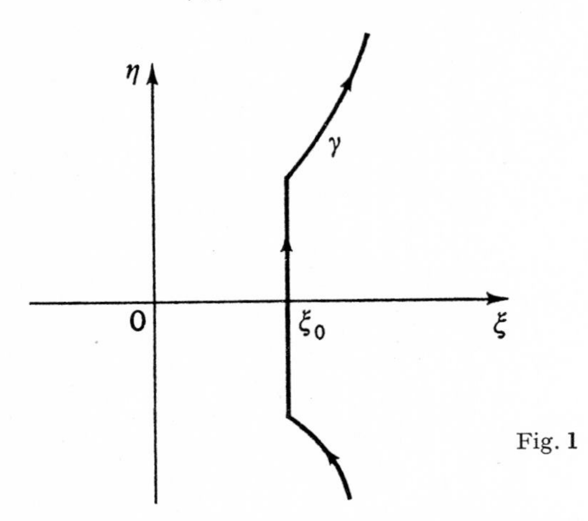

Fig. 1

Then, $\forall \varphi \in \mathcal{D}(\mathbf{R})$, and *subject to verification of convergence*, we define:

(11.26) $$\mathcal{G}(\varphi) = \frac{1}{2\pi i} \int_{\gamma} R(\lambda)\, \Phi(\lambda)\, d\lambda.$$

But, by integration by parts:

(11.27) $$\Phi(\lambda) = \lambda^{-q} \int e^{\lambda t} \varphi^{(q)}(t)\, dt, \quad \forall q,$$

and if φ has support in $[\tau_0, \tau_1]$, it follows that

(11.28) $$|\Phi(\lambda)| \leq c_q(\varphi)\, e^{\tau_1 \xi}\, |\lambda|^{-q}, \quad c_q(\varphi) = \sup_t |\varphi^{(q)}(t)|.$$

But then, according to (11.7), we have

(11.29) $$\| R(\lambda)\, \Phi(\lambda) \|_{\mathcal{L}(E;E)} \leq c_q(\varphi)\, (1 + |\lambda|)^m\, |\lambda|^{-q}\, e^{\tau_1 \xi}, \; \lambda \in \mathcal{R}.$$

Since, on γ, $e^{\tau_1 \xi} \sim |\lambda|^{\varrho}$, we see that (11.29) yields:

$$\| R(\lambda)\, \Phi(\lambda) \|_{\mathcal{L}(E;E)} \leq c_q(\varphi)\, (1 + \lambda)^m\, |\lambda|^{\varrho - q}.$$

Therefore we can choose q so that the integral in (11.26) converges and we have

(11.30) $$\| \mathcal{G}(\varphi) \| \leq c\, c_q(\varphi).$$

It follows that (11.26) defines $\mathcal{G}$ with

(11.31) $$\mathcal{G} \in \mathcal{L}\big(\mathcal{D}(\mathbf{R});\, \mathcal{L}(E; E)\big).$$

But we also have

$$\mathcal{G} \in \mathcal{L}(\mathcal{D}(\mathbf{R}); \mathcal{L}(E; \mathrm{D}(A))), \tag{11.32}$$

since $AR(\lambda) = I - \lambda R(\lambda)$, therefore

$$\| R(\lambda) \|_{\mathcal{L}(E;\mathrm{D}(A))} \leq (1 + |\lambda|) \, \| R(\lambda) \|_{\mathcal{L}(E;E)}$$

and it is sufficient to replace q with $q + 1$ in the above argument to obtain (11.32).

We still need to verify the properties:

$$\mathcal{G} = 0 \quad \text{for} \quad t < 0, \tag{11.33}$$

$$\begin{cases} \left(\dfrac{\mathrm{d}}{\mathrm{d}t} + A\right) * \mathcal{G} = \delta \otimes I_E, \\ \mathcal{G} * \left(\dfrac{\mathrm{d}}{\mathrm{d}t} + A\right) = \delta \otimes I_{\mathrm{D}(A)}. \end{cases} \tag{11.34}$$

Proof of (11.33). Let

$$\gamma_N = \text{translate of } \gamma \text{ by } +N \text{ in the direction of } \xi \geq 0.$$

By Cauchy's Theorem, we have

$$\mathcal{G}(\varphi) = \frac{1}{2\pi \mathrm{i}} \int_{\gamma_N} R(\lambda) \, \Phi(\lambda) \, \mathrm{d}\lambda. \tag{11.35}$$

Let $\varphi \in \mathcal{D}(\mathbf{R})$, with support in $]-\infty, -\varepsilon[$, $\varepsilon > 0$. It suffices to show that $\mathcal{G}(\varphi) = 0$. Now, if $\xi \geq N + \xi_0$, we have:

$$|\Phi(\lambda)| \leq c_q(\varphi) \, \mathrm{e}^{-\varepsilon N} \, |\lambda|^{-q},$$

from which, in the same way as above, using (11.35), we deduce that

$$\| \mathcal{G}(\varphi) \|_{\mathcal{L}(E;E)} \leq c c_q(\varphi) \, \mathrm{e}^{-\varepsilon N}.$$

Letting $N \to \infty$, it follows that $\mathcal{G}(\varphi) = 0$.

Proof of (11.34). For $\varphi \in \mathcal{D}(\mathbf{R})$, we have

$$\frac{\mathrm{d}}{\mathrm{d}t} \mathcal{G}(\varphi) = \mathcal{G}(-\varphi') = \frac{1}{2\pi \mathrm{i}} \int_{\gamma} \lambda R(\lambda) \, \Phi(\lambda) \, \mathrm{d}\lambda$$

(since $\int \mathrm{e}^{\lambda t}(-\varphi') \, \mathrm{d}\lambda = \lambda \Phi(\lambda)$ and the integral $\int_{\gamma} \lambda R(\lambda) \, \Phi(\lambda) \, \mathrm{d}\lambda$ converges (after a suitable choice for q)). As $\lambda R(\lambda) = I_E - AR(\lambda)$, it follows that

$$\frac{\mathrm{d}}{\mathrm{d}t} \mathcal{G}(\varphi) = A \mathcal{G}(\varphi) + I_E \frac{1}{2\pi \mathrm{i}} \int_{\gamma} \Phi(\lambda) \, \mathrm{d}\lambda.$$

Since $\frac{1}{2\pi i}\int_\gamma \Phi(\lambda)\,d\lambda = \varphi(0)$, the first equality in (11.34) follows. The second equality is verified in the same way. ☐

Remark 11.1. The distribution $\mathscr{G}$ introduced in (11.8) has, in particular, the property:

$$\mathscr{G}(\varphi * \psi) = \mathscr{G}(\varphi)\cdot\mathscr{G}(\psi),\ \forall \varphi, \psi \in \mathscr{D}(\mathbf{R}), \tag{11.36}$$

which is the basis for the definition of *distribution semi-groups.* ☐

Remark 11.2. Theorem 11.1 is a variant of the Hille-Yosida Theorem (see Yosida [1]), characterizing the infinitesimal generators of the usual semi-groups. It can be shown that, in an appropriate sense, $-A$ is the *"infinitesimal generator" of the distribution semi-group* $\mathscr{G}$. See Lions [4]. ☐

Orientation

The criterion (11.6), (11.7), yields the most general conditions for the "abstract Cauchy problem" (11.3) to be well-posed, *in the spaces* $\mathscr{L}(\mathscr{D}_-(\mathbf{R}); E)$. We can pose a more general problem, by replacing $\mathscr{L}(\mathscr{D}_-(\mathbf{R}); E)$ *with spaces of ultra-distributions.* This is the aim of the next section. ☐

11.2 Ultra-Distribution Semi-Groups

The data are the same as in Section 11.1.

We further consider a non-quasi-analytic sequence M_k, more precisely satisfying (1.22), (1.23), (1.24).

In the same way as in Section 11.1, we seek the necessary and sufficient condition on A for the existence of a unique ultra-distribution u having the following properties:

$$u \in \mathscr{L}(\mathscr{D}_{-,M_k}(\mathbf{R}); D(A)), \tag{11.37}$$

$$Au + u' = f,\ f \text{ given in } \mathscr{L}(\mathscr{D}_{-,M_k}(\mathbf{R}), E), \tag{11.38}$$

$$\inf_t \{\text{support of } u \text{ in } t\} \le \inf_t \{\text{support of } f \text{ in } t\}, \tag{11.39}$$

$$\left\{\begin{array}{l} f \to u \text{ is a continuous mapping of} \\ \mathscr{L}(\mathscr{D}_{-,M_k}(\mathbf{R}); E) \to \mathscr{L}(\mathscr{D}_{-,M_k}(\mathbf{R}); D(A)). \end{array}\right. \tag{11.40}$$ ☐

In order to state the main result, we introduce the function $\lambda \to M(\lambda)$ by:

$$\exp M(\lambda) = \sup_{k\ge 0}\left(|\lambda|^k \frac{M_0}{M_k}\right). \tag{11.41}$$

For $h, \alpha, \beta \in \mathbf{R}$, we define:

$$\mathscr{R}^{\alpha,\beta}_{M,h} = \{\lambda \mid \lambda = \xi + i\eta,\ \xi \ge \alpha M(h^{-1}\eta) + \beta\}. \tag{11.42}$$

Then we have:

Theorem 11.2 (Chazarain [2]). *In order that conditions* (11.37), ..., (11.40) *be satisfied, it is necessary and sufficient that* A *has the following properties*:

$$\text{(11.43)} \quad \begin{cases} \forall h > 0, \text{ there exists } \mathscr{R}^{\alpha,\beta}_{M,h} \text{ such that } (A+\lambda) \text{ is} \\ \text{an isomorphism of } \mathrm{D}(A) \text{ onto } E,\ \forall \lambda \in \mathscr{R}^{\alpha,\beta}_{M,h}; \end{cases}$$

$$\text{(11.44)} \quad \begin{cases} \forall L, h, \forall \varepsilon > 0, \text{ there exists } \mathscr{R}^{\alpha,\beta}_{M,h} \text{ with (11.43) and} \\ \text{there exists a constant } c \text{ such that} \\ \|(A+\lambda)^{-1}\|_{\mathscr{L}(E;E)} \leq c \exp\left(M\left(\dfrac{\lambda}{L}\right) + \varepsilon\xi\right), \quad \lambda \in \mathscr{R}^{\alpha,\beta}_{M,h}. \end{cases} \quad \square$$

The principle of the proof is analogous to the one used for the proof of Theorem 11.1. We refer the reader to Chazarain [2], for technical details.

Remark 11.3. The same type of problem can be posed by replacing the spaces $\mathscr{L}(\mathscr{D}_{-,M_k}(\mathbf{R}); F)$ with the Beurling spaces $\mathscr{L}(B_{-,M_k}(\mathbf{R}); F)$.

We obtain the same type of result, replacing the conditions $\forall L$, $\forall h$ with: "*there exist* L_0 *and* h_0" such that the conditions contained in (11.43), (11.44) are satisfied. See Chazarain [2]. $\square$

Remark 11.4. If we take $M_k = (k!)^s$, $s > 1$, the regions $\mathscr{R}^{\alpha,\beta}_{M,h}$ decrease in size as s decreases—i.e. the hypotheses on A become more and more general (as would be expected!). $\square$

12. A General Local Existence Result

12.1 Statement of the Result

Let s be a parameter $\in [0, 1]$. Let E_s be a *family of Banach spaces*, with norm denoted by $\| \|_s$. Assume that

$$\text{(12.1)} \quad E_1 \subset \cdots \subset E_s \subset E_{s'} \subset \cdots E_0, \quad 1 \geq s \geq s' \geq 0,$$

with

$$\text{(12.2)} \quad \begin{cases} \text{the injection of } E_s \to E_{s'} \text{ is continuous with norm} \leq 1, \text{ i.e.} \\ \|e\|_{s'} \leq \|e\|_s, \quad \forall e \in E_s,\ s \geq s'. \end{cases}$$

Let t be a real or complex parameter which varies in a closed neighborhood $\mathcal{O}$ of the origin, thus:

$$\text{(12.3)} \quad t \in \mathcal{O},\ \mathcal{O} \subset \mathbf{R} \text{ or } \mathcal{O} \subset \mathbf{C},\ \mathcal{O} \text{ neighborhood of the origin}.$$

Consider a family of operators $A(t)$, $t \in \mathcal{O}$, with the following properties:

$$(12.4)\qquad \begin{cases} \forall t \in \mathcal{O},\ \forall s, s',\ 0 \leq s' < s \leq 1, \text{ we have:} \\ A(t) \in \mathcal{L}(E_s; E_{s'}), \\ \| A(t) \|_{\mathcal{L}(E_s, E_{s'})} \leq c_1 \dfrac{1}{s - s'}, \\ (c_1 = \text{constant independent of } t, s, s'); \end{cases}$$

$$(12.5)\qquad \begin{cases} \text{the function } t \to A(t) \text{ of } \mathcal{O} \to \mathcal{L}(E_s; E_{s'}) \text{ is:} \\ \text{continuous if } \mathcal{O} \subset \mathbf{R}, \\ \text{holomorphic if } \mathcal{O} \in \mathbf{C}. \quad \square \end{cases}$$

Remark 12.2. Examples of this situation will be given in Section 12.2. □

Theorem 12.1 (Ovciannikov [1], Treves [4], [2]). *Let hypotheses* (12.1), ..., (12.5) *be satisfied. Let $t \to f(t)$ be a given function of $\mathcal{O} \to E_1$, with:*

$$(12.6)\qquad \begin{cases} t \to f(t) \textit{ is continuous (resp. holomorphic) from} \\ \mathcal{O} \in \mathbf{R} \text{ (resp. } \mathcal{O} \in \mathbf{C}) \to E_1. \end{cases}$$

Let u_0 be given with

$$(12.7)\qquad u_0 \in E_1.$$

Then, for every $s \in]0, 1[$, there exists one and only one function $t \to u(t)$ having the following properties:

$$(12.8)\qquad \begin{cases} u \textit{ is defined in } \mathcal{O} \cap \{t \mid |t| \leq K(1-s)\} = \mathcal{O}_s{}^{((1))}, \\ t \to u(t) \textit{ is continuously differentiable if } \mathcal{O} \subset \mathbf{R} \\ \text{(resp. } \textit{holomorphic if } \mathcal{O} \subset \mathbf{C}) \textit{ from } \mathcal{O}_s \to E_s, \end{cases}$$

$$(12.9)\qquad \frac{du}{dt} + A(t)\, u(t) = f(t),\ t \in \mathcal{O}_s,$$

$$(12.10)\qquad u(0) = u_0.$$

(Of course, u is "independent of s").

Proof. The proof is an application of the successive approximation method. Equations (12.9), (12.10) are equivalent to

$$(12.11)\qquad u(t) = \int_0^t f(\sigma)\, d\sigma + u_0 - \int_0^t A(\sigma)\, u(\sigma)\, d\sigma,$$

((1)) K = constant to be specified in the course of the proof.

which by recurrence leads to defining

$$u_k(t) = \int_0^t f(\sigma)\, d\sigma + u_0 - \int_0^t A(\sigma)\, u_{k-1}(\sigma)\, d\sigma, \tag{12.12}$$

with $u_0(t) = u_0$.

Let us introduce constants c_2 and L satisfying

$$c_2 \geq \sup_{t \in \mathcal{O}} \| f(t) \|_1 + c_1 \| u_0 \|_1, \tag{12.13}$$

$$L \geq c_1 e. \tag{12.14}$$

We shall show that

$$\| u_{k+1}(t) - u_k(t) \|_s \leq c_2 L^k \frac{|t|^{k+1}}{(1-s)^{k+1}}. \tag{12.15}$$

Let us first verify (12.15) for $k = 0$. By (12.12) for $k = 1$, we have:

$$\begin{cases} \| u_1(t) - u_0 \|_s \leq \left\| \int_0^t f(\sigma)\, d\sigma \right\|_s + \left\| \int_0^t A(\sigma)\, u_0\, d\sigma \right\|_s \leq \\ \qquad \leq |t| \sup_{t \in \mathcal{O}} \| f(t) \|_s + |t| \sup_{t \in \mathcal{O}} \| A(t)\, u_0 \|_s. \end{cases} \tag{12.16}$$

But by (12.4), $\| A(t)\, u_0 \|_s \leq \dfrac{c_1}{1-s} \| u_0 \|_1$ and it follows a fortiori from (12.16) that

$$\| u_1(t) - u_0 \|_s \leq \frac{|t|}{1-s} \left[\sup_{t \in \mathcal{O}} \| f(t) \|_1 + c_1 \| u_0 \|_1 \right] \leq \frac{c_2 |t|}{1-s}.$$

Let us admit (12.15) up to $k - 1$ and prove it for k. It follows from (12.12) that

$$u_{k+1}(t) - u_k(t) = - \int_0^t A(\sigma) \left(u_k(\sigma) - u_{k-1}(\sigma) \right) d\sigma,$$

whence

$$\begin{cases} \| u_{k+1}(t) - u_k(t) \|_s \leq \dfrac{c_1}{\varepsilon} \left\| \int_0^t \left(u_k(\sigma) - u_{k-1}(\sigma) \right) d\sigma \right\|_{s+\varepsilon}, \\ \varepsilon > 0 \quad \text{such that} \quad s + \varepsilon \leq 1. \end{cases} \tag{12.17}$$

By induction, (12.17) yields

$$\| u_{k+1}(t) - u_k(t) \|_s \leq \frac{c_1}{\varepsilon} c_2 L^{k-1} \frac{1}{(1-s-\varepsilon)^k} \frac{|t|^{k+1}}{k+1}. \tag{12.18}$$

Choose ε such that

$$\varepsilon(k+1) = 1 - s$$

(and then $s + \varepsilon \leq 1$); (12.18) then yields

$$\| u_{k+1}(t) - u_k(t) \|_s \leq \frac{c_1 c_2}{L} \frac{|t|^{k+1}}{(1-s)^{k+1}} L^k \left(1 + \frac{1}{k}\right)^k \leq$$
$$\leq c_2 \left(\frac{c_1 \mathrm{e}}{L}\right) \frac{|t|^{k+1} L^k}{(1-s)^{k+1}},$$

from which we obtain (12.15) thanks to (12.14).

It follows that

$$u(t) = \sum_{k=0}^{\infty} (u_{k+1}(t) - u_k(t))$$

converges in E_s and yields a solution of the problem, if

$$\frac{|t|\, L}{1-s} < 1,$$

i.e., taking into account (12.14), if

$$|t| < \frac{1}{c_1 \mathrm{e}} (1-s). \tag{12.19}$$

We have thus shown the existence in Theorem 12.1, in which the constant K must be chosen $< 1/c_1\mathrm{e}$. The proof of uniqueness is standard. ▯

12.2 Examples

Example 12.1. In a Banach space E, let A be a closed unbounded operator. In the notation of Section 7.2, for $s \in [0, 1]$, we define

$$E_s = \mathrm{D}^{1/s}(A^\infty; k!), \tag{12.20}$$

i.e.

$$E_s = \{\mathrm{e} \mid \mathrm{e} \in \mathrm{D}(A^\infty),\ \|\mathrm{e}\|_s = \sup_{k \geq 0} \frac{s^k}{k!} \|A^k \mathrm{e}\| < \infty\}. \tag{12.21}$$

We assume that E_s does not reduce to $\{0\}$ (see Section 7). Then it is easily verified that A satisfies (12.4). ▯

Example 12.2 (Treves [4]). Let $r \to \varphi(r)$ be a given function such that

$$r \to \varphi(r) \text{ is non-decreasing } > 0 \text{ on } r > 0. \tag{12.22}$$

Let X be a Banach space. Define:

$$\begin{cases} E(\varphi; X) = \text{space of entire functions } z \to f(z) \text{ of } \mathbf{C} \to X, \\ \text{such that } \sup\limits_{z \in \mathbf{C}} (\exp(-\varphi(|z|))\, \|f(z)\|_X) = \|f\|_{E(\varphi;X)} < \infty, \end{cases} \tag{12.23}$$

which is a Banach space for the norm $\|f\|_{E(\varphi;X)}$.

For $s \in [0, 1]$, set

$$\varphi_s(r) = \varphi\left(\left(1 - \frac{s}{2}\right)r\right), \tag{12.24}$$

and

$$E_s = E(\varphi; X). \tag{12.25}$$

Then we have

Proposition 12.1. *The operator $A = \partial/\partial z$ satisfies* (12.4), (12.5) *in the family E_s defined by* (12.25). *In* (12.4), *we may take*

$$c_1 = 2 \exp \varphi(1). \tag{12.26}$$

Proof. Let us first verify that

$$\begin{cases} \text{if } r \geq 1, \text{ then, for } |z| = r, \text{ we have} \\ \exp\left(-\varphi_{s'}(r)\right) \left\| \dfrac{\partial f}{\partial z}(z) \right\|_X \leq \dfrac{2 - s}{s - s'} \|f\|_{E_s}, \quad 0 \leq s' < s \leq 1. \end{cases} \tag{12.27}$$

Indeed, according to the Cauchy theorem:

$$\left\| \frac{\partial f}{\partial z}(z) \right\|_X \leq \varepsilon^{-1} \sup_{|\xi| = r + \varepsilon} \|f(\xi)\|_X \leq \varepsilon^{-1} \exp \varphi_s(r - \varepsilon) \|f\|_{E_s}.$$

Therefore

$$\exp\left(-\varphi_{s'}(r)\right) \left\| \frac{\partial f}{\partial z}(z) \right\|_X \leq \varepsilon^{-1} \exp\left(\varphi_s(r + \varepsilon) - \varphi_{s'}(r)\right) \|f\|_{E_s}. \tag{12.28}$$

Choosing $\varepsilon = \dfrac{(s - s')\, r}{2 - s}$, we have: $\varphi_s(r + \varepsilon) = \varphi_{s'}(r)$ and (12.28) yields (12.27) (since $r \geq 1$).

For $|z| = r < 1$, we have

$$\exp\left(-\varphi_{s'}(r)\right) \left\| \frac{\partial f}{\partial z}(z) \right\|_X \leq \exp \varphi_{s'}(1) \sup_{|\xi| = 1} \left(\exp -\varphi_{s'}(|\xi|) \left\| \frac{\partial f}{\partial \xi}(\xi) \right\|_X \right) \leq$$

$$\leq \left(\text{by } (12.27)\right) \frac{2 - s}{s - s'} \exp \varphi_{s'}(1) \|f\|_{E_s},$$

from which the desired result follows. ∎

A particular case

Letting

$$\varphi(r) = \tau r^k, \ \tau > 0, \ \text{integer } k > 0, \tag{12.29}$$

we obtain for $E(\varphi, X)$, as τ varies, the space of entire functions f with values in X such that $\|f(z)\|_X \leq M \exp(M' |z|^k)$ for suitable constants M and M'. ∎

Extension to n complex variables

We introduce

$$(12.30)\quad \begin{cases} \varphi(r_1, \ldots, r_n) = \sum_{j=1}^{n} \varphi_j(r_j), \\ \text{each } \varphi_j \text{ satisfying the analogue to (12.22)}. \end{cases}$$

We define

$$(12.31)\quad \begin{cases} E(\varphi; X) = \text{spaces of entire function } z \to f(z) \text{ of } \mathbf{C}^n \to X \\ \text{such that} \\ \sup_{z \in \mathbf{C}^n} \left(\exp\left(- \sum_{j=1}^{n} \varphi_j(|z_j|) \, \|f(z)\| \right) \right) = \|f\|_{E(\varphi; X)} < \infty, \end{cases}$$

and then we define E_s as in (12.24), (12.25).

As for Proposition 12.1, we verify

Proposition 12.2. *For every* j, *the operator* $\partial/\partial z_j$ *satisfies the analogues to* (12.4), (12.5), *with* $c_1 = 2 \exp(\varphi_j(1))$. □

Then we define

$$(12.32)\quad A(t) = \sum_{j=1}^{n} a_j(t) \frac{\partial}{\partial z_j} + a_0(t),$$

where

$$(12.33)\quad \begin{cases} t \to a_j(t),\ 0 \leq i \leq n, \text{ is a continuous (resp.} \\ \text{holomorphic) function of } \mathcal{O} \to \mathcal{L}(X; X). \end{cases}$$

We are within the conditions of Theorem 12.1. Thus we obtain the existence and uniqueness of a function $u(z, t)$, $z \in \mathbf{C}^n$, $t \in \mathcal{O}_s$, solution of

$$(12.34)\quad \frac{\partial u}{\partial t} + \sum_{j=1}^{n} a_j(t) \frac{\partial u}{\partial z_j} + a_0(t)\, u = f(z, t),$$

$$(12.35)\quad u(z, 0) = u_0(z),\ z \in \mathbf{C}^n.$$

If $t \to f(., t)$ is a continuous (resp. holomorphic) function of $\mathcal{O} \to E_1(\varphi; X)$ and if $u_0 \in E_1(\varphi; X)$, then the solution $u(0, t)$ belongs to $E_s(\varphi; X)$ for $t \in \mathcal{O}_s$ and depends continuously (resp. holomorphically) on t. □

13. Comments

The results of Section 1 have been given in Lions-Magenes [3]. The results of Sections 2 and 4 have been obtained in Lions-Magenes [4], but under much stronger hypotheses on $A(t)$. C. Baiocchi [2] observed that it is possible to obtain "the good estimates" when $(A'(t)\, v, v) \leq -\gamma \|v\|^2$ and that the problem can be reduced to this case by a simple change of variables (this change of variables was introduced independently by J. Lieutaud (personal communication) in order to reduce nonlinear problems to the case $(A'(t)\, v, v) \leq 0$).

The results of Sections 3 and 5 are given here for the first time.

The results of Section 6 are given in Lions-Magenes [3, 4].

Sections 7.1 through 7.4 develop certain results contained in the author's notes [6]. See also Lions [11], Magenes [5].

As stated in the text, the results of Section 7.5 are due to Tanabe [1]. For the case $M_k = k!$, see Agmon-Nirenberg [1], A. Friedman [7], Kato-Tanabe [1], Kato [1], Komatsu [6], Kotake [1], S. Krein [1], Shirota [1], Sobolevski [1], Yosida [1]. *For the interpolation of the spaces* $\mathrm{D}(A^\infty; M_k)$, see Goulaouic [1, 2].

For the regularity of solutions of evolution equations by the method of semi-groups, see P. Suryanarayana [1].

The results of Sections 8, 9, 10 are given here for the first time. Other results of this type should exist, relative to other classes of operator equations (see Problem 14.10).

The results of Section 11 are due to Chazarain [1, 2]. They extend previous results of C. Foias [1] pertaining to the Hilbert case, the operator A being normal (Foias used spectral decomposition). The distribution semi-groups, mentioned in Section 10, were introduced by Lions [4], numerous additional properties being given by Peetre [1], Da Prato [1, 2], Da Prato-Mosco [1], D. Fujiwara [1], Fattorini [1], I. Cioranescu [1].

Numerous other properties can be found in Chazarain [3, 4]. In particular, he shows that certain mixed hyperbolic problems are well-posed in certain spaces of ultra-distributions, while the corresponding problems in the "usual" spaces of distributions are not (cf. Ikawa, Add. Bibliography).

The results of Section 12 are due to Ovciannikov [1], Treves [4, 2] and Treves-Steinberg [1]. We have followed the presentation of Treves [4], where numerous other results can be found; Theorem 12.1 extends to certain nonlinear problems, which Treves applies, in particular, to the Cauchy-Kowaleska theorem.

14. Problems

14.1. The results of Section 1 most probably extend to the case where the spaces V are replaced by families of spaces $V(t)$ *depending on* t in a "regular of class M_k" fashion (for example if the $V(t)$'s are closed subspaces of a Hilbert space K, the orthogonal projection operator $P(t)$ of K onto $V(t)$ satisfying

$$t \to P(t) \text{ belongs to } \mathscr{E}_{M_k}(\mathbf{R}; \mathscr{L}(K; K))\big).$$

But we have not attempted to treat this point in detail. The analog extension for operators of the second order in t or for Schroedinger operators is certainly much more delicate.

14.2. We have not systematically studied what happens to the results of this chapter if the classes $\mathscr{E}_{M_k}$ are replaced with the Beurling classes $\mathscr{B}_{M_k}$ (except in Section 1.4).

14.3. It would be of interest to examine the stability properties in M_k classes by elliptic regularization (see Chapter 3, Section 7).

14.4. In relation to Section 7, can one find examples where $\mathrm{D}(A^\infty; M_k)$ does not reduce to $\{0\}$ without being dense in E?

14.5. Do the results given in Amerio-Prouse [1] pertaining to *almost periodic* solutions in t of partial differential equations extend to *almost periodic solutions of class M_k*?

14.6. A number of considerations pertaining to the spaces $\mathrm{D}(A^\infty; M_k)$ have meaning *without $-A$ being the infinitesimal generator of a semi-group*; how far can one go in this direction?

14.7. "*Perturbation*" problems: for which operators B "small relative to A" do we have

$$\mathrm{D}((A + B)^\infty; M_k) = \mathrm{D}(A^\infty; M_k)\,?$$

If we assume that E is a space of local type on an open set Ω of $\mathbf{R}^n$ and if A and B are differential operators, when do the spaces $\mathrm{D}((A + B)^\infty; M_k)$ and $\mathrm{D}(A^\infty; M_k)$ coincide on every open set $\mathcal{O}$ with $\bar{\mathcal{O}} \subset \Omega$?

14.8. What are the necessary and sufficient conditions on $f^{(j)}(0)$ in order to have

$$\left\| \frac{\mathrm{d}^k}{\mathrm{d}t^k} u(t) \right\| \leq c L^k M_k, \ \forall k, \forall t \in [0, T], \tag{14.1}$$

instead of (7.94)? (See Remarks 7.13 and 7.14).

14.9. More generally, consider problem (7.94), (7.96). What "compatibility relations" must u_0 and $f(0), f'(0), \ldots$ satisfy in order to have (14.1)? (See Remark 7.14).

14.10. Is it possible to extend Theorem 9.3, under suitable hypotheses, to the operator equations

$$\Lambda u + \mathscr{A} u = f, \tag{14.2}$$

studied by P. Grisvard [1]? In the text, we have given M_k-regularity results *in Λ*. It should be possible to give "abstract" M_k-regularity in Λ and M_k^*-regularity in $\mathscr{A}$ results pertaining to (14.2).

14.11. The following is an example of an equation which fits Section 9, but where hypotheses (9.20) or (9.25) *are not satisfied*. Let

$$\mathscr{V} = L^2(0, T; V) \times L^2(0, T; V), \quad \mathscr{H} = L^2(0, T; H) \times L^2(0, T; H),$$

$$(14.3) \qquad \begin{cases} \Lambda = \begin{pmatrix} \mathrm{d}/\mathrm{d}t & 0 \\ 0 & -\mathrm{d}/\mathrm{d}t \end{pmatrix} \\ \mathrm{D}(\Lambda; \mathscr{H}) = \{f \mid f = \{f_1, f_2\}, f_i, f_i' \in L^2(0, T; H), \\ f_1(0) = 0, f_2(T) = 0\}, \end{cases}$$

$$(14.4) \qquad \mathscr{A} = \begin{pmatrix} A(t) & +I \\ -I & A(t)^* \end{pmatrix}.$$

Assume that

$$(14.5) \qquad \mathrm{Re}\,\big(A(t)\, v, v\big) \geq \alpha \, \|v\|^2, \; \alpha > 0, \; \forall v \in V, \; t \in [0, T].$$

Then hypotheses (9.2) and (9.3) are satisfied. In fact, we could more generally consider $\mathscr{A}$ to be given by

$$\mathscr{A} = \begin{pmatrix} A_1(t) & +I \\ -I & A_2(t) \end{pmatrix},$$

$$A_i(t) \in \mathscr{L}(V; V'), \; i = 1, 2$$

$$\mathrm{Re}\big(A_i(t)\, v, v\big) \geq \alpha \, \|v\|^2, \; i = 1, 2,$$

but (14.4) is the case which comes up in the applications to the optimal control of systems governed by parabolic partial differential equations (see Lions [5.])

Equation (9.5) becomes:

$$\frac{\partial u_1}{\partial t} + A(t)\, u_1 + u_2 = f_1,$$

$$-\frac{\partial u_2}{\partial t} + A^*(t)\, u_2 - u_1 = f_2,$$

$$u_1(0) = 0, \; u_2(T) = 0.$$

But hypothesis (9.9) *is not satisfied*. Can one still obtain regularity results —and in particular M_k-regularity in t results?

14.12. Characterization of the space described by $v(0)$ as v describes X, outside the cases given in Sections 10.2.2 and 10.2.3.

14.13. Problems analogous to Problem 6.4 of Chapter 8 come up in connection with Section 10.3.

14.14. One can probably extend most of the results given in this chapter to the case of equations *with delay*, for example of the type

$$\frac{\partial u}{\partial t} + A(t)\, u + u(t - \omega(t)) = f(t), \tag{14.6}$$

where the function $t \to \omega(t)$ is of class M_k. For the study of (14.6) in L^2-spaces, see Artola [1], Baiocchi [3].

14.15. Do the results of Section 11 extend to *families* of operators $A(t)$, suitably dependent on t?

14.16. The results of Sections 1 and 7 apply to the operator

$$u \to \frac{\partial u}{\partial t} + \text{p.v.} \int_{-1}^{1} \frac{\frac{\partial u}{\partial x}(y, t)}{x - y}\, dy$$

(take $V = H_{00}^{1/2}(\Omega)$, $\Omega =]-1, 1[$, $H = L^2(\Omega)$). Thus if A is defined by

$$Au = \text{p.v.} \int_{-1}^{1} \frac{\frac{du}{dx}(y)}{x - y}\, dy,$$

one is led to study the space $D(A^\infty; M_k)$. Probably

$$D(A^\infty; M_k) = \{v \mid \|v^{(k)}\|_{L^2} \leq c L^k M_k,\ \forall k,\ v^{(k)}(\pm 1) = 0\}, \tag{14.7}$$

but this has not been shown.

Chapter 10

Parabolic Boundary Value Problems in Spaces of Ultra-Distributions

This Chapter develops boundary value problems in spaces of ultra-distributions for parabolic partial differential equations.

From the point of view of partial differential equations, we assume the knowledge of the essential part of Chapter 4. Chapter 7 is also used.

1. Regularity in the Interior of Solutions of Parabolic Equations

1.1 The Hypoellipticity of Parabolic Equations

The application of the results of Chapter 9 to the concrete case of parabolic partial differential equations of evolution has given us mostly information about the regularity of solutions with respect to the variable t in Gevrey type spaces. We shall now study the *global* regularity, either with respect to t or with respect to the space variables $x_1, \ldots, x_n$; and to start, we examine the regularity in the *interior*.

The first question which arises is the *hypoellipticity* of parabolic operators (see Chapter 2, Section 3.3 for elliptic operators). We shall of course restrict ourselves to the parabolic evolution operators which we have considered in this text (see the Comments for the general case) and more precisely to the operators P introduced in Section 1.1 of Chapter 4.

Therefore, let us again consider the notation of Chapter 4. Let Ω be a bounded open set in $\mathbf{R}^n$ and Γ its boundary (in this Section we do not make any regularity hypothesis on Γ, since we are only concerned with regularity in the *interior*). In the space $\mathbf{R}^{n+1} = \mathbf{R}^n_x \times \mathbf{R}^1_t$, we consider the cylinder $Q = \Omega \times]0, T[$, $0 < T < \infty$ and set $\Sigma = \Gamma \times]0, T[$. In Q, set

$$Pu = A\left(x, t, \frac{\partial}{\partial x}\right) + \frac{\partial}{\partial t}, \tag{1.1}$$

where $A = A(x, t, \partial/\partial x)$ is given by

$$Au = \sum_{|p|,|q| \leq m} (-1)^{|p|} \mathrm{D}_x^p(a_{pq}(x, t) \mathrm{D}_x^q u) \tag{1.2}$$

and the coefficients a_{pq} are *infinitely differentiable* in $\overline{Q}$, i.e.

$$a_{pq} \in \mathscr{D}(\overline{Q}). \tag{1.3}$$

We assume that

$$\begin{cases} \textit{for every } \theta \in \left[-\dfrac{\pi}{2}, \dfrac{\pi}{2}\right] \textit{ and every } t_0 \in [0, T], \textit{ the operator} \\ A\left(x, t_0, \dfrac{\partial}{\partial x}\right) + (-1)^m \mathrm{e}^{i\theta} \mathrm{D}_y^{2m} \\ \textit{is properly elliptic in } \overline{\Omega} \times \mathbf{R}_y \text{ (see Chapter 2, Section 1).} \end{cases} \tag{1.4}$$

Note that, under hypothesis (1.4), the operator P is *parabolic* in the sense of Petrowski [2].

Then we have

Theorem 1.1. *If P is given by* (1.1), (1.2) *with* (1.3), (1.4), *then P is hypoelliptic in Q*, i.e. *if $u \in \mathscr{D}'(\Omega)$ and $Pu \in \mathscr{E}(Q)$, then $u \in \mathscr{E}(Q)$.*

Proof. We shall give an indirect proof of the theorem, using the results on boundary value problems for the operator P given in Chapter 4.

Let us first prove

Lemma 1.1. *The operator*

$$u \to \frac{\partial u}{\partial x_i} \quad i = 1, \ldots, n.$$

is a continuous linear mapping of $H^{2mr,r}(Q)$ into $H^{2mr-1,r-1/2m}(Q)$, for every real r with $2mr \neq 1/2$.

Proof. a) If $r \geq 1/2m$, then the result is a particular case of Proposition 2.3 of Chapter 4.

b) If $r \leq 0$, the result follows by duality of a), since, if $u \in H^{2mr,r}(Q)$, for every $\varphi \in \mathscr{D}(\Omega)$, we have

$$\left|\left\langle \frac{\partial u}{\partial x_i}, \varphi \right\rangle\right| = \left|\left\langle u, \frac{\partial \varphi}{\partial x_i} \right\rangle\right| \leq \|u\|_{H^{2mr,r}(Q)} \left\|\frac{\partial \varphi}{\partial x_i}\right\|_{H^{-2mr,-r}(Q)}$$

(we recall that by definition $H^{2mr,r}(Q)$ is the dual of $H_{0,0}^{-2mr,-r}(Q)$) and therefore we deduce from a) (since $-2mr + 1 \geq 1$) that

$$\left|\left\langle \frac{\partial u}{\partial x_i}, \varphi \right\rangle\right| \leq \|u\|_{H^{2mr,r}(Q)} \|\varphi\|_{H^{-2mr+1,-r+1/2m}(Q)}$$

and therefore

$$\frac{\partial u}{\partial x_i} \in H^{2mr-1, r-1/2m}(Q).$$

c) If $0 < r < 1/2m$, we first note that, thanks to a), $\partial/\partial x_i$ is a continuous linear mapping of

$$H^{1,1/2m}(Q) \text{ into } H^{0,0}(Q) = L^2(Q) \tag{1.5}$$

and, thanks to b), of

$$H^{0,0}(Q) \text{ into } H^{-1,-1/2m}(Q) \tag{1.6}$$

and therefore by interpolation of

$$[H^{1,1/2m}(Q), H^{0,0}(Q)]_\theta \text{ into } [H^{0,0}(Q), H^{-1,-1/2m}(Q)]_\theta, \quad 0 < \theta < 1. \tag{1.7}$$

But (see Proposition 2.1, Chapter 4) we have

$$[H^{1,1/2m}(Q), H^{0,0}(Q)]_\theta = H^{1-\theta,(1-\theta)/2m}(Q)$$

and (duality theorem, Section 6.2, Chapter 1)

$$[H^{0,0}(Q), H^{-1,-1/2m}(Q)]_\theta = [H_{0,0}^{1,1/2m}(Q), H^{0,0}(Q)]'_{1-\theta}. \tag{1.8}$$

Now, by an analogous proof to the one of Theorem 11.6, Chapter 1, we have

$$[H_{0,0}^{1,1/2m}(Q), H^{0,0}(Q)]_{1-\theta} = H_{0,0}^{\theta,\theta/2m}(Q) \text{ if } \theta \neq 1/2. \tag{1.9}$$

Thus the right-hand side of (1.8) is equal to $H^{-\theta,-\theta/2m}(Q)$ and then (1.7) proves the lemma for the case $0 < r < 1/2m$ (set $1 - \theta = 2mr$), under the condition $2mr \neq 1/2$. ☐

Now let (x_0, t_0) be an interior point of Q and let ϱ_0 be the distance from this point to the boundary of Q. For $0 < \varrho \leq \varrho_0$, let Q_ϱ denote the cylinder $\{(x, t) \mid |x - x_0| < \varrho, |t - t_0| < \varrho\}$. Theorem 1.1 will be proven if we can show that $u \in \mathcal{D}(\overline{Q}_\varrho)$, $0 < \varrho < \varrho_0$.

Choose ϱ_1 and ϱ_2 such that $\varrho < \varrho_2 < \varrho_1 < \varrho_0$. The distributions on Q being locally of finite order, we can find a real r such that the restriction of u to Q_{ϱ_1} (which we still denite by u) belongs to $H^{2mr,r}(Q_{\varrho_1})$; in order to avoid the difficulties of "exceptional parameters" we also assume that r is *irrational*, which is always possible. Let us show that

$$\begin{cases} \textit{if } u \in H^{2mr,r}(Q_{\varrho_1}) \textit{ and } Pu = f \in \mathcal{E}(Q), \\ \textit{then } u \in H^{2m(r+1/2m), r+1/2m}(Q_{\varrho_2}). \end{cases} \tag{1.10}$$

Let $\varphi \in \mathcal{D}(Q_{\varrho_1})$ be fixed with $\varphi = 1$ in Q_{ϱ_2}; and set $v = \varphi u$. Then $v \in H^{2mr,r}(Q_{\varrho_1})$ and has compact support in Q_{ϱ_1}. Furthermore, if Σ_ϱ and

σ_ϱ denote respectively the lateral surface of Q_ϱ and the lower base of Q_ϱ, we have

$$\begin{cases} Pv = \psi \text{ in } Q_{\varrho_1}, \\ \left.\dfrac{\partial^j v}{\partial \nu^j}\right|_{\Sigma_{\varrho_1}} = 0, \; j = 0, \ldots, m-1, \; \nu \text{ normal to } \Sigma_{\varrho_1}, \\ v\,|_{\sigma_{\varrho_1}} = 0, \end{cases} \tag{1.11}$$

with $\psi = \varphi f + [P(\varphi u) - \varphi P u]$. Thanks to Lemma 1.1 and (1.1), (1.2), we have

$$\begin{cases} \psi \in H^{2m(r-1+1/2m), r-1+1/2m}(Q_{\varrho_1}), \\ \text{with compact support in } Q_{\varrho_1}. \end{cases} \tag{1.12}$$

But then v appears as a solution of the boundary value problem (1.11), to which we can apply the results of Chapter 4, for the system of Dirichlet conditions $\{\partial^j/\partial \nu^j\}_{j=0}^{n-1}$ *covers* every properly elliptic operator and therefore (1.11) satisfies the conditions of Section 1.1, Chapter 4.

The application of these results yields

$$v \in H^{2m(r+1/2m), r+1/2m}(Q_{\varrho_1}). \tag{1.13}$$

Indeed:

a) If $r - 1 + 1/2m \geq 0$, we can apply Theorem 6.2, Chapter 4 (note that r is assumed to be irrational and thus the "exceptional" cases of the parameter do not interfere, see (6.28) in Chapter 4);

b) if $r - 1 + 1/2m \leq -1$, we can apply Theorem 12.1 of Chapter 4, since, thanks to (1.12), we have

$$\psi \in \Xi^{2m(r-1+1/2m), r-1+1/2m}(Q_{\varrho_1})$$

(and the exceptional cases of the parameter are excluded);

c) there only remains the case

$$-\frac{1}{2m} < r < -\frac{1}{2m} + 1, \tag{1.14}$$

which fits the problems studied in Sections 13, 14, 15 of Chapter 4: in the notation of these Sections (see Section 13.1) it corresponds to the case where the data of problem (13.1) of Section 13.1, Chapter 4, are of "regularity" s, with $-1 < s < 0$, and furthermore

$$g_j = 0, \; j = 0, \ldots, m-1, \text{ and } u_0 = 0. \tag{1.15}$$

Now it is easy to see that, under conditions (1.15), the mapping

$$\mathscr{G}: \{f, 0, 0\} \to u = \mathscr{G}(f, 0, 0), \tag{1.16}$$

defined by (13.7) of Chapter 4, is a continuous mapping of

$$[H^{0,0}(Q), (H^{2m,1}(Q))']_\theta \text{ into } [H^{2m,1}(Q), H^{0,0}(Q)]_\theta, \ 0 < \theta < 1,$$

and therefore, in the notation of Section 15.2, Chapter 4, of

$$\mathscr{H}^{-2\theta m, -\theta}(Q) \text{ into } H^{2m(1-\theta),1-\theta}(Q), \tag{1.17}$$

without exception for θ, $0 < \theta < 1$. Therefore it follows that (see (iii) of Section 15.1, Chapter 4):

$$\begin{cases} \text{if } f \in \mathscr{H}^{2ms,s}(Q),\ g_j = 0,\ u_0 = 0, \text{ then the solution } u \\ \text{of problem (13.1), Chapter 4, belongs to } H^{2m(s+1),s+1}(Q) \\ \text{for every } s, \text{ with } -1 < s < 0. \end{cases} \tag{1.18}$$

We can now easily apply this result to problem (1.11) under conditions (1.14), since, thanks to (1.12), we have

$$\psi \in \mathscr{H}^{2m(r-1+1/2m),r-1+1/2m}(Q_{\varrho_1}).$$

Thus we finally obtain the fact that (1.13) holds, even for the case (1.14).

But then it follows from (1.13), since $\varphi = 1$ in Q_{ϱ_2}, that $u \in H^{2m(r+1/2m),r+1/2m}(Q_{\varrho_2})$ and this proves (1.10).

Let us now choose a sequence of numbers $\{\varrho_\nu\}$, $\nu = 3, 4, \ldots$ such that $\varrho < \varrho_{\nu+1} < \varrho_\nu < \varrho_2$; for each ν, a proposition analogous to (1.10) applies and it follows that

$$u \in H^{2m(r+\nu/2m),r+\nu/2m}(Q_{\varrho_{\nu+1}}),\ \forall \nu$$

and therefore $u \in \bigcap_\nu H^{2m(r+\nu/2m),r+\nu/2m}(Q_\varrho) = \mathscr{D}(\overline{Q}_\varrho)$ and the theorem is proved. ☐

Remark 1.1. Hypothesis (1.3) may of course be replaced with the hypothesis $a_{pq} \in \mathscr{E}(Q)$. ☐

1.2 The Regularity in the Interior in Gevrey Spaces

Since the operator P is hypoelliptic, there exists a "characteristic" Gevrey class for P (see for example Hörmander [1]): this is the Gevrey class to which each solution of the equation $Pu = 0$ belongs, under the assumption that the coefficients of P are constants. In the present case, it is the Gevrey class of order $2m$ in t and of order 1 (i.e. analytic) in x, that is $\mathscr{E}_{2m,1}(Q)$ (notation of Chapter 7). As for the elliptic case (see Chapter 8), we shall prove a more general theorem, by assuming variable coefficients of P ($a_{pq} \in \mathscr{D}_{2m,1}(\overline{Q})$) and by studying the regularity of the solutions of the equation $Pu = f$ in the spaces $\mathscr{E}_{s,r}(Q)$, with $s \geq 2m$ and $r > 1$ (see also Remark 2.3 below and problem 6.2).

Thus let P be given by (1.1), (1.2) with (1.4) and

$$a_{pq} \in \mathscr{D}_{2m,1}(\overline{Q}). \tag{1.19}$$

We aim to show that

(1.20)

if $u \in \mathscr{E}(Q)$ *and* $Pu \in \mathscr{E}_{s,r}(Q)$, *with* $s \geq 2m$, $r > 1$, *then* $u \in \mathscr{E}_{s,r}(Q)$.

To this end, we first prove some lemmas.

Lemma 1.2. *There exists a constant* $c' > 0$ *such that for every* $\varepsilon > 0$ *and every function* $v \in \mathscr{D}(\mathbf{R}^{n+1})$ *we have*

$$\varepsilon^{|\alpha|} \| D_x^\alpha v \|_{L^2(\mathbf{R}^{n+1})} \leq c' \Big\{ \varepsilon^{2m} \sum_{|\beta|=2m} \| D_x^\beta v \|_{L^2(\mathbf{R}^{n+1})} + \| v \|_{L^2(\mathbf{R}^{n+1})} \Big\} \tag{1.21}$$

for every α *with* $|\alpha| \leq 2m$.

Proof. Applying the Fourier transform, the result is an immediate consequence of the inequality

$$\varepsilon^{|\alpha|} \, |\xi|^{|\alpha|} \leq c' \varepsilon^{2m} \, (|\xi|^{2m} + 1), \quad \xi = (\xi_1, \ldots, \xi_n), \quad |\alpha| \leq 2m. \quad \square$$

Now let (x_0, t_0) be an interior point of Q; let us again use the cylinders Q_ϱ, $0 < \varrho < \varrho_0$, as in Section 1.1; and assume, which is permissible, that $\varrho_0 < 1$. Denote by $\|u\|_\varrho$ the norm of u in $L^2(Q_\varrho)$.

Then we have

Lemma 1.3. *Let* P *be given by* (1.1), (1.2) *with* (1.3) *and* (1.4); *there exists a constant* $C_1 > 0$, *such that if* $0 < \varrho < \varrho + \delta < \varrho_0$ *and if* $u \in \mathscr{D}(\overline{Q}_{\varrho_0})$, *we have*:

$$\begin{aligned} \sum_{|\alpha| \leq 2m} \varepsilon^{|\alpha|} \| D_x^\alpha u \|_\varrho + \varepsilon^{2m} \| D_t u \|_\varrho \leq C_1 \\ \times \Big\{ \varepsilon^{2m} \| Pu \|_{\varrho+\delta} + \varepsilon^{2m} \sum_{|\alpha| < 2m} \frac{1}{\delta^{2m-|\alpha|}} \| D_x^\alpha u \|_{\varrho+\delta} + \| u \|_{\varrho+\delta} \Big\}. \end{aligned} \tag{1.22}$$

for every $\varepsilon > 0$.

Proof. Let v be an arbitrary functions of $\mathscr{D}(Q_{\varrho_0})$; then setting $Pv = w$, we may consider v as a solution of the boundary value problem

$$\begin{cases} Pv = w \\ \left. \dfrac{\partial^j v}{\partial \nu^j} \right|_{\Sigma_\varrho} = 0, \quad j = 0, \ldots, m-1, \\ v \,|_{\sigma_\varrho} = 0, \end{cases} \tag{1.23}$$

to which, thanks to hypothesis (1.4) on P and to the fact that the Dirichlet conditions *cover* every properly elliptic operator, we may apply the results of Chapter 4; in particular Theorem 6.1, Chapter 4 yields the existence of a constant c independent of v such that

$$\| v \|_{H^{2m,1}(Q,\varrho_0)} \leq c \, \| Pv \|_{\varrho_0}. \tag{1.24}$$

Applying Lemma 1.2, we obtain the estimate

$$\sum_{|\alpha| \leq 2m} \varepsilon^{|\alpha|} \, \| D_x^\alpha v \|_{\varrho_0} + \varepsilon^{2m} \, \| D_t v \|_{\varrho_0} \leq C \{ \varepsilon^{2m} \, \| Pv \|_{\varrho_0} + \| v \|_{\varrho_0} \} \tag{1.25}$$

for every $\varepsilon > 0$ and every $v \in \mathscr{D}(Q_{\varrho_0})$, with C independent of ε and v.

Also note (see Chapter 8, Section 2.2) that we can construct a function $\varphi_{\varrho,\delta}(x, t) \in \mathscr{D}(\overline{Q}_{\varrho_0})$ such that

$$\begin{cases} \varphi_{\varrho,\delta}(x, t) = 1 \text{ in } \overline{Q}_\varrho, \text{ with support contained in } \overline{Q}_{\varrho+\delta} \\ \text{and satisfying} \\ \sup\limits_{x,t} | D_x^\alpha D_t^k \varphi_{\varrho,\delta}(x, t) | \leq \gamma_{\alpha,k} \delta^{-(|\alpha| + 2mk)}, \\ \gamma_{\alpha,k} \text{ depending only on } \alpha \text{ and } k. \end{cases} \tag{1.26}$$

Now, if $u \in \mathscr{D}(\overline{Q}_{\varrho_0})$, let us apply (1.25) to the function v given by $v = \varphi_{\varrho,\delta} u$. Then we have

$$\sum_{|\alpha| \leq 2m} \varepsilon^{|\alpha|} \, \| D_x^\alpha u \|_\varrho + \varepsilon^{2m} \, \| D_t u \|_\varrho \leq C \{ \varepsilon^{2m} \, \| P(\varphi_{\varrho,\delta} u) \|_{\varrho+\delta} + \| \varphi_{\varrho,\delta} u \|_{\varrho,\delta} \}. \tag{1.27}$$

P may be written in the form

$$P = \sum_{|p| \leq 2m} a_p \, D_x^p + D_t. \tag{1.28}$$

Then

$$P(\varphi_{\varrho,\delta} u) = \varphi_{\varrho,\delta} \, Pu + \sum_{|p| \leq 2m} \sum_{\substack{q_i \leq p_i \\ q \neq 0}} a_p c_{p,q} \, D_x^{p-q} u \, D_x^q \varphi_{\varrho,\delta} + u \, D_t \varphi_{\varrho,\delta},$$

where the $c_{p,q}$'s denote suitable constants; therefore, by (1.26),

(1.29)

$$\| P(\varphi_{\varrho,\delta} u) \|_{\varrho+\delta} \leq \| Pu \|_{\varrho+\delta} + c_1 \sum_{|p| \leq 2m} \sum_{\substack{q_i \leq p_i \\ q \neq 0}} \frac{1}{\delta^{|q|}} \| D_x^{p-q} u \|_{\varrho+\delta} + \frac{1}{\delta^{2m}} \| u \|_{\varrho+\delta} \leq$$

$$\leq \| Pu \|_{\varrho+\delta} + c_2 \sum_{|\alpha| \leq 2m} \frac{1}{\delta^{2m-|\alpha|}} \| D_x^\alpha u \|_{\varrho+\delta},$$

with suitable constants c_1 and c_2. Then (1.22) follows from (1.27) and (1.29). □

Now, for fixed real μ and R with $\mu > 0$, $0 < R \leq \varrho_0$, let us introduce the following notation:

$$N_{\mu,R}(u) = \sup_{0<\varrho<R} (R - \varrho)^\mu \, \|u\|_\varrho . \tag{1.30}$$

The function $u \to N_{\mu,R}(u)$ is a norm which is analogous to the various norms $\sigma^k(u, \lambda, R)$, $\sigma_{k,q}(u, \lambda, \theta, R)$ introduced in Chapter 8 for the study of the regularity of solutions of elliptic equations.

Note that $N_{\mu,R}(u)$ (u fixed) is monotonically increasing in R and monotonically decreasing in μ.

Lemma 1.4. *Under the hypotheses of Lemma* 1.3, *there exists a constant* K *such that for every* $u \in \mathscr{D}(\overline{Q}_{\varrho_0})$ *and every* $R \leq \varrho_0$, *we have*

$$\sum_{|\alpha|\leq 2m} N_{|\alpha|,R}(\mathrm{D}_x^\alpha u) + N_{2m,R}(\mathrm{D}_t u) \leq K\{N_{2m,R}(Pu) + \|u\|_R\}. \tag{1.31}$$

Proof. Let us apply formula (1.22) with $\varepsilon = \eta\delta$, where η is a positive number to be chosen later on and $\delta = (R - \varrho)/2$. Then

$$\sum_{|\alpha|\leq 2m} \eta^{|\alpha|}\, 2^{-|\alpha|}(R-\varrho)^{|\alpha|} \, \|\mathrm{D}_x^\alpha u\|_\varrho + \eta^{2m}\, 2^{-2m}(R-\varrho)^{2m} \, \|\mathrm{D}_t u\|_\varrho \leq$$

$$\leq C_1 \Big\{\eta^{2m}(R-\varrho-\delta)^{2m} \, \|Pu\|_{\varrho+\delta} + \eta^{2m} \sum_{|\alpha|<2m} (R-\varrho-\delta)^{|\alpha|} \, \|\mathrm{D}_x^\alpha u\|_{\varrho+\delta} +$$

$$+ \|u\|_{\varrho+\delta}\Big\} \leq \eta^{2m} C_2 \Big\{N_{2m,R}(Pu) + \sum_{|\alpha|<2m} N_{|\alpha|,R}(\mathrm{D}^\alpha u) + \frac{1}{\eta^{2m}} \|u\|_R\Big\}$$

and, taking the sup over $\varrho \in \,]0, R[$ on the left-hand side:

$$\sum_{|\alpha|\leq 2m} \left(\frac{\eta}{2}\right)^{|\alpha|} N_{|\alpha|,R}(\mathrm{D}_x^\alpha u) + \left(\frac{\eta}{2}\right)^{m} N_{2m,R}(\mathrm{D}_t u) \leq \tag{1.32}$$

$$\leq \eta^{2m} C_2 N_{2m,R}(Pu) + C_2 \, \|u\|_R + \eta^{2m} C_2 \sum_{|\alpha|<2m} N_{|\alpha|,R}(\mathrm{D}_x^\alpha u).$$

If we now *fix* η such that

$$\left(\frac{\eta}{2}\right)^{|\alpha|} - \eta^{2m} C_2 \geq \frac{1}{2} \quad \text{for} \quad |\alpha| < 2m,$$

(1.31) follows from (1.32), by moving the term

$$\eta^{2m} C_2 \sum_{|\alpha|<2m} N_{|\alpha|,R}(\mathrm{D}_x^\alpha u)$$

to the left-hand side. $\square$

Let us now make use of hypothesis (1.19) on the coefficients of the operator P. Writing P in the form (1.28), it follows from hypothesis (1.19) that

$$\text{(1.33)}\quad \begin{cases} \sum_{|p|\le 2m} \sup_{Q_R} |D_x^\alpha D_t^h a_p(x,t)| \le cL^{|\alpha|+h}(|\alpha|+h)^{|\alpha|+2mh}, \\ \text{for every } \alpha, h \text{ and for } R \le \varrho_0, \text{ with suitable constants } c \text{ and } L. \end{cases}$$

Analogously, having *fixed* $R < \varrho_0$, it follows from the hypothesis

$$f \in \mathscr{E}_{s,r}(Q) \quad (f = Pu)$$

that

$$\text{(1.34)}\qquad \sup_{Q_R} |D_x^\alpha D_t^h f(x,t)| \le cL^{|\alpha|+h}(|\alpha|+h)^{r|\alpha|+sh}, \quad \forall \alpha, h,$$

where, R *being fixed,* we may assume that c and L are the same constants as in (1.33).

It follows from (1.33), if $\delta > 0$ and $R - j\delta > 0$, that

$$\text{(1.35)}\quad \delta^{r|\alpha|+sh} \sum_{|p|\le 2m} \sup_{Q_{R-j\delta}} |D_x^\alpha D_t^h a_p(x,t)| \le cL^{|\alpha|+h}\left(\frac{|\alpha|+h}{j}\right)^{|\alpha|+h},$$

for $|\alpha| + h \le j$.

Indeed it suffices to note that $j\delta < R$, since $R < 1$ and to multiply the left-hand side of (1.33) with $(j\delta)^{r|\alpha|+sh}$; it follows that

$$\delta^{r|\alpha|+sh} \sum_{|p|\le 2m} \sup_{Q_{R-j\delta}} |D_x^\alpha D_t^h a_p(x,t)| \le cL^{|\alpha|+h}\frac{(|\alpha|+h)^{\alpha+2hm}}{j^{r|\alpha|+sh}}$$

from which, thanks to the hypothesis $r > 1$, $s \ge 2m$, we obtain (1.35) for $|\alpha| + h \le j$.

Similarly, it follows from (1.34) that

$$\text{(1.36)}\quad \delta^{r|\alpha|+sh} \sup_{Q_R} |D_x^\alpha D_t^h f(x,t)| \le cL^{|\alpha|+h} \text{ for } (|\alpha|+h)\,\delta \le 1.$$

Lemma 1.5. *Under the hypotheses of Lemma* 1.3 *and with* (1.33) *and* (1.34), *if* $u \in \mathscr{D}(\overline{Q}_R)$ *with* $Pu = f$, *then there exist two positive constants* c_0 *and* B *such that*

$$\text{(1.37)}\quad \begin{cases} \delta^{r|\beta|+sh}\left(\sum_{|\alpha|\le 2m} N_{|\alpha|,R-(|\beta|+h)\delta}(D_x^{\alpha+\beta}D_t^h u) + N_{2m,R-(|\beta|+h)\delta}(D_x^\beta D_t^{h+1}u)\right) \\ \le c_0 B^{|\beta|+h}, \end{cases}$$

for every β, h *and* δ *with* $R - (|\beta| + h)\,\delta > 0$.

Proof. We prove the lemma by induction on $|\beta| + h$. The inequality (1.37) is evidently true for $|\beta| + h = 0$; let us assume that it holds for every β and h such that $|\beta| + h < j$ and prove it for β and h such that $|\beta| + h = j > 0$.

Using Lemma 1.4 applied to $D_x^\beta D_t^h u$, we obtain

$$(1.38)\quad \begin{cases} \delta^{r|\beta|+sh}\left\{\sum\limits_{|\alpha|\le 2m} N_{|\alpha|,R-j\delta}(D_x^{\alpha+\beta}D_t^h u) + N_{2m,R-j\delta}(D_x^\beta D_t^{h+1}u)\right\} \le \\ \le K\{\delta^{r|\beta|+sh}N_{2m,R-j\delta}\big(P(D_x^\beta D_t^h u)\big) + \delta^{r|\beta|+sh}\,\|D_x^\beta D_t^h u\|_{R-j\delta}\}. \end{cases}$$

There are two cases: either $h \ge 1$ or $h = 0$ and $\beta = (\beta_1, \ldots, \beta_n)$ with at least one $\beta_i \ge 1$.

In the first case ($h \ge 1$), we have, thanks to the induction hypothesis,

$$\delta^{r|\beta|+s(h-1)}N_{2m,R-(j-1)\delta}\,(D_x^\beta D_t^h u) \le c_0 B^{|\beta|+h-1} = c_0 B^{j-1}$$

and therefore, by definition (1.30):

$$(1.39)\qquad \delta^{r|\beta|+s(h-1)}\big(R-(j-1)\,\delta-\varrho\big)^{2m}\,\|D_x^\beta D_t^h u\|_\varrho \le c_0 B^{j-1}$$

for $0 < \varrho < R-(j-1)\,\delta$. Take $\varrho = R - j\delta$; then we have

$$\delta^{r|\beta|+sh-s}\delta^{2m}\,\|D_x^\beta D_t^h u\|_{R-j\delta} \le c_0 B^{j-1};$$

thus, recalling that $s \ge 2m$ and $\delta < 1$, we have

$$(1.40)\qquad \delta^{r|\beta|+sh}\,\|D_x^\beta D_t^h u\|_{R-j\delta} \le c_0 B^{j-1}.$$

In the second case ($h = 0$, $\beta_i \ge 1$ for at least one i), set $\beta' = (\beta_1, \ldots, \beta_{i-1}, \beta_i - 1, \beta_{i+1}, \ldots, \beta_n)$; thus $\beta = \beta' + e_i$, $e_i = (0, \ldots, 0, 1, 0, \ldots, 0)$ and $|\beta'| = j - 1$; still by the induction hypothesis, we have

$$\delta^{r|\beta'|}N_{1,R-(j-1)\delta}(D_x^{e_i+\beta'}\,u) \le c_0 B^{j-1}$$

and therefore

$$\delta^{r(|\beta|-1)}\big(R-(j-1)\,\delta-\varrho\big)\,\|D_x^\beta u\|_\varrho \le c_0 B^{j-1},\quad 0 < \varrho < R-(j-1)\,\delta$$

whence, taking $\varrho = R - j\delta$ and recalling that $r \ge 1$ and $\delta < 1$,

$$(1.41)\qquad \delta^{r|\beta|}\,\|D_x^\beta u\|_{R-j\delta} \le c_0 B^{j-1},$$

which is precisely inequality (1.40) in the case $h = 0$; therefore (1.40) is still valid.

Now note that

$$(1.42)\quad P(D_x^\beta D_t^h u) = D_x^\beta D_t^h Pu + [P(D_x^\beta D_t^h u) - D_x^\beta D_t^h Pu] = D_x^\beta D_t^h f + g.$$

First, thanks to (1.36), we have

$$(1.43)\qquad \delta^{r|\beta|+sh}N_{2m,R-j\delta}(D_x^\beta D_t^h f) =$$

$$= \delta^{r|\beta|+sh}\sup_{0<\varrho<R-j\delta}(R-j\,\delta-\varrho)^{2m}\,\|D_x^\beta D_t^h f\|_\varrho \le$$

$$\le \delta^{r|\beta|+sh}\sup_{Q_{R-j\delta}}|D_x^\beta D_t^h f(x,t)|\sup_{0<\varrho<R-j\delta}(R-j\,\delta-\varrho)^{2m}\,(\text{meas. } Q_\varrho)^{1/2} \le$$

$$\le \delta^{r|\beta|+sh}\sup_{Q_{R-j\delta}}|D_x^\beta D_t^h f(x,t)| \le cL^{|\beta|+h} = cL^j.$$

Next, taking into account (1.28), we have:

$$(1.44)\qquad \delta^{r|\beta|+sh} N_{2m,R-j\delta}(g) \le \sum_{|p|\le 2m;\ } \sum_{\substack{l=0,\dots,h;\eta\le\beta\\ l+|\eta|>0}} \binom{h}{l}\binom{\beta_1}{\eta_1}\cdots\binom{\beta_n}{\eta_n}\times$$

$$\times\delta^{r|\eta|+sh} \sup_{Q_{R-j\delta}} |D_x^\eta D_t^l a_p(x,t)|\, \delta^{r(|\beta|-|\eta|)+s(h-l)} N_{2m,R-j\delta}(D_x^{p+\beta-\eta} D_t^{h-l} u).$$

But

$$N_{2m,R-j\delta}(D_x^{p+\beta-\eta} D_t^{h-l} u) \le N_{|p|,R-(|\beta-\eta|+h-l)\delta}(D_x^{p+\beta-\eta} D_t^{h-l} u)$$

since $|\beta-\eta|+h-l<j$, thanks to the fact that $l+|\eta|>0$.

Therefore, using (1.35) and the induction hypothesis, we have

$$\delta^{r|\beta|+sh} N_{2m,R-j\delta}(g) \le \sum_{\substack{l=0,\dots,h;\eta\le\beta\\ l+|\eta|>0}} \binom{h}{l}\binom{\beta_1}{\eta_1}\cdots\binom{\beta_n}{\eta_n} cL^{|\eta|+l}\left(\frac{|\eta|+l}{j}\right)^{|\eta|+l}\times$$

$$\times\, \delta^{r(|\beta-\eta|)+s(h-l)} \sum_{|p|\le 2m} N_{|p|,R-(|\beta-\eta|+h-l)\delta}(D_x^{p+\beta-\eta} D_t^{h-l} u) \le$$

$$\le \sum_{\substack{l=0,\dots,h;\eta\le\beta\\ l+|\eta|>0}} \binom{h}{l}\binom{\beta_1}{\eta_1}\cdots\binom{\beta_n}{\eta_n} cL^{|\eta|+l}\left(\frac{|\eta|+l}{j}\right)^{|\eta|+l} c_0 B^{|\beta-\eta|+h-l}.$$

Noting that

$$\sum_{\substack{\eta\le\beta\\ |\eta|=k}} \binom{\beta_1}{\eta_1}\cdots\binom{\beta_n}{\eta_n} \le \binom{|\beta|}{k},\quad 0\le k\le|\beta|,$$

we obtain (recalling that $|\beta|+h=j$ and that if $\eta\le\beta$, $|\beta-\eta|=|\beta|-|\eta|$)

$$(1.45)\qquad \delta^{r|\beta|+sh} N_{2m,R-j\delta}(g) \le \sum_{i=1}^{j}\binom{j}{i}\left(\frac{i}{j}\right)^i cL^i c_0 B^{j-i} \le$$

$$\le \left(\text{we have} \binom{j}{i}\left(\frac{i}{j}\right)^i \le e^{i-1} \text{ if } 1\le i\le j\right) \le$$

$$\le cc_0 LB^{i-1} \sum_{i=1}^{j}\left(\frac{eL}{B}\right)^{i-1} \le cc_0 LB^{j-1}$$

for $B\ge 2eL$.

Let us now fix (which is permissible) c_0 and B so as to satisfy

$$(1.46)\qquad B\ge 2eL,\quad B\ge K+LK+cLK,\quad c_0\ge c.$$

It then follows from (1.38), (1.40), (1.42), (1.43) and (1.45) that

$$\delta^{r|\beta|+sh}\left\{\sum_{|\alpha|\le 2m} N_{|\alpha|,R-j\delta}(\mathrm{D}_x^{\alpha+\beta}\mathrm{D}_t^h u) + N_{2m,R-j\delta}(\mathrm{D}_x^{\beta}\mathrm{D}_t^{h+1}u)\right\} \le$$

$$\le K(c_0 B^{j-1} + cL^j + cc_0 LB^{j-1}) \le c_0 B^j\left(\frac{K}{B} + \frac{LK}{B} + \frac{cLK}{B}\right) \le c_0 B_j$$

and we thus obtain (1.37), for $|\beta| + h = j$. Q.E.D. ☐

Finally, we have

Lemma 1.6. *Under the hypotheses of Lemma* 1.5, *for every* $R_1 < R$, *there exist* c_* *and* L_* *such that*

$$\sup_{Q_{R_1}} |\mathrm{D}_x^{\beta}\mathrm{D}_t^h u(x, t)| \le c_* L_*^{|\beta|+h}(|\beta| + h)^{r|\beta|+sh}, \ \forall \beta \ \textit{and} \ \forall h, \tag{1.47}$$

and therefore $u \in \mathscr{D}_{s,r}(\overline{Q}_{R_1})$.

Proof. Indeed, it follows from Lemma 1.5 that

$$\delta^{r|\beta|+sh}\,\|\mathrm{D}_x^{\beta}\mathrm{D}_t^h u\|_{R-(|\beta|+h)\delta} \le c_0 B^{|\beta|+h} \tag{1.48}$$

for $R - (|\beta| + h)\,\delta > 0$.

Let $\delta = \dfrac{R - R_1}{|\beta| + h}$ in (1.48); we obtain

$$\|\mathrm{D}_x^{\beta}\mathrm{D}_t^h u\|_{R_1} \le c_0 \frac{B^{|\beta|+h}}{(R - R_1)^{r|\beta|+sh}}\,(|\beta| + h)^{r|\beta|+sh}$$

whence, applying the Sobolev inequalities (see Chapter 1, Section 9.4), we deduce (1.47), with c_* and L_* depending on c_0, B and R_1. And we know that (1.47) is equivalent to saying that $u \in \mathscr{D}_{s.r}(\overline{Q}_{R_1})$. ☐

In summary, we have

Theorem 1.2. *Let* P *be a parabolic operator given by* (1.1), (1.2) *with* (1.4) *and* (1.19); *then if* $u \in \mathscr{E}(Q)$ *and if*

$$Pu \in \mathscr{E}_{s,r}(Q) \ \textit{with} \ s \ge 2m \ \textit{and} \ r > 1,$$

we have

$$u \in \mathscr{E}_{s,r}(Q). \quad ☐$$

Remark 1.2. Hypothesis (1.19) may of course be replaced with the hypothesis: $a_{pq} \in \mathscr{E}_{2m,1}(Q)$. ☐

Remark 1.3. If $u \in \mathscr{D}'(Q)$ and $Pu \in \mathscr{E}_{s,r}(Q)$, it still follows that $u \in \mathscr{E}_{s,}(Q)$, thanks to Theorems 1.1 and 1.2. ☐

2. The Regularity at the Boundary of Solutions of Parabolic Boundary Value Problems

2.1 The Regularity in the Space $\mathscr{D}(\overline{Q})$

In this Section we shall study the regularity up to the boundary of solutions of parabolic boundary value problems in the *non-variational* formulation which we have already studied in the spaces $H^{2mr,r}(Q)$ in Chapter 4. In this manner, we obtain a very extensive class of boundary value problems. We note that many of the results to be obtained are still valid for other problems, which fit different formulations (variational theory, semi-groups, ...) and which we have discussed in the preceding Chapters. □

Let Q and P be the cylinder and the operator introduced in Section 1.1, P being given by (1.1) and (1.2), with (1.3) and (1.4). Furthermore we shall assume that

$$\text{(2.1)} \qquad \begin{cases} \textit{the boundary } \Gamma \textit{ of } \Omega \textit{ is an } (n-1)\textit{-dimensional,} \\ \textit{infinitely differentiable variety, } \Omega \textit{ being locally} \\ \textit{on only one side of } \Gamma. \end{cases}$$

We shall also assume (see Chapter 4, Section 1.1) that a system $\{B_j\}_{j=0}^{m-1}$ of "boundary" operators, defined by

$$\text{(2.2)} \qquad B_j u = \sum_{|h| \leq m_j} b_{jh} \mathrm{D}_x^h u,$$

is given, where the functions $b_{jh} = b_{jh}(x, t)$ satisfy

$$\text{(2.3)} \qquad b_{jh} \in \mathscr{D}(\overline{\Sigma}) \quad (\Sigma = \text{lateral boundary of } Q),$$

with the hypotheses:

$$\text{(2.4)} \qquad 0 \leq m_j \leq 2m-1;$$

$$\text{(2.5)} \qquad \begin{cases} \textit{for every } t_0 \in [0, T], \textit{ the system } \{B_j(x, t_0, \mathrm{D}_x)\}_{j=0}^{m-1} \\ \textit{is normal on } \Gamma \text{ (see Chapter 2, Section 1.4)}; \end{cases}$$

$$\text{(2.6)} \qquad \begin{cases} \textit{for every } \theta \in [-\pi/2, \pi/2] \textit{ and every } t_0 \in [0, T], \textit{ the} \\ \textit{system } \{B_j(x, t_0, \mathrm{D}_x)\}_{j=0}^{m-1} \textit{ "covers" the operator (considered} \\ \textit{in the space } \mathbf{R}_x^n \times \mathbf{R}_y^1) \ A(x, t_0, \mathrm{D}_x) + (-1)^m e^{i\theta} \mathrm{D}_y^{2m} \textit{ on } \Gamma \times \mathbf{R}_y^1. \end{cases}$$

Then Theorem 6.2 of Chapter 4 applies for *every* integer $r \geq 0$. But $\bigcap_{r=0}^{\infty} H^{2mr,r}(Q) = \mathscr{D}(\overline{Q})$, thanks to the Sobolev inequalities (see Chapter 1, Section 9.4) which easily extend to the case of the cylinder Q. Furthermore, we are given f, u_0 and g_j, $j = 0, \ldots, m-1$, with

$$f \in \mathcal{D}(\overline{Q}), \; u_0 \in \mathcal{D}(\overline{\Omega}), \; g_j \in \mathcal{D}(\overline{\Sigma}) \tag{2.7}$$

and with the *compatibility relations* $(\mathcal{R} \cdot \mathcal{C})$ such that there exists a $w \in \mathcal{D}(\overline{Q})$ *with*

$$\begin{cases} B_j w = g_j, \; j = 0, \ldots, m-1, \text{ on } \Sigma, \; w(x, 0) = u_0(x) \text{ on } \Omega, \\ D_t^k [A(x, t, D_x)\, w + D_t w]_{t=0} = D_t^k f(x, 0) \text{ on } \Omega, \; k = 0, 1, 2, \ldots \end{cases} \tag{2.8}$$

Thanks to Whitney's theorem (see Malgrange [2]), the $(\mathcal{R} \cdot \mathcal{C})$ are the conditions for linking g_j, u_0 and f on Γ so that (2.8) holds.

Thus we have

Theorem 2.1. *Let P be given by* (1.1), (1.2) *and* $\{B_j\}_{j=0}^{m-1}$ *by* (2.2) *under the hypotheses* (1.3), (1.4), (2.1), (2.3), (2.4), (2.5), (2.6); *and let f, u_0 and g_j, $j = 0, \ldots, m-1$, be given with* (2.7) *and the compatibility relations* $(\mathcal{R} \cdot \mathcal{C})$; *then the problem*

$$\begin{cases} Pu = f \text{ in } Q, \\ B_j u = g_j \text{ on } \Sigma, \; j = 0, \ldots, m-1, \\ u(x, 0) = u_0(x) \text{ in } \Omega, \end{cases} \tag{2.9}$$

admits a unique solution u belonging to $\mathcal{D}(\overline{Q})$.

2.2 The Regularity in Gevrey Spaces

Let us now study the regularity of the solution of problem (2.9) when the data $(f, g_j, u_0, \Omega, a_{pq})$ belong to Gevrey classes. As we have seen for the problem of regularity in the interior, the characteristic Gevrey classes for the parabolic operators of the type of operator P are analytic classes in x and of order $2m$ in t. Therefore, to the hypotheses on Ω, on P and on the B_j's used in Section 2.1, we add the following hypotheses:

$$\Gamma \text{ is an analytic variety}, \tag{2.10}$$

$$a_{pq} \in \mathcal{D}_{2m,1}(\overline{Q}), \tag{2.11}$$

$$b_{jh} \in \mathcal{D}_{2m,1}(\overline{\Sigma}). \tag{2.12}$$

With these conditions, we have

Theorem 2.2. *Let P and B_j, $j = 0, \ldots, m-1$, be given by* (1.1), (1.2) *and* (2.2), *under the hypotheses* (2.1), (2.10), (2.11), (2.12), (2.4), (2.5), (2.6). *Let f, u_0, g_j, $j = 0, \ldots, m-1$, be given with*

$$f \in \mathcal{D}_{2m,1}(\overline{Q}), \; g_j \in \mathcal{D}_{2m,1}(\overline{\Sigma}), \; u_0 \in \mathcal{H}(\overline{\Omega}) \; (= \mathcal{D}_1(\overline{\Omega})) \tag{2.13}$$

and with the compatibility relations $(\mathcal{R}\cdot\mathcal{C})$ *such that there exists a* w *with*

$$(2.14)\quad \begin{cases} w \in \mathcal{D}_{2m,1}(\overline{Q}),\ B_j w = g_j \ \textit{on}\ \Sigma,\ j = 0, \dots, m-1, \\ w(x, 0) = u_0(x) \ \textit{on}\ \Omega, \\ \mathrm{D}_t^k(A(x, t, \mathrm{D}_x)\, w + \mathrm{D}_t w)\,|_{t=0} = \mathrm{D}_t^k f(x, 0) \ \textit{on}\ \Omega, \forall k. \end{cases}$$

Then there exists one and only one solution u *of problem* (2.9), *belonging to the space* $\mathcal{D}_{2m,1}(\overline{Q})$. ☐

Remark 2.1. The problem of specifying the compatibility relations (2.14) still seems to be open; it would require a Whitney theorem in the space $\mathcal{D}_{2m,1}(\overline{Q})$, that is to show the existence of $w \in \mathcal{D}_{2m,1}(\overline{Q})$, satisfying (2.14), f, g_j and u_0 being given with (2.13) and the relations $(\mathcal{R}\cdot\mathcal{C})$, of Section 2.1. ☐

Proof. Let us introduce the function $v = u - w$, where w satisfies (2.14). Then

$$(2.15)\quad \begin{cases} Pv = \varphi \ \text{ in } Q \text{ (where } \varphi = f - Pw), \\ B_j v = 0 \ \text{ on } \Sigma,\ j = 0, \dots, m-1, \\ v(x, 0) = 0 \ \text{ on } \Omega, \end{cases}$$

with

$$(2.16)\quad \varphi \in \mathcal{D}_{2m,1}(\overline{Q}),$$

$$(2.16')\quad \mathrm{D}_t^k \varphi(x, 0) = 0,\ k = 0, 1, \dots$$

Then it is sufficient to show that $v \in \mathcal{D}_{2m,1}(\overline{Q})$. We may assume[1] to have *extended* the coefficients a_{pq} and b_{jh} in a cylinder $Q' = \Omega \times]t', T[$, with $t' < 0$ and sufficiently small so that the operators A and B_j, thus extended to $\overline{Q}'$, still satisfy the same hypotheses in $\overline{Q}'$ as in $\overline{Q}$. Then, if we also extend the functions φ and v to $\overline{Q}'$, setting $\varphi(x, t) = v(x, t) = 0$ for $t < 0$, we see that $v \in \mathcal{D}(\overline{Q}')$, thanks to Theorem 2.1, and that

$$(2.17)\quad \begin{cases} Pv = \varphi \ \text{ in } Q', \text{ with } \varphi \in \mathcal{D}_{2m,1}(\overline{Q}'), \\ B_j v = 0 \ \text{ on } \Gamma \times]t', T[, \\ v(x, t') = 0 \ \text{ on } \Omega, \end{cases}$$

and the regularity problem for v in $\overline{Q}$ thus reduces to two problems: that of the regularity of v in the *interior* of Q', which is already solved by Theorem 1.2, and that of the regularity of v in the neighborhood of the lateral boundary $\Gamma \times]t', T[$ of Q'

---

[1] For example using Carleson [1], pages 197—201.

This second problem is solved by

Theorem 2.3. *Let P and B_j, $j = 0, \ldots, m-1$, be given by* (1.1), (1.2) *and* (2.2), *under the hypotheses* (1.4), (2.1), (2.10), (2.11), (2.12), (2.4), (2.5), (2.6); *let (x_0, t_0) be a point of Σ ($x_0 \in \Gamma$, $0 < t_0 < T$) and V_0 be a neighborhood of (x_0, t_0) in $\mathbf{R}^{n+1}$. If $v \in \mathscr{D}(V_0 \cap \overline{Q})$ and satisfies the conditions*

$$\text{(2.18)} \qquad Pv = \varphi \text{ in } V_0 \cap Q, \text{ with } \varphi \in \mathscr{D}_{2m,1}(\overline{V_0 \cap Q}),$$

$$\text{(2.19)} \qquad B_j v = \psi_j \text{ on } \Sigma \cap V_0,\ j = 0, \ldots, m-1, \text{ with } \psi_j \in \mathscr{D}_{2m,1}(\overline{\Sigma \cap V_0}),$$

then there exists another neighborhood V_0' of (x_0, t_0), $V_0' \subset V_0$, such that $v \in \mathscr{D}_{2m,1}(\overline{V_0' \cap Q})$.

In order not to burden our presentation unnecessarily, we shall not give the proof of this theorem here and refer the reader to the original work of Cavallucci [1] for the case of Dirichlet conditions ($B_j = \gamma_j$) and of Matsuzawa [2] for the case of general B_j; in these studies, Cavallucci and Matsuzawa more generally examined the regularity at the boundary of solutions of quasi-elliptic equations in Gevrey spaces and Theorem 2.2 is a particular case of their results.

Let us just note that the methods of proof make use of techniques analogous to the ones we have used in Chapter 8, for the regularity of solutions of elliptic equations, and in this Chapter, for the proof of Theorem 1.2.

Finally, Theorem 2.2 follows from Theorems 1.2 and 2.3. ☐

Remark 2.2. A simple proof of Theorem 2.2 can be given in the particular case of *operators A and B_j with coefficients independent of t.*

Indeed, the problem reduces to (2.15) with (2.16) and (2.16') and one can show that $v \in \mathscr{D}_{2m,1}(\overline{Q})$ in the following way. Under the given hypotheses we can apply Theorem 8.1 of Chapter 9, from which it follows that $v \in \mathscr{D}_{2m}([0, T]; L^2(\Omega))$.

Then, differentiating the equation $Pv = \varphi$ with respect to t and applying the operator A, it follows that

$$\text{(2.20)} \qquad A^i v = (-1)^i \mathrm{D}_t^i v + \sum_{h=0}^{i-1} (-1)^{i-h-1} A^h(\mathrm{D}_t^{i-h-1} \varphi), \quad i = 1, 2, \ldots$$

and

$$\text{(2.21)} \qquad B_j(A^i v) = \sum_{h=0}^{i-1} (-1)^{i-h-1} B_j(A^h(\mathrm{D}_t^{i-h-1} \varphi)), \quad i = 1, 2, \ldots, \quad j = 0, \ldots, m-1.$$

Thanks to the fact that $v \in \mathscr{D}_{2m}([0, T]; L^2(\Omega))$ and to (2.16), it follows that there exist c_0 and L_0 such that, for $0 \leq t \leq T$,

$$\| A^i v(x, t) \|_{L^2(\Omega)} \leq c_0 L_0^i (2mi)!, \quad i = 0, 1, 2, \ldots, \tag{2.22}$$

$$\sum_{j=0}^{m-1} \| B_j(A^i v) \|_{H^{2m+2mk-m_j-1/2}(\Gamma)} \leq c_0 L_0^{k+i+1} (2m(i + k + 1))! \qquad i, k = 0, 1, 2, \ldots; \tag{2.23}$$

thus, applying the theorem on *elliptic iterates* (Theorem 1.2, Chapter 8), we obtain that $v(x, t)$ remains in a bounded set of $\mathscr{H}(\bar{\Omega})$ $(= \mathscr{D}_1(\bar{\Omega}))$ for every t of $[0, T]$.

But from (2.20) we obtain

$$|\mathrm{D}_x^p \mathrm{D}_t^i v(x, t)| \leq |\mathrm{D}_x^p A^i v(x, t)| + \sum_{h=0}^{i-1} |\mathrm{D}_x^p A^h (\mathrm{D}_t^{i-h-1} \varphi(x, t))|$$

and then it can be seen that there exist c_* and L_* such that

$$|\mathrm{D}_x^p \mathrm{D}_t^i v(x, t)| \leq c_* L_*^{|p|+i} (2mi)! \, |p|!, \ (x, t) \in \bar{Q}$$

and consequently we have

$$v \in \mathscr{D}_{2m,1}(\bar{Q}). \quad \square$$

Remark 2.3. The above proof, for the case of *coefficients independent of t*, can easily be extended to Gevrey spaces of the type $\mathscr{D}_{N_h, M_k}(\bar{Q})$, more general than $\mathscr{D}_{2m,1}(\bar{Q})$. Let us again start from problem (2.15), assuming that

$$\varphi \in \mathscr{D}_{N_h, M_k}(\bar{Q}), \quad \text{with} \quad (2.16'), \tag{2.24}$$

instead of (2.16).

And let us assume that the sequence $\{N_h\}$ satisfies hypotheses (1.22), (1.23), (1.24) of Chapter 3 and that the sequence $\{M_k\}$ satisfies conditions (1.6), ..., (1.11) of Chapter 8.

Then Theorem 8.1 of Chapter 9 again applies to v and it follows that $v \in \mathscr{D}_{N_h}([0, T]; L^2(\Omega))$. We continue as in the preceding Remark and find (2.20) and (2.21).

Thanks to (2.24) and to the fact that $v \in \mathscr{D}_{N_h}([0, T]; L^2(\Omega))$, we deduce from (2.20) that

$$\| A^i v(x, t) \|_{L^2(\Omega)} \leq c_0 L_0^i N_i + \sum_{h=0}^{i-1} c_0 L_0^{i-1} M_{2hm} N_{i-h-1} \tag{2.25}$$

with suitable constants c_0 and L_0.

Assume that

$$N_h \leq M_{2mh}, \ \forall h. \tag{2.26}$$

Then it follows that

$$\|A^i v\|_{L^2(\Omega)} \leq c_0 L_0^i M_{2mi} + c_0 L^{i-1} \sum_{h=0}^{i-1} M_{2hm} M_{2m(i-h-1)}.$$

But, thanks to the hypotheses on $\{M_k\}$, we have

$$M_{2hm} M_{2m(h-i-1)} \leq M_{2hm} M_{2m(h-i)} \leq c_1 M_{2mi}$$

and therefore

$$\|A^i v\|_{L^2(\Omega)} \leq c_* L_*^i M_{2mi}, \tag{2.27}$$

with suitable c_* and L_* and for every $i = 0, 1, 2, \ldots$

Similarly, we have

(2.28)

$$\sum_{j=0}^{m-1} \| B_j(A^i v) \|_{H^{2m+2mk-m_j-1/2}(\Gamma)} \leq c_* L_*^{k+i+1} M_{2m(k+i+1)}, \quad k, i = 0, 1, 2, \ldots$$

and therefore we can apply the theorem on elliptic iterates of Chapter 8, and continuing as in the preceding Remark, we find that

$$v \in \mathscr{D}_{N_h, M_k}(\overline{Q}).$$

In particular, we can take

$$N_h = (h!)^s, \ M_k = (k!)^r, \ \text{with } r \geq 1, \ 1 < s \leq 2mr. \tag{2.29}$$

Thus this Remark poses the problem of finding under what conditions on the sequences $\{M_k\}$ and $\{N_h\}$ and on the coefficients of P and of the B_j's, Theorem 2.2 is still valid in the space $\mathscr{D}_{N_h, M_k}(\overline{Q})$; the conditions used here, in particular (2.26), are not necessary (compare for example (2.29) with the conditions $s \geq 2m$, $r \geq 1$ of Theorem 1.2 and with the result of Friedman [4] on the regularity "in the interior"). □

3. Application of Transposition: The Finite Cylinder Case

3.1 The Existence of Solutions in the Space $\mathscr{D}'(Q)$: Generalities, the Spaces X and Y

We shall now apply the regularity results obtained in the preceding Sections in order to study, by the transposition method, the boundary value problem

$$\begin{cases} Pu = f \ \text{ in } \ Q = \Omega \times]0, T[, \ T \text{ finite}, \\ B_j u = g_j \ \text{ on } \ \Sigma, \ j = 0, \ldots, m-1, \\ u(x, 0) = u_0 \ \text{ on } \ \Omega, \end{cases} \tag{3.1}$$

in spaces of distributions, and of ultra-distributions on Q. We shall first examine the case of distributions.

The notation and the hypotheses on Ω, P and $\{B_j\}_{j=0}^{m-1}$ are those of Section 2.1 *of this Chapter.*

We further introduce (see Section 1.3 of Chapter 4) the operator P^*, *formal adjoint of* P:

$$(3.2) \qquad P^* = A^* - D_t = \sum_{|p|,|q|\le m} (-1)^{|p|} D_x^p\big(\overline{a_{pq}(x,t)}\, D_x^p\big) - D_t.$$

We recall Green's formula:

$$(3.3)$$

$$\int_Q (Pu)\,\bar{v}\,dx\,dt - \int_Q u\overline{P^*v}\,dx\,dt = \sum_{j=0}^{m-1} \int_\Sigma S_j u \overline{C_j v}\,d\sigma - \sum_{j=0}^{m-1} \int_\Sigma B_j u \overline{T_j v}\,d\sigma +$$
$$+ \int_\Omega u(x,T)\,\overline{v(x,T)}\,dx - \int_\Omega u(x,0)\,\overline{v(x,0)}\,dx, \quad u, v \in \mathscr{D}(\overline{Q}),$$

where C_j, S_j, T_j, $j = 0, \ldots, m-1$, are suitable "boundary" operators (see Chapter 2, Section 1 and Chapter 4, Section 1.3); we note that the coefficients of C_j, S_j and T_j will also belong to $\mathscr{D}_{2m,1}(\overline{\Sigma})$, thanks to the hypotheses on a_{pq} and b_{jh}. ☐

We start from the adjoint problem:

$$(3.4) \qquad \begin{cases} P^*v = \varphi \text{ in } Q,\ \varphi \in \mathscr{D}(Q), \\ C_j v = 0 \text{ on } \Sigma,\ j = 0, \ldots, m-1, \\ v(x,T) = 0 \text{ on } \Omega, \end{cases}$$

to which we can apply Theorem 2.1 (with the obvious changes of operators and of time direction). We obtain the following result.

Let

$$X = \{v \mid v \in \mathscr{D}(\overline{Q}),\ v(x,T) = 0,\ C_j v = 0,\ j = 0, \ldots, m-1,\ P^*v \in \mathscr{D}(Q)\}$$

provided with the inductive limit topology of the spaces

$$X^{(\nu)} = \{v \mid v \in \mathscr{D}(\overline{Q}),\ v(x,T) = 0,\ C_j v = 0,\ j = 0, \ldots, m-1,\ P^*v \in \mathscr{D}_{\mathscr{K}_\nu}(Q)\},$$

where $\{\mathscr{K}_\nu\}$ is an increasing sequence of compact sets contained in Q, of union Q, ($\mathscr{D}_{\mathscr{K}_\nu}(Q)$ being the Frechet space of functions of $\mathscr{D}(Q)$ with support contained in $\mathscr{K}_\nu$, $X^{(\nu)}$ being provided with the natural Frechet space topology); thus, X is a strict $(\mathscr{L}\mathscr{F})$-space.

Then it is easy, by an application of Theorem 2.1 and of the closed graph theorem, to obtain

Proposition 3.1. *The operator P^* defines an (algebraic and topological) isomorphism of X onto $\mathscr{D}(Q)$.* ☐

By transposition of Proposition 3.1, we deduce

Proposition 3.2. *For every continuous antilinear form $v \to L(v)$ on X, there exists a unique distribution $u \in \mathscr{D}'(Q)$ such that*

$$\langle u, \overline{P^* v}\rangle = L(v), \quad \forall v \in X \tag{3.5}$$

and u depends continuously on L (for the strong dual topologies). ☐

It is now required to choose the form L in an appropriate way and to *interpret* (3.5); *formally* L should be in the form (see Green's formula (3.3))

$$L(v) = \langle f, \bar{v}\rangle + \langle u_0, \overline{v(x, 0)}\rangle + \sum_{j=0}^{m-1} \langle g_j, \overline{T_j v}\rangle, \tag{3.6}$$

where f, u_0, g_j are "functions" given on Q, Ω and Σ in such a way as to be able to deduce the solution to problem (3.1) from (3.5). ☐

Concerning the choice of f, we can proceed along considerations analogous to those of Section 3.2 of Chapter 8 for the elliptic case. Choose a space $K(Q)$ of distributions on Q, such that

$$\begin{cases} X \subset K(Q) \subset L^2(Q), \text{ with continuous injection}, \\ K(Q) \text{ is reflexive}, \\ \mathscr{D}(Q) \text{ is dense in } K(Q). \end{cases} \tag{3.7}$$

Then, if we pick f in $K'(Q)$, dual of $K(Q)$, we easily see (write (3.5) with (3.6) for every $v \in \mathscr{D}(Q)$) that

$$Pu = f \text{ in the sense of } \mathscr{D}'(Q). \ ☐ \tag{3.8}$$

Among the possibilities for $K(Q)$, we shall choose one, which seems reasonable and is valid for all boundary conditions of the problem (i.e. for all B_j). Similarly as for the elliptic case (see Section 3.2, Chapter 8) we shall take for $K(Q)$, the space $\Xi(Q)$ defined in the following way. Let $\varrho(x, t)$ be a function of $\mathscr{D}(\overline{Q})$, positive in Q, zero on the boundary of Q of the same order as the distance from (x, t) to this boundary, then

$$\Xi(Q) = \{u \mid \varrho^{|\alpha|} D^\alpha u \in L^2(Q), \ \forall \alpha\} \tag{3.9}$$

is provided with the Frechet space topology given by the semi-norms

$$\|\varrho^{|\alpha|} D^\alpha u\|_{L^2(\Omega)}, \ \forall \alpha.$$

The space $\Xi(Q)$ satisfies conditions (3.7); by the same proofs as for the space $\Xi(\Omega)$ (see Propositions 3.3, 3.4, 3.5 of Chapter 8) we show that $\mathscr{D}(Q)$ is dense in $\Xi(Q)$ and that $\Xi(Q)$ is reflexive, and we obtain a representation theorem for the elements f of the dual $\Xi'(Q)$ of $\Xi(Q)$. ☐

Having fixed the choice of $\Xi(Q)$, we now introduce the space

$$Y = \{u \mid u \in \mathscr{D}'(Q),\ Pu \in \Xi'(Q)\},$$

provided with the coarsetst locally convex topology which makes the mappings $u \to u$ and $u \to Pu$ of Y into $\mathscr{D}'(Q)$ and $\Xi'(Q)$ respectively, continuous.

As for the elliptic case, we have the *density theorem*:

Theorem 3.1. *The space $\mathscr{D}(\overline{Q})$ is dense in Y.*

Proof. Let $u \to M(u)$ be a continuous antilinear form on Y; it may be written, all the spaces introduced being reflexive,

$$M(u) = \langle f, \bar{u}\rangle + \langle g, \overline{Pu}\rangle \text{ with } f \in \mathscr{D}(Q) \text{ and } g \in \Xi(Q). \tag{3.10}$$

Assume that we have $M(\varphi) = 0$, $\forall \varphi \in \mathscr{D}(\overline{Q})$ and let us show that $M(u) = 0$, $\forall u \in Y$.

Note that, thanks to the hypotheses on A, there exists an $\varepsilon > 0$ and an open set Ω_ε such that $\overline{\Omega} \subset \Omega_\varepsilon$ and a linear operator $\mathscr{A}(x, t, \mathrm{D}_x)$ extension of A to $Q_\varepsilon = \Omega_\varepsilon \times]-\varepsilon, T + \varepsilon[$, with coefficients belonging to $\mathscr{D}_{2m,1}(\overline{Q}_\varepsilon)$ and such that the operator

$$\mathscr{A}(x, t_0, \mathrm{D}_x) + (-1)^m\, \mathrm{e}^{i\theta}\, \mathrm{D}_y^{2m}$$

is properly elliptic in $\overline{\Omega}_\varepsilon \times \mathbf{R}_y$ for every $\theta \in [-\pi/2, \pi/2]$ and every $t_0 \in [-\varepsilon, T + \varepsilon]$ (see also the footnote to the proof of Theorem 2.2 of this Chapter).

Let $\tilde{f}$ and $\tilde{g}$ denote the extensions of f and g to Q_ε by 0 outside Q. Let Φ be arbitrary in $\mathscr{D}(Q_\varepsilon)$; then we have

$$\langle \tilde{f}, \overline{\Phi}\rangle + \langle \tilde{g}, \overline{\mathscr{A}\Phi + \mathrm{D}_t\Phi}\rangle = 0, \tag{3.11}$$

the brackets being taken in the duality between $\mathscr{D}'(Q_\varepsilon)$ and $\mathscr{D}(Q_\varepsilon)$; indeed, this expression is equivalent to $M(\varphi)$, $\varphi =$ restriction of Φ to $\overline{Q}$, element of $\mathscr{D}(\overline{Q})$. It follows, $\mathscr{A}^*$ denoting the formal adjoint of $\mathscr{A}$, that

$$\mathscr{A}^*\tilde{g} - \mathrm{D}_t\tilde{g} = -\tilde{f} \text{ in the sense of } \mathscr{D}'(Q_\varepsilon). \tag{3.12}$$

But then, thanks to the fact that $\tilde{f} \in \mathscr{D}(Q_\varepsilon)$ and to the hypoellipticity of $\mathscr{A}^* - \mathrm{D}_t$ (see Theorem 1.1), it follows that $\tilde{g}$ is infinitely differentiable in Q_ε; furthermore, it follows from Theorem 1.2 that for each t, $\tilde{g}$ is analytic in x in $\Omega_\varepsilon - K_t$ (K_t support of $x \to f(x, t)$); now, by definition $\tilde{g}$ vanishes in $Q_\varepsilon - Q$, therefore it vanishes for *each* t in a neighborhood of Γ and for every $x \in \Omega_\varepsilon$, if $t > t_1$ or $t < t_0$ (where $0 < t_0 < t_1 < T$, in such a way that $f(x, t) = 0$ for $t \le t_0$ and for $t \ge t_1$, $x \in \overline{\Omega}$). Therefore $g \in \mathscr{D}(Q)$.

Then (3.12) implies

$$P^*g = -f \text{ in } Q$$

and therefore in (3.10) we have

$$M(u) = -\langle P^*g, \bar{u}\rangle + \langle g, \overline{Pu}\rangle$$

and so $M(u) = 0$, $\forall u \in Y$, since $g \in \mathscr{D}(Q)$. ☐

Remark 3.1. For Theorem 3.1 to be valid, it is sufficient that P satisfies the hypotheses of Theorem 1.2 of this Chapter.

3.2 Space described by $\mathscr{T}v$ as v Describes X

We now have to choose the *boundary* data in (3.7) and prove a trace theorem for the elements of Y. To this end, the essential point is the study of the space described by

$$\mathscr{T}v = \{v(x, 0), T_0 v, \ldots, T_{m-1} v\}$$

as v describes X.

We have completely solved this problem only under more restrictive hypotheses on A and the B_j's, the problem being already technically very complicated in that case. We have met a similar problem in Chapter 9, Section 9.2; similarly as for that Section (see the hypotheses (9.11), (9.13)) we shall make the following hypotheses:

$$(3.13)\quad \begin{cases} \textit{the operator } A \textit{ and } B_j,\ j = 0, \ldots, m-1, \textit{ do not depend} \\ \textit{on } t \textit{ in a neighborhood of } t = 0 \text{ (i.e. } a_{pq}(x, t) = a_{pq}(x), \\ b_{jh}(x, t) = b_{jh}(x) \text{ for } 0 \le t \le t_0); \end{cases}$$

$$(3.14)\quad \begin{cases} -A(0) = -A(x, 0, \mathrm{D}_x), \textit{ considered as an operator in } L^2(\Omega), \\ \textit{unbounded, with domain} \\ \mathrm{D}(A(0)) = \{v \mid v \in H^{2m}(\Omega),\ B_j(x, 0, \mathrm{D}_x)\, v = 0,\ j = 0, \ldots, m-1\}, \\ \textit{is the infinitesimal generator of an analytic semi-group.} \end{cases}$$

Note that (3.13) (choosing, which is permissible, the operators S_j independent of t) implies that C_j and T_j are independent of t for $0 \le t \le t_0$.

Also note that (3.14) implies that $-A^*(0) = -A^*(x, 0, \mathrm{D}_x)$, considered as an unbounded operator in $L^2(\Omega)$, with domain

$$\mathrm{D}(A^*(0)) = \{v \mid v \in H^{2m}(\Omega),\ C_j v = 0,\ j = 0, \ldots, m-1\},$$

is also an infinitesimal generators of an analytic semi-groups.

Finally, let us note that (3.14) is surely satisfied if the problem $\{A(0), B_j(x, 0, \mathrm{D}_x)\}$ is *variational and coercive* (see Chapter 9, Section 9). ▯

Thus, let us study the image of X under $\mathscr{T}$. First of all, in a neighborhood of T, $v(x, t)$ vanishes, for in this neighborhood we have $P^*v = 0$ (thanks to $P^*v \in \mathscr{D}(Q)$) and $v(x, T) = 0$, $C_jv = 0$, $j = 0, \ldots, m-1$; therefore

$$(3.15)\qquad T_jv = 0 \ \textit{for}\ x \in \Gamma \ \textit{and}\ t\ \textit{in a neighborhood of}\ T.$$

Next, applying Theorem 2.3 and the fact that $P^*v = 0$ in a neighborhood of Σ (still because of $P^*v \in \mathscr{D}(Q)$) and $C_jv = 0$ on Σ, $j = 0, \ldots, m-1$, we see that in a neighborhood of Σ, $v(x, t)$ is a Gevrey function of order $2m$ in t and of order 1 (i.e. analytic) in x; and therefore

$$(3.16)\qquad \begin{cases} \textit{there exist } c \textit{ and } L, \textit{ depending on } v, \textit{ such that} \\ \left\| \dfrac{\mathrm{d}^k(T_jv)}{\mathrm{d}t^k} \right\|_{\mathscr{H}_L(\Gamma)} \leq cL^k(k!)^{2m},\ 0 \leq t \leq T,\ j = 0, \ldots, m-1, \\ \qquad\qquad\qquad\qquad\qquad\qquad\qquad\qquad k = 0, 1, 2, \ldots \end{cases}$$

Next, using Remark 7.11 of Chapter 9, thanks to the fact that $P^*v \in \mathscr{D}(Q)$, $C_jv = 0$, $j = 0, \ldots, m-1$, and thanks to (3.13) and (3.14), we find that v is analytic in t with values in $\mathrm{D}(A^{*\infty}(0); k!)$, in a neighborhood of $t = 0$ and therefore *there exist* c *and* L *such that*

(3.17)

$$\left\| \frac{\mathrm{d}^k(T_jv)}{\mathrm{d}t^k} \right\|_{\mathscr{H}_L(\Gamma)} \leq cL^kk!,\ 0 \leq t \leq 1/L,\ j = 0, \ldots, m-1,\ k = 0, 1, 2, \ldots$$

and

$$(3.18)\qquad v(x, 0) \in \mathrm{D}^L\big(A^{*\infty}(0); k!\big).$$

Finally, we have the compatibility conditions on Γ, still thanks to the fact that $P^*v \in \mathscr{D}(Q)$:

$$(3.19)\qquad \left.\frac{\mathrm{d}^k(T_jv)}{\mathrm{d}t^k}\right|_{t=0} = T_j(x, 0)\big(A^{*k}(0)\, v(x, 0)\big),\ x \in \Gamma,\ j = 0, \ldots, m-1,\quad k = 0, 1, \ldots$$ ▯

It now seems natural to introduce the following space: for every fixed $L > 0$, let

$$\mathscr{V}_L = \big\{\{\psi; \varphi_0, \ldots, \varphi_{m-1}\} \mid \psi \in D^L\big(A^{*\infty}(0); k!\big),$$

$$\varphi_j \in \mathscr{D}\big([0, T], \mathscr{H}_L(\Gamma)\big),\ \varphi_j(t) = 0 \ \text{if}\ T - 1/L \leq t \leq T,$$

and there exists c, depending on φ_j, such that

$$\|\varphi_j^{(k)}(t)\|_{\mathscr{H}_L(\Gamma)} \leq cL^k(k!)^{2m},\ 1/L \leq t \leq T,\ \forall k;$$
$$\|\varphi_j^{(k)}(t)\|_{\mathscr{H}_L(\Gamma)} \leq cL^k k!,\ 0 \leq t \leq 1/L,\ \forall k;$$
$$\varphi_j^{(k)}(0) = T_j(x, 0)\left(A^{*k}(0)\,\psi(x)\right),\ \forall k\Big\}.$$

Provided with the norm

$$\|\{\cdots\}\|_{\mathscr{V}_L} = \|\psi\|_{D^L(A^{*\infty}(0);k!)} + \sum_{j=0}^{m-1} \sup_k \times$$
$$\times \left\{\sup_{1/L \leq t \leq T} \frac{\|\varphi_j^{(k)}(t)\|_{\mathscr{H}_L(\Gamma)}}{L^k(k!)^{2m}} + \sup_{0 \leq t \leq 1/L} \frac{\|\varphi_j^{(k)}(t)\|_{\mathscr{H}_L(\Gamma)}}{L^k k!}\right\}$$

$\mathscr{V}_L$ is a Banach space.

By the same method as for Theorem 7.2, Chapter 9, it can be also shown that

(3.20) $\qquad \mathscr{V}_L \subset \mathscr{V}_{L'}$ if $L < L'$, *with compact injection*.

Finally set

$$\mathscr{V} = \operatorname*{ind\,lim}_{L \to +\infty} \mathscr{V}_L.$$

In this manner, we obtain an *inductive limit space of a regular sequence of Banach spaces* (see Chapter 7, Section 1.2).

From (3.15), ..., (3.19) and from the closed graph theorem (see Grothendieck [1]), we then deduce:

Theorem 3.2. *Under the hypotheses of Section 2.1 and furthermore with* (3.13) *and* (3.14), $v \to \mathscr{T}v$ *is a continuous linear mapping of* X *into* $\mathscr{V}$. ☐

We shall now see that this mapping is also *surjective*. More precisely:

Theorem 3.3. *Let the hypotheses of Theorem* 3.2 *be satisfied. Let* $L > 0$ *be fixed; there exists a "right-inverse" of* $\mathscr{T}$, *continuous from* $\mathscr{V}_L$ *into* X, *i.e. there exists a continuous linear mapping* $\{\psi, \varphi_0, \ldots, \varphi_{m-1}\} \to v = \mathscr{R}(\psi, \varphi_0, \ldots, \varphi_{m-1})$ *of* $\mathscr{V}_L$ *into* X *such that*

(3.21) $\qquad v(x, 0) = \psi(x),\ T_j v = \varphi_j,\ j = 0, \ldots, m-1.$

Proof. Let $\{\psi; \varphi_0, \ldots, \varphi_{m-1}\}$ be given in $\mathscr{V}_L$. We first consider the function

(3.22) $$w(x, t) = \sum_{k=0}^{\infty} \left(A^{*k}(0)\,\psi(x)\right)\frac{t^k}{k!}$$

and show that it is analytic in t with values in $\mathrm{D}(A^{*\infty}(0); k!)$ in a neighborhood of 0. Indeed the series (3.22) converges uniformly on every interval $0 \leq t \leq \delta$ with $\delta < 1/L$, as a series of functions with values in $L^2(\Omega)$, since $\psi \in \mathrm{D}^L(A^{*\infty}(0); k!)$. Furthermore, we have

$$\frac{\partial^h}{\partial t^h}(A^{*i}(0)\, w) = \sum_{k=h}^{\infty} A^{*(k+i)}(0)\, \psi(x)\, \frac{t^{k-h}}{(k-h)!} = \sum_{l=0}^{\infty} A^{*(l+h+i)}(0)\, \psi(x)\, \frac{t^l}{l!}$$

and therefore

$$\left\| \frac{\partial^h}{\partial t^h} A^{*i}(0)\, w \right\|_{L^2(\Omega)} \leq \sum_{l=0}^{\infty} c L^{l+h+i} (l+h+i)!\, \frac{\delta^l}{l!} \leq c_0 L_0^{h+i} h! i!,\ 0 \leq t \leq \delta$$

with suitable c_0 and L_0; and therefore w is analytic in t with values in $\mathrm{D}^{L_0}(A^{*\infty}(0); k!)$.

By differentiation, we immediately verify that

$$P^* w = 0,\ x \in \Omega,\ 0 \leq t \leq \delta \tag{3.23}$$

and

$$w(x, 0) = \psi(x),\ x \in \Omega. \tag{3.24}$$

Next, we consider the *Cauchy problem*:

$$\begin{cases} P^* z = 0 \text{ in a neighborhood of } \Sigma, \\ C_j z \;= 0 \text{ on } \Sigma,\ j = 0, \ldots, m-1, \\ T_j z \;= \varphi_j \text{ on } \Sigma,\ j = 0, \ldots, m-1. \end{cases} \tag{3.25}$$

This is indeed a Cauchy problem, for $\{C_0, \ldots, C_{m-1}, T_0, \ldots, T_{m-1}\}$ is a Dirichlet system of order $2m$ on Σ. Then we can apply the results of Talenti [3] on the Cauchy problem in Gevrey classes and find that there exists a unique solution z of (3.25) in the space $\mathcal{D}_{2m}([0, T]; \mathcal{H}(\overline{\varrho(\Gamma)}))$, $\varrho(\Gamma)$ being a suitable neighborhood of Γ (which depends on L).

Furthermore, thanks to the fact that φ_j is analytic in (x, t) for $x \in \Gamma$, $0 \leq t \leq 1/L$, we find that z is also analytic for $x \in \overline{\Omega}$, $0 \leq t \leq 1/L$.

Also note that, thanks to the compatibility relations between ψ and φ_j (i.e. $\varphi_j^{(k)}(0) = T_j(x, 0)\, (A^{*k}(0)\, \psi(x))$, $\forall k$), and to the definition (3.22) of w, we have, in a suitable neighborhood of 0, say for $0 \leq t \leq \delta$:

$$\begin{cases} T_j(x, t)\, w(x, t) = \displaystyle\sum_{k=0}^{\infty} T_j(x, 0)\, \big(A^{*k}(0)\, \psi(x)\big) \frac{t^k}{k!} \\ \qquad\qquad\qquad = \displaystyle\sum_{k=0}^{\infty} \varphi_j^{(k)}(0)\, \frac{t^k}{k!} = \varphi_j(t), \\ C_j(x, t)\, w(x, t) = 0,\ j = 0, 1, \ldots, m-1; \end{cases} \tag{3.26}$$

therefore, by uniqueness of the Cauchy problem

$$w(x, t) = z(x, t) \tag{3.27}$$

in a suitable neighborhood of $\Gamma \times [0, \delta]$ (we may assume in $\varrho(\Gamma) \times [0, \delta]$).

Finally we note that

$$z(x, t) = 0, \ x \in \Omega, \ T - 1/L \leq t \leq T, \tag{3.28}$$

since $\varphi_j(t) = 0$ for $T - 1/L < t \leq T$.

We can now construct the "right-inverse" $v = \mathscr{R}(\psi; \varphi_0, \ldots, \varphi_{m-1})$ by setting:

$$\begin{cases} v(x, t) = 0, \ \text{in} \ \overline{\Omega} \times [T - 1/L, T], \\ v(x, t) = w(x, t)\, \alpha(x, t) \ \text{in} \ \Omega \times [0, \delta], \\ v(x, t) = z(x, t)\, \alpha(x, t) \ \text{in} \ \overline{\varrho(\Gamma)} \times [0, T], \\ v(x, t) = 0 \ \text{elsewhere in} \ Q, \end{cases} \tag{3.29}$$

where α is a fixed function of $\mathscr{D}(\overline{Q})$, with $\alpha(x, t) = 0$ in $(\overline{\Omega} - \varrho(\Gamma)) \times [\delta, T]$ and $\alpha(x, t) = 1$ in $\overline{\Omega} \times [0, \delta/2]$ and in $\dfrac{\varrho(\Gamma)}{2} \times [0, T]$.

Finally, using for example the closed graph theorem, it is easy to see that the right-inverse $\mathscr{R}$ is *continuous* from $\mathscr{V}_L$ into X. □

Corollary 3.1. *The image of X under the mapping $\mathscr{C}$ is the space $\mathscr{V}$.*

3.3 Trace and Existence Theorems in the Space Y

Let us now see how it is possible to define the traces $u(x, 0)$ and $B_j u$ for the elements u of Y.

Here we meet a similar difficulty as in Section 12.3 of Chapter 4 for the case of the spaces $D_{\bar{P}}^{-(r-1)}(Q)$, and which does not come up in the elliptic case (see Section 3.5 of Chapter 8): we shall extend to the space Y the operator $u \to \sigma u = \{u(x, 0); B_0 u, \ldots, B_{m-1} u\}$, but not the operators $u \to u(x, 0)$ and $u \to B_j u$ separately (see Remark 3.2). □

Let $\mathscr{V}'$ denote the (strong) dual of $\mathscr{V}$; then $\mathscr{V}'$ is the dual space of an inductive limit space of a regular sequence of Banach spaces (see Chapter 7, Section 1.2) and therefore $\mathscr{V}'$ is a Frechet-Schwartz space.

We first note the following: let $u_0 \in \mathscr{D}(\overline{\Omega})$ and $g_j \in \mathscr{D}(\overline{\Sigma})$, and verify the compatibility relations (2.8); then the form

$$\{\psi; \varphi_0, \ldots, \varphi_{m-1}\} \to \int_\Omega u_0(x)\, \psi(x)\, dx + \sum_{j=0}^{m-1} \int_\Sigma g_j \varphi_j \, d\sigma \tag{3.30}$$

is continuous linear on $\mathscr{V}$, therefore it may be written in the form:

$$\langle L_{\{u_0;g_0,\ldots,g_{m-1}\}}, \{\psi; \varphi_0, \ldots, \varphi_{m-1}\}\rangle$$

the bracket denoting the duality between $\mathscr{V}'$ and $\mathscr{V}$, and $L_{\{u_0;g_0,\ldots,g_{m-1}\}}$ belonging to $\mathscr{V}'$. Later on, we shall *identify* $\{u_0; g_0, \ldots, g_{m-1}\}$ with $L_{\{u_0,g_0,\ldots,g_{m-1}\}}$, which is legitimate, because, as we shall see, the mapping $\{u_0; g_0, \ldots, g_{m-1}\} \to L_{\{u_0;g_0,\ldots,g_{m-1}\}}$ is one-to-one from a subspace of $\mathscr{D}(\overline{\Omega}) \times [\mathscr{D}(\overline{\Sigma})]^m$ of elements satisfying the $(\mathscr{R} \cdot \mathscr{C})$ into $\mathscr{V}'$. Indeed, if $L_{\{u_0;g_0,\ldots,g_{m-1}\}} = 0$, then

$$\int_\Omega u_0 \psi \,\mathrm{d}x + \sum_{j=0}^{m-1} \int_\Sigma g_j \varphi_j \,\mathrm{d}\sigma = 0, \quad \forall \{\psi; \varphi_0, \ldots, \varphi_{m-1}\}. \tag{3.31}$$

Let $u \in \mathscr{D}(\overline{Q})$ be the solution of

$$Pu = 0, \ u(x, 0) = u_0(x), \ B_j u = g_j, \ j = 0, \ldots, m-1,$$

which exists thanks to Theorem 2.1. Then for arbitrary v in X, applying Green's formula (3.3), we have

$$-\int_Q u\overline{P^* v} \,\mathrm{d}x \,\mathrm{d}t = -\int_\Omega u_0(x)\, \overline{v(x, 0)} \,\mathrm{d}x - \sum_{j=0}^{m-1} \int_\Sigma g_j \overline{T_j v} \,\mathrm{d}\sigma$$

and therefore, thanks to (3.31) and Theorem 3.2,

$$\int_Q u P^* v \,\mathrm{d}x \,\mathrm{d}t = 0, \ \forall v \in X.$$

But for arbitrary θ in $\mathscr{D}(Q)$ there exists a $v \in X$ with $P^* v = \theta$ and therefore

$$\int_Q u\theta \,\mathrm{d}x \,\mathrm{d}t = 0, \ \forall \theta \in \mathscr{D}(Q) \text{ and therefore } u = 0 \text{ in } Q$$

$$\text{and } u_0 = 0, \ g_j = 0, \ j = 0, \ldots, m-1,$$

which proves our assertion.

Therefore in the sequel, we shall always assume that this identification has been made. □

Theorem 3.4. *Under the hypotheses of Theorem 3.2, the mapping* $u \to \sigma u = \{u(x, 0); B_0 u, \ldots, B_{m-1} u\}$ *of* $\mathscr{D}(\overline{Q})$ *into* $\mathscr{D}(\overline{\Omega}) \times [\mathscr{D}(\overline{\Sigma})]^m$ *extends by continuity to a continuous linear mapping, still denoted by* $u \to \sigma u$, *of* Y *into* $\mathscr{V}'$; *furthermore, for* $u \in Y$ *and* $v \in X$, *we have the "Green's formula"*

$$\langle Pu, \bar{v}\rangle - \langle u, \overline{P^* v}\rangle = -\langle \sigma u, \overline{\mathscr{T} v}\rangle, \tag{3.32}$$

where the first bracket denotes the duality between $\Xi'(Q)$ *and* $\Xi(Q)$, *the second between* $\mathscr{D}'(Q)$ *and* $\mathscr{D}(Q)$ *and the third between* $\mathscr{V}'$ *and* $\mathscr{V}$.

Proof. 1) Let u be given in Y and $\{\psi; \varphi_0, \ldots, \varphi_{m-1}\}$ in $\mathscr{V}$; then $\{\psi; \varphi_0, \ldots, \varphi_{m-1}\} \in \mathscr{V}_L$ for suitable L; choose $v = \mathscr{R}\{\psi; \varphi_0, \ldots, \varphi_{m-1}\}$ as in Theorem 3.3 and introduce the form:

$$\mathscr{L}(v) = \langle u, \overline{P^* v}\rangle - \langle Pu, \bar{v}\rangle, \tag{3.33}$$

the first bracket denoting the duality between $\mathscr{D}'(Q)$ and $\mathscr{D}(Q)$ and the second between $\Xi'(\Omega)$ and $\Xi(Q)$ (note that $X \subset \Xi(Q)$).

Let us first verify that $\mathscr{L}(v)$ is independent of the "right-inverse" used and only depends on $\{\psi; \varphi_0, \ldots, \varphi_{m-1}\}$. Indeed, if v_1 and v_2 are such right-inverses, then $\chi = v_1 - v_2$ satisfies the conditions

$$\chi(x, T) = 0, \; \chi(x, 0) = 0, \; C_j\chi = 0, \; T_j\chi = 0, \; j = 0, \ldots, m-1$$

and since $P^*\chi \in \mathscr{D}(Q)$ (and therefore $= 0$ in a neighborhood of the boundary of Q) it follows that

$$\begin{cases} \chi \text{ vanishes in a neighborhood of } \Sigma \text{ (thanks to} \\ \text{the uniqueness of the Cauchy problem);} \end{cases} \tag{3.34}$$

$$\begin{cases} \chi \text{ vanishes in a neighborhood of } t = 0 \text{ for } x \in \bar{\Omega} \\ \text{(thanks to (3.34) and the analyticity in } x \text{ for} \\ \text{sufficiently small } t \geq 0) \end{cases} \tag{3.35}$$

We already know that χ vanishes in a neighborhood of $t = T$ for $x \in \bar{\Omega}$, since $\chi \in X$, therefore

$$\chi \in \mathscr{D}(Q).$$

But then we have

$$\langle u, \overline{P^*\chi}\rangle = \langle Pu, \bar{\chi}\rangle$$

and therefore

$$\mathscr{L}(v_1) = \mathscr{L}(v_2).$$

Thus the form (3.33) depends only on $\{\psi; \varphi_0, \ldots, \varphi_{m-1}\}$ and may be written $\mathscr{L}(\psi, \vec{\varphi})$.

2) It is easy to see that $\mathscr{L}(\psi, \vec{\varphi})$ is continuous and antilinear on $\mathscr{V}$; it is sufficient to prove it in $\mathscr{V}_L$ for every fixed L; but then we can use the expression (3.33) for $\mathscr{L}$ and the proof results from Theorem 3.3.

3) Consequently, we have

$$\mathscr{L}(\psi, \vec{\varphi}) = \langle \sigma u, \overline{\{\psi, \vec{\varphi}\}}\rangle, \tag{3.36}$$

where $\sigma u \in \mathscr{V}'$.

Thus, we have defined a mapping $u \to \sigma u$ of Y into $\mathscr{V}'$ and we immediately see, still using expression (3.33) for $\mathscr{L}$, that σu is antilinear.

Let us now show the continuity of σu. It is sufficient to show that, given a bounded set $\mathscr{B}$ in $\mathscr{V}$ (and therefore in $\mathscr{V}_L$ for suitable L), there exists a neighborhood $\mathscr{U}$ of zero in Y such that

$$|\overline{\langle \sigma u, \{\psi, \vec{\varphi}\}\rangle}| \leq 1, \forall u \in \mathscr{U}, \{\psi, \vec{\varphi}\} \in \mathscr{B}.$$

Let us choose the right-inverse v of $\{\psi, \vec{\varphi}\}$ given by Theorem 3.3; we have

$$\langle \sigma u, \{\psi, \vec{\varphi}\}\rangle = -\langle Pu, \bar{v}\rangle + \langle u, \overline{P^* v}\rangle$$

and v belongs to a bounded set of X and therefore to a bounded set of $X^{(\nu)}$, fixed ν, and therefore in particular v belongs to a bounded set $\mathscr{B}_1$ of $\mathscr{D}(\overline{Q})$ (which is also a bounded set in $\Xi(Q)$) and $P^* v$ to a bounded set of $\mathscr{D}_{\mathscr{K}_\nu}(Q)$. Then we may take

$$\mathscr{U} = \left\{u \mid u \in \frac{1}{2} \mathscr{B}_2^0, \, Pu \in \frac{1}{2} \mathscr{B}_1^0\right\},$$

where E^0 denotes the polar of E, and the result follows.

4) Now if $u \in \mathscr{D}(\overline{Q})$, using Green's formula (3.3) and the *identification* given earlier in this Section, we see that

$$\sigma u = \{u(x, 0); B_0 u, \ldots, B_{m-1} u\}.$$

5) Finally Green's formula (3.32) results from the above, for if $v \in X$, then $\mathscr{C} v \in \mathscr{V}$ and in (3.33) we may take exactly this v and then (3.32) follows from (3.33) and (3.36). ☐

We now come to the existence theorem. In Proposition 3.2, we can choose the form $L(v)$ in the following way:

$$L(v) = \langle f, \bar{v}\rangle + \langle g^*, \overline{\mathscr{C} v}\rangle, \tag{3.37}$$

where $f \in \Xi'(Q)$ and $g^* \in \mathscr{V}'$, the brackets denoting the duality between $\Xi'(Q)$ and $\Xi(Q)$ and $\mathscr{V}'$ and $\mathscr{V}$ respectively. Therefore, for every $f \in \Xi'(Q)$ and $g^* \in \mathscr{V}'$, there exists a unique u in $\mathscr{D}'(Q)$ such that

$$\langle u, \overline{P^* v}\rangle = \langle f, \bar{v}\rangle + \langle g^*, \overline{\mathscr{C} v}\rangle, \quad \forall v \in X. \tag{3.38}$$

It follows that

$$Pu = f \quad \text{in the sense of } \mathscr{D}'(Q) \tag{3.39}$$

and therefore $u \in Y$; but then, thanks to Green's formula (3.32), we see that

$$\langle \sigma u - g^*, \overline{\mathscr{C} v}\rangle = 0, \; \forall v \in X.$$

Therefore, the image of X under $\mathscr{C}$ being $\mathscr{V}$ (see Corollary 3.1), we have

$$\sigma u = g^*. \tag{3.40}$$

Thus we have shown

Theorem 3.5. *Under the hypotheses of Section 2.1 and if furthermore* (3.13) *and* (3.14) *are satisfied, the mapping* $u \to \{Pu, \sigma u\}$ *is an (algebraic and topological) isomorphism of* Y *onto* $\Xi'(Q) \times \mathscr{V}'$.

Therefore the problem

$$Pu = f \text{ in the sense of } \mathscr{D}'(Q), \tag{3.41}$$

$$\sigma u = g^* \text{ in the sense of Theorem 3.4} \tag{3.42}$$

admits a unique solution u in Y, for every $f \in \Xi'(Q)$ and $g^* \in \mathscr{V}'$, u depending continuously on f and g^*. □

Remark 3.2. As we have already, pointed out in Theorems 3.4 and 3.5, the boundary conditions, $B_j u = g_j$ and $u(x, 0) = u_0$, are "united" in condition (3.42). Now, a natural problem would be to separate (if possible) in σu the operators $B_j u$ on Σ and $u(x, 0)$ on Ω. We have met a similar problem in the case of the spaces $D_p^{-(r-1)}(Q)$ in Section 12.3 of Chapter 4; in this case the problem can be resolved either by interpreting the space analogous to $\mathscr{V}'$ as a product of distribution spaces on $\Omega \times \Sigma \times \Gamma$ (see Baiocchi [1]), or as we have done in Chapter 4, by giving meaning to the operators $B_j u$ and $u(x, 0)$ *separately* in *larger* spaces than the "optimal spaces" and by taking, for the existence theorem, the data g_j and u in *smaller* spaces than the "optimal spaces".

Here, the problem is appreciably more complicated; we shall restrict outselves to showing how the operators $B_j u$ for $u \in Y$ can be defined in a space of ultra-distributions on Σ; we have the following Proposition:

$$\begin{cases} \textit{the mapping } u \to Bu = \{B_0 u, \ldots, B_{m-1} u\} \textit{ of } \mathscr{D}(\overline{Q}) \\ \textit{into } [\mathscr{D}(\overline{\Sigma})]^m \textit{ extends by continuity to a continuous} \\ \textit{linear mapping, still denoted by } u \to Bu, \textit{ of } Y \\ \textit{into } [\mathscr{D}'_{2m}(]0, T[; \mathscr{H}'(\Gamma))]^m. \end{cases} \tag{3.43}$$

The proof of this Proposition is similar to the proof of Theorem 3.4, taking the following two remarks into consideration:

a) thanks to the results of Geymonat [2] recapitulated in Chapter 7, Section 4.4, the space $\mathscr{D}_{2m}(]0, T[; \mathscr{H}(\Gamma))$ may be considered as an inductive limit space of a regular sequence of Banach spaces (see Chapter 7, Section 1.2):

(3.44)

$$\mathscr{D}_{2m}(]0, T[; \mathscr{H}(\Gamma)) = \underset{\mathscr{K} \to]0,T[, L \to +\infty, M \to +\infty}{\text{inductive lim}} \mathscr{D}_{2m}(]0, T[; \mathscr{K}, L; \mathscr{H}_M(\Gamma)),$$

where $\mathscr{K}$ = compact set contained in $]0, T[$;

b) The same construction as in Theorem 3.3 (if one takes $\psi = 0$ and $\varphi_j \in \mathscr{D}_{2m}(]0, T[; \mathscr{H}(\Gamma))$, therefore in a $\mathscr{D}_{2m}(]0, T[; \mathscr{K}, L; \mathscr{H}_M(\Gamma)))$ yields a right-inverse $v = \mathscr{R}(0, \varphi_0, \ldots, \varphi_{m-1})$, continuous from $\mathscr{D}_{2m}(]0, T[; \mathscr{K}, L; \mathscr{H}_M(\Gamma))$ into X, with $v(x, t) = 0$ in a neighborhood of $t = 0$, for $x \in \bar{\Omega}$. □

Remark 3.3. It would be of interest to give a structure theorem for the elements of the space $\mathscr{V}'$ (see Problem 6.7).

In any case, they are rather general ultra-distributions on $\Sigma \cup \bar{\Omega}$; for example, we easily see that the form

$$\langle u_0, \psi \rangle + \sum_{j=0}^{m-1} \langle g_j, \varphi_j \rangle,$$

where $u_0 \in (\mathrm{D}(A^{*\infty}(0), k!))'$ and $g_j \in \mathscr{D}'_{+,2m}(]0, T[; \mathscr{H}'(\Gamma))$, the brackets denoting the duality between $(\mathrm{D}(A^{*\infty}(0), k!))'$ and $\mathrm{D}(A^{*\infty}(0), k!)$ and $\mathscr{D}'_{+,2m}(]0, T[; \mathscr{H}'(\Gamma))$ and $\mathscr{D}_{-,2m}(]0, T[; \mathscr{H}(\Gamma))$ respectively, defines an element of $\mathscr{V}'$.

In this regard, note the differences between the preceding results and those obtained in Chapter 9, Section 10.6, Examples 10.1 and 10.2.

In Chapter 9, we could take f and g_j in spaces

$$\Xi'_{M_k}(]0, T[; \Xi^{-1}(\Omega)) \quad \text{and} \quad \Xi'_{M_k}(]0, T[; H^{-s}(\Gamma))^{((1))},$$

with an "arbitrary" sequence $\{M_k\}$ (i.e. only subject to the "general" conditions (1.22), ..., (1.25), Chapter 9); thus, the ultra-distributions considered in Chapter 9 *are more general in* t; on the other hand, they are *less general in the space variables,* since here, we can take the g_j's with values in $\mathscr{H}'(\Gamma)$. Concerning the *initial conditions,* the generality of the result is the same in both cases.

Remark 3.4. For all results of this Section, one can take, instead of the space $\Xi'(\Omega)$, a general space $K'(Q)$, dual of a space $K(Q)$ satisfying (3.7); indeed, only the properties (3.7) intervened in the proofs; the space Y then being defined by

$$Y = \{u \mid u \in \mathscr{D}'(Q),\ Pu \in K'(Q)\}.$$

3.4 The Existence of Solutions in the Spaces $\mathscr{D}'_{s,r}(Q)$ of Gevrey Ultra-Distributions, with $r > 1$, $s \geqq 2m$

In this Section, we shall investigate how the same methods as in the preceding Section can be used to study problem (3.1) in the spaces $\mathscr{D}'_{s,r}(Q)$ of Gevrey ultra-distributions on Q, of order s in t and r in x, with

$$s \geq 2m, \quad r > 1. \tag{3.45}$$

((1)) We have restricted our discussion to s "integer $+1/2$", but s may in fact be chosen arbitrarily.

Still under the hypotheses of Section 2.1 of this Chapter and furthermore (3.13) *and* (3.14), *we consider the spaces* $\mathcal{D}_{s,r}(Q)$ *and* $\mathcal{D}'_{s,r}(Q)$ *instead of* $\mathcal{D}(Q)$ *and* $\mathcal{D}'(Q)$ *respectively*; and we follow the same reasoning as before.

We thus start from the adjoint problem (3.4) with $\varphi \in \mathcal{D}_{s,r}(Q)$ and introduce the space

(3.46)

$$X_{s,r} = \{v \mid v \in \mathcal{D}(\bar{Q}), v(x, T) = 0, C_j v = 0, j = 0, \ldots, m-1, P^* v \in \mathcal{D}_{s,r}(Q)\}.$$

The operator P^* defines a one-to-one mapping of $X_{s,r}$ onto $\mathcal{D}_{s,r}(Q)$, thanks to Theorem 2.1, we can therefore define on $X_{s,r}$ the image topology of $\mathcal{D}_{s,r}(Q)$ under $(P^*)^{-1}$. Then $X_{s,r}$ is an inductive limit of a regular sequence of Banach spaces $X_{s,r}^{(\nu)}$ (Chapter 7, Section 1.2) and

(3.47) $\quad$ *P^* is an isomorphism of $X_{s,r}$ onto $\mathcal{D}_{s,r}(Q)$.*

Therefore by transposition:

$$\text{(3.48)} \quad \begin{cases} \textit{for every continuous antilinear form } v \to L(v) \textit{ on } X_{s,r}, \\ \textit{there exists a unique } u \textit{ in } \mathcal{D}'_{s,r}(Q) \textit{ such that} \\ \langle u, \overline{P^* v} \rangle = L(v), \ \forall v \in X_{s,r}. \end{cases}$$

Still following the method of Sections 3.1—3.3, we introduce a space $K_{s,r}(Q)$ of ultra-distributions on Q such that

$$\text{(3.49)} \quad \begin{cases} X_{s,r} \subset K_{s,r}(Q) \subset L^2(Q), \textit{ with continuous injection,} \\ K_{s,r}(Q) \textit{ is reflexive,} \\ \mathcal{D}_{s,r}(Q) \textit{ is dense in } K_{s,r}(Q). \end{cases}$$

In order to fix our ideas and taking as a pattern the spaces Ξ introduced previously, we shall take, for the space $K_{s,r}(Q)$, the space $\Xi_{s,r}(Q)$ defined in the following way: let $\varrho(x)$ be the function introduced in Section 3.2 of Chapter 8 and let $\mathrm{d}(t)$ be an analogous function defined on $[0, T]$ (continuous, positive on $]0, T[$, $\mathrm{d}(t) = t$ (resp. $T - t$) in the neighborhood of 0 (resp. T)).

For every $L > 0$, we define:

$$\Xi^L_{s,r}(Q) = \{u \mid u \in \mathcal{E}(Q), \text{ such that there exists a } c$$

$$\text{with } \sum_{|\alpha|=k} \|\varrho^k \, \mathrm{d}^h \mathrm{D}_x^\alpha \mathrm{D}_t^h u\|_{L^2(\Omega)} \leq c L^{k+h} (k!)^r \, (h!)^s, \ \forall k \text{ and } h\};$$

it is a Banach space for the norm

$$\|u\| = \sup_{k,h} \sum_{|\alpha|=k} \frac{\|\varrho^k \, \mathrm{d}^h \mathrm{D}_x^\alpha \mathrm{D}_t^h u\|_{L^2(Q)}}{L^{h+k} (k!)^r \, (h!)^s};$$

then we define

$$\Xi_{s,r}(Q) = \operatorname*{ind\,lim}_{L\to+\infty} \Xi^L_{s,r}(Q),$$

L increasing monotonically to $+\infty$.

Through methods entirely analogous to those used for the space $\Xi_{M_k}(\Omega)$ in Section 4.2 of Chapter 8, it can be shown that $\Xi_{s,r}(Q)$ is *reflexive* and that $\mathcal{D}_{s,r}(Q)$ *is dense in* $\Xi_{s,r}(Q)$. Concerning the condition

$$X_{s,r} \subset \Xi_{s,r}(Q),$$

we note that the elements of $X_{s,r}$ belong to $\mathcal{E}_{s,r}(Q)$, thanks to Theorem 1.2.

Therefore $\Xi_{s,r}(Q)$ satisfies conditions (3.49). □

We then introduce the space

$$Y_{s,r} = \{u \mid u \in \mathcal{D}'_{s,r}(Q),\ \ Pu \in \Xi'_{s,r}(Q)\},$$

provided with the coarsest locally convex topology which makes the mappings $u \to u$ and $u \to Pu$ of Y into $\mathcal{D}'_{s,r}(Q)$ and $\Xi'_{s,r}(Q)$ respectively, continuous. □

Next, we show that

$$\mathcal{D}(\bar{Q}) \text{ is dense in } Y_{s,r}. \tag{3.50}$$

We follow the same steps as for Theorem 3.1; this time we have $f \in \mathcal{D}_{s,r}(Q)$ and $g \in \Xi_{s,r}(Q)$. Formulas (3.11) and (3.12) are again taken in the sense of $\mathcal{D}'(Q_\varepsilon)$ and it follows again that $g \in \mathcal{D}(Q)$; finally using Theorem 1.2 of this Chapter, it follows that $g \in \mathcal{D}_{s,r}(Q)$ and therefore we have

$$M(u) = -\langle P^*g, \bar{u}\rangle + \langle g, \overline{Pu}\rangle$$

in the sense of $\mathcal{D}'_{s,r}(Q)$, and therefore again $M(u) = 0$. □

The study of the image of $X_{s,r}$ under the mapping $\mathcal{T}$ is the same as for X and the image of $X_{s,r}$ *is again given by* $\mathcal{V}$, Theorems 3.2 and 3.1 being still valid with $X_{s,r}$ replacing X; indeed the only point which requires modification in the proofs of Section 3.2 is the function $\alpha(x, t)$ (used in formula (3.29)). It now has to belong to $\mathcal{D}_{s,r}(\bar{Q})$ as well, which is possible since $r > 1$ and $s \geq 2m$.

We note that it is precisely at this point that the hypothesis "$s \geq 2m$" intervenes, for the function z belongs to $\mathcal{D}_{2m}([0, T]; \mathcal{H}(\overline{\varrho(\Gamma)}))$ and therefore cannot be extended to the entire cylinder Q in such a way as to belong to $\mathcal{D}_{s,r}(\bar{Q})$ if $s < 2m$. □

Finally the trace and existence theorems (Theorem 3.4 and 3.5) extend to the present case; thus, we have:

Theorem 3.6. *Under the hypotheses of Section 2.1* and *if furthermore* (3.13), (3.14) *and* (3.45) *hold, the mapping* $u \to \sigma u = \{u(x, 0); B_0 u, \ldots, B_{m-1} u\}$

of $\mathscr{D}(\bar{Q})$ *into* $\mathscr{D}(\bar{\Omega})\times[\Omega(\bar{\Sigma})]^m$ *extends by continuity to a continuous linear mapping of* $Y_{s,r}$ *into* $\mathscr{V}'$, *and we have*

$$(3.51) \qquad \langle Pu, \bar{v}\rangle - \langle u, \overline{P^*v}\rangle = -\langle \sigma u, \overline{\mathscr{C}v}\rangle, \ \forall u \in Y_{s,r}, \ v \in X_{s,r},$$

the first bracket denoting the duality between $\Xi'_{s,r}(Q)$ *and* $\Xi_{s,r}(Q)$, *the second between* $\mathscr{D}'_{s,r}(Q)$ *and* $\mathscr{D}_{s,r}(Q)$ *and the third between* $\mathscr{V}'$ *and* $\mathscr{V}$, *the space* $\mathscr{V}$ *being defined in Section 3.2.*

Theorem 3.7. *Under the hypothesis of Theorem 3.6, the mapping* $u \to \{Pu, \sigma u\}$ *is an isomorphism of* $Y_{s,r}$ *onto* $\Xi'_{s,r}(Q)\times\mathscr{V}'$. □

Remark 3.5. Remarks entirely analogous to Remarks 3.2, 3.4 are valid for the spaces $Y_{s,r}$. □

Remark 3.6. It is also possible to apply the results of this Section to the *regularity of Gevrey ultra-distribution solutions* of the equation $Pu = f$ in a similar way as was done for elliptic equations in Section 4.4 of Chapter 8. Indeed, the spaces Y and $Y_{s,r}$ have the *same* space $\mathscr{V}'$ of traces on $\Sigma \cup \bar{\Omega}$.

Thus let u be an ultra-distribution belonging to $\mathscr{D}'_{s,r}(Q)$ with $s \geq 2m$, $r > 1$ and $Pu = f$, $f \in L^2(Q)$ (for example) and let us assume that there exists a system of operators $\{B_j\}_{j=0}^{m-1}$ such that problem (3.1) satisfies the hypotheses of Section 2.1 and (3.13), (3.14); then, thanks to Theorem 3.6, u admits a trace σu which belongs to $\mathscr{V}'$ and therefore, applying Theorem 3.5, it follows that $u \in \mathscr{D}'(Q)$. This situation occurs, for example, if $A(x, t, D_x) = A(x, D_x)$ *does not depend on* t *and if it is strongly uniformly elliptic in* $\bar{\Omega}$ *with analytic coefficients in* $\bar{\Omega}$; then the hypotheses of Section 2.1 and (3.13), (3.14) are satisfied with the Dirichlet system $\{\gamma_j\}_{j=0}^{m-1}$.

Therefore in this case the problem of regularity in the "interior" of Gevrey ultra-distribution solutions of order $s \geq 2m$ and $r > 1$ of equations $Pu = f$, reduces to that of the regularity of distribution solutions, which we have already studied in Section 1. □

4. Application of Transposition: The Infinite Cylinder Case

4.1 The Existence of Solutions in the Space $\mathscr{D}'_+(\mathbf{R}; \mathscr{D}'(\Omega))$; the Space X_-

In this Section, we shall study non-homogeneous boundary value problems for the parabolic operator P in an infinite cylinder. We shall still use the transposition method, starting from regularity results which can be deduced from Sections 1 and 2.

We again denote by Ω a bounded open set of $\mathbf{R}^n$ whose boundary Γ is *an* $(n-1)$*-dimensional analytic variety,* Ω *being locally on only one side of* Γ. In the space $\mathbf{R}^{n+1} = \mathbf{R}^n_x \times \mathbf{R}^1_t$, we consider the infinite cylinder $Q = \Omega \times \mathbf{R}^1_t$ and its boundary $\Sigma = \Gamma \times \mathbf{R}^1_t$. The operator P is given by $P = A + \dfrac{\partial}{\partial t}$, where

$$Au = A(x, t; \mathrm{D}_x) = \sum_{|p|,|q| \le m} (-1)^{|p|} \mathrm{D}^p_x \big(a_{pq}(x, t)\, \mathrm{D}^q_x u\big), \tag{4.1}$$

under the hypotheses:

$$a_{pq} \in \mathscr{E}_{2m}\big(\mathbf{R}; \mathscr{H}(\bar{\Omega})\big). \tag{4.2}$$

We are also given a system of *boundary operators*:

$$B_j u = B_j(x, t; \mathrm{D}_x)\, u = \sum_{|h| \le m_j} b_{jh}(x, t)\, \mathrm{D}^h_x u, \quad \text{order } B_j = m_j, \tag{4.3}$$

with

$$0 \le m_j \le 2m - 1, \quad (j = 0, \ldots, m-1); \tag{4.4}$$

$$b_{jh} \in \mathscr{E}_{2m}\big(\mathbf{R}; \mathscr{H}(\Gamma)\big); \tag{4.5}$$

$$\begin{cases} \textit{for every } \theta \in [-\pi/2, \pi/2] \textit{ and every } t_0 \in \mathbf{R}, \textit{ the} \\ \textit{operator } A(x, t_0; \mathrm{D}_x) + (-1)^m \mathrm{e}^{i\theta} \mathrm{D}^{2m}_y \\ \textit{is properly elliptic in } \bar{\Omega} \times \mathbf{R}^1_y; \end{cases} \tag{4.6}$$

$$\begin{cases} \textit{for every } t_0 \in \mathbf{R}, \textit{ the system } \{B_j(x, t_0; \mathrm{D}_x)\}_{j=0}^{m-1} \\ \textit{is normal on } \Gamma; \textit{ for every } \theta \in [-\pi/2, \pi/2] \textit{ and} \\ t_0 \in \mathbf{R}, \textit{ the system } \{B_j(x, t_0; \mathrm{D}_x)\}_{j=0}^{m-1} \textit{ covers the} \\ \textit{operator } A(x, t_0; \mathrm{D}_x) + (-1)^m \mathrm{e}^{i\theta} \mathrm{D}^{2m}_y \textit{ on } \Gamma \times \mathbf{R}^1_y. \end{cases} \tag{4.7}$$

We consider the boundary value problem:

$$Pu = f \text{ in } Q, \tag{4.8}$$

$$\begin{cases} B_j u = g_j \text{ on } \Sigma, \ j = 0, 1, \ldots, m-1, \\ f, g_j, u \text{ with support in } t \text{ bounded on the left.} \end{cases} \tag{4.9}$$

It seems natural to consider the problem in spaces of vector-valued distributions or ultra-distributions on $\mathbf{R}^1_t$. Thus we shall first seek the solution u in the space $\mathscr{D}'_+(\mathbf{R}; \mathscr{D}'(\Omega))$.

We start from the *adjoint problem*:

$$\begin{cases} P^* v = \varphi \text{ in } Q, \\ C_j v = 0 \text{ on } \Sigma, \ j = 0, \ldots, m-1, \end{cases} \tag{4.10}$$

with $\varphi \in \mathscr{D}_-(\mathbf{R}; \mathscr{D}(\Omega))$,(we recall that by definition $\mathscr{D}'_+(\mathbf{R}; \mathscr{D}'(\Omega))$ is the dual of $\mathscr{D}_-(\mathbf{R}; \mathscr{D}(\Omega))$, see Section 5.1, Chapter 7), the C_j's still being given by Green's formula (see (3.3) for the finite cylinder case)

$$\int_Q (Pu)\,\bar{v}\,dx\,dt - \int_Q u\overline{P^*v}\,dx\,dt = \sum_{j=0}^{m-1} \int_\Sigma S_j u \overline{C_j v}\,d\sigma - \tag{4.11}$$

$$- \sum_{j=1}^{m-1} \int_\Sigma B_j u \overline{T_j v}\,d\sigma, \quad u, v \in \mathscr{D}\big(\mathbf{R}; \mathscr{D}(\bar{\Omega})\big)$$

(note that, thanks to (4.2) and (4.5), the coefficients of S_j, C_j, T_j also belong to $\mathscr{E}_{2m}(\mathbf{R}; \mathscr{H}(\Gamma))$).

We introduce the space

$$X_- = \{v \mid v \in \mathscr{D}_-\big(\mathbf{R}; \mathscr{D}(\bar{\Omega})\big),\ C_j v = 0,\ j = 0, \ldots, m-1,$$
$$P^*v \in \mathscr{D}_-\big(\mathbf{R}; \mathscr{D}(\Omega)\big)\}.$$

Thanks to Theorem 2.1, we can easily see that P^* defines a one-to-one mapping of X_- onto $\mathscr{D}_-(\mathbf{R}; \mathscr{D}(\Omega))$; we can therefore define on X_- the image topology of $\mathscr{D}_-(\mathbf{R}; \mathscr{D}(\Omega))$ under $(P^*)^{-1}$. We thus obtain:

$$\begin{cases} P^* \textit{ is an (algebraic and topological) isomorphism} \\ \textit{of } X_- \textit{ onto } \mathscr{D}_-\big(\mathbf{R}; \mathscr{D}(\Omega)\big). \quad \square \end{cases} \tag{4.12}$$

We now have to study the space described by

$$\mathscr{T}v = \{T_0 v, \ldots, T_{m-1} v\}$$

as v describes X_-; we have

Theorem 4.1. *$v \to \mathscr{T}v$ is a continuous linear mapping of X_- into*

$$[\mathscr{D}_{-,2m}\big(\mathbf{R}; \mathscr{H}(\Gamma)\big)]^m.$$

Proof. We recall that by definition (see Remark 4.1 of Chapter 7) we have

$$\mathscr{D}_-\big(\mathbf{R}; \mathscr{D}(\Omega)\big) = \operatorname*{ind\,lim}_{b \to +\infty} \mathscr{D}_b\big(]-\infty, b]; \mathscr{D}(\Omega)\big), \tag{4.13}$$

where $\mathscr{D}_b(]-\infty, b]; \mathscr{D}(\Omega)) = \operatorname*{proj\,lim}_{a \to -\infty} \mathscr{D}_b([a, b]; \mathscr{D}(\Omega))$ and where $\mathscr{D}_b([a, b]; \mathscr{D}(\Omega))$ is the space of infinitely differentiable functions φ on $[a, b]$ such that $\varphi^{(k)}(b) = 0$, $\forall k$, space provided with the topology of uniform convergence on $[a, b]$ for φ and each of its derivatives.

We also recall that

$$\mathscr{D}_b\big([a, b]; \mathscr{D}(\Omega)\big) = \operatorname*{ind\,lim}_{i \to \infty} \mathscr{D}_b\big([a, b]; \mathscr{D}_{\mathscr{K}_i}(\Omega)\big), \tag{4.14}$$

where $\{\mathscr{K}_i\}$ is an increasing sequence of compact sets contained in Ω, of union Ω, and $\mathscr{D}_{\mathscr{K}_i}(\Omega)$ is the (Frechet) space of functions of $\mathscr{D}(\Omega)$ with support in $\mathscr{K}_i$; the proof of (4.14) follows, for example, from the Corollary in Chapter 1, page 47, of Grothendieck [1] and from the Proposition in Chapter 2, pages 84—85, of the same work.

Let us introduce

$$(4.15)\quad \begin{cases} X_-([a\ b]; \mathscr{K}_i) = \{v \mid v \in \mathscr{D}_b([a\ b]; \mathscr{D}(\bar{\Omega})) \\ C_j v = 0 \text{ on } \Gamma \times [a, b],\ j = 0, \ldots, m-1, \\ P^* v \in \mathscr{D}_b([a, b]; \mathscr{D}_{\mathscr{K}_i}(\Omega))\}, \\ X_-([a, b]) = \{v \mid v \in \mathscr{D}_b([a, b]; \mathscr{D}(\bar{\Omega})),\ C_j v = 0 \text{ on} \\ \Gamma \times [a, b],\ j = 0, \ldots, m-1,\ P^* v \in \mathscr{D}_b([a, b]; \mathscr{D}(\Omega))\}, \\ X_-(b) = \{v \mid v \in \mathscr{D}_b(]-\infty, b]; \mathscr{D}(\bar{\Omega})),\ C_j v = 0 \text{ on} \\ \Gamma \times]-\infty, b],\ j = 0, \ldots, m-1,\ P^* v \in \mathscr{D}_b(]-\infty, b]; \mathscr{D}(\Omega))\}, \end{cases}$$

each space being provided with the image topology of $\mathscr{D}_b([a, b]; \mathscr{D}_{\mathscr{K}_i}(\Omega))$, resp. $\mathscr{D}_b([a, b]; \mathscr{D}(\Omega))$, resp. $\mathscr{D}_b(]-\infty, b]; \mathscr{D}(\Omega))$ under $(P^*)^{-1}$, which is possible since P^*, thanks to Theorem 2.1, defines a one-to-one mapping of $X_-([a, b]; \mathscr{K}_i)$, resp. $X_-([a, b])$, resp. $X_-(b)$, onto $\mathscr{D}_b([a, b]; \mathscr{D}_{\mathscr{K}_i}(\Omega))$, resp. $\mathscr{D}_b([a, b]; \mathscr{D}(\Omega))$, resp. $\mathscr{D}_b(]-\infty, b]; \mathscr{D}(\Omega))$. Also note that:

$$(4.16)\quad \begin{cases} X_-([a, b]) = \operatorname*{ind\,lim}_{i \to +\infty} X_-([a, b]; \mathscr{K}_i) \\ \text{and therefore, thanks to (4.14) and to the fact} \\ \text{that } \mathscr{D}_b([a, b]; \mathscr{D}_{\mathscr{K}_i}(\Omega)) \text{ is a Frechet space,} \\ X_-([a, b]) \text{ is an } (\mathscr{L}\mathscr{F})\text{-space.} \end{cases}$$

Furthermore, we have

$$X_-(b) = \operatorname*{proj\,lim}_{a \to -\infty} X_-([a, b]);\ X_- = \operatorname*{ind\,lim}_{b \to -\infty} X_-(b).$$

Let us show that

$$(4.17)\quad \begin{cases} v \to \mathscr{C} v \textit{ is a continuous linear mapping of} \\ X_-([a, b]) \textit{ into } [\mathscr{D}_{b,2m}([a, b]; \mathscr{H}(\Gamma))]^m \text{ (notation} \\ \text{of Section 4.3 of Chapter 7).} \end{cases}$$

Indeed, we note that if $v \in X_-([a, b])$, then $P^* v$ vanishes in a neighborhood of $\Gamma \times [a, b]$ and $C_j v = 0$, $j = 0, \ldots, m-1$, on $\Gamma \times ([a, b]$; therefore we can apply Theorem 2.3 and deduce from it that v is of

Gevrey class of order $2m$ in t and analytic in x in a neighborhood of $\Gamma \times [a, b]$ and therefore $T_j v$ belongs to $\mathscr{D}_{b,2m}([a, b]; \mathscr{H}(\Gamma))$ for $j = 0, \ldots, m - 1$. (4.17) then follows from the closed graph theorem of L. Schwartz [4], thanks to (4.16) ($X_-([a, b])$ then is a space of type (β) according to Grothendieck [1], Chapter 1, page 17) and to the fact that $\mathscr{D}_{b,2m}([a, b]; \mathscr{H}(\Gamma))$ is a Suslin space (see Chapter 7, Section 4.4, Remark 4.4).

We can now pass to the projective limits with respect to $a \to -\infty$, in (4.17); we obtain that $v \to \mathscr{T} v$ is a continuous linear mapping of $X_-(b)$ into

$$[\mathscr{D}_{b,2m}(]-\infty, b]; \mathscr{H}(\Gamma))]^m,$$

and finally, by passage to the inductive limits with respect to $b \to +\infty$, the Theorem follows. ☐

We now have to show that $\mathscr{T}$ is surjective. To this end, let us prove

Theorem 4.2. *For every fixed positive L and M and every fixed interval* $[a, b]$, *there exists a continuous linear mapping ("right-inverse" of* $\mathscr{T}$) $\vec{\varphi} = \{\varphi_0, \ldots, \varphi_{m-1}\} \to v = \mathscr{R}(\vec{\varphi})$ *of*

$$[\mathscr{D}_{b,2m}([a, b], L; \mathscr{H}_M(\Gamma))]^m$$

into $X_-([a, b]; \mathscr{K}_i)$, *where the compact set* $\mathscr{K}_i$ *depends on* L, M *and* $[a, b]$, *such that*

$$T_j v = \varphi_j, \; j = 0, \ldots, m - 1. \tag{4.18}$$

Proof. Let $\varphi_j \in \mathscr{D}_{b,2m}([a, b], L; \mathscr{H}_M(\Gamma))$, $j = 0, \ldots, m - 1$. Consider the Cauchy problem

$$\begin{cases} P^* w = 0 \text{ in a neighborhood of } \Gamma \times [a, b], \\ C_j w = 0, \; T_j w = \varphi_j, \; j = 0, \ldots, m - 1, \text{ on } \Gamma \times [a, b]. \end{cases} \tag{4.19}$$

Thanks to the results of Talenti [2, 3], there exists a neighborhood $\mathscr{O}$ of Γ, depending on L and M, and in this neighborhood one and only one function $w \in \mathscr{D}_{b,2m}([a, b]; \mathscr{H}(\overline{\mathscr{O}}))$ (and more precisely, $w \in \mathscr{D}_{2b,m}([a, b]; L', \mathscr{H}_{M'}(\overline{\mathscr{O}}))$ with L' and M' depending on L and M) solution of (4.19); furthermore w depends continuously in $\mathscr{D}_{b,2m}([a, b]; L', \mathscr{H}_{M'}(\mathscr{O}))$ on φ_j, as φ_j varies in $\mathscr{D}_{b,2m}([a, b]; L, \mathscr{H}_M(\Gamma))$.

Next, let $\alpha(x, t)$ be a function of $\mathscr{D}(\overline{\Omega} \times [a, b])$ with:

$$\begin{cases} \alpha(x, t) = 1 \text{ in a suitable neighborhood } \mathscr{O}' \times [a, b] \text{ of } \Gamma \times [a, b], \; \overline{\mathscr{O}'} \subset \mathscr{O}, \\ \alpha(x, t) = 0 \text{ in } (\overline{\Omega} - \mathscr{O}) \times [a, b]. \end{cases}$$

Then, the function v defined by

$$(4.20) \qquad v(x,t) = \begin{cases} \alpha(x,t)\, w(x,t) \text{ in } \bar{\mathcal{O}} \times [a,b], \\ 0 \text{ in } (\bar{\Omega} - \mathcal{O}) \times [a,b], \end{cases}$$

provides the required right-inverse $\mathscr{R}(\vec{\varphi})$. ☐

Remark 4.1. If $\vec{\varphi}(x,t) = 0$ for $x \in \Gamma$ and $t \in [a', b']$, $a \leq a' \leq b' < b$, then the given construction of v guarantees that $v(x,t) = 0$ in $\bar{\Omega} \times [a', b']$. ☐

Corollary. *$v \to \mathscr{T} v$ is a surjective mapping of X_- onto*

$$[\mathscr{D}_{-,2m}(\mathbf{R}; \mathscr{H}(\Gamma))]^m.$$

Proof. It is sufficient to slighlty modify the given construction for the right-inverse of Theorem 4.2: let $\vec{\varphi} = (\varphi_0, \ldots, \varphi_{m-1})$ belong to $[\mathscr{D}_{-,2m}(\mathbf{R}; \mathscr{H}(\Gamma))]^m$; there exists a b such that φ_j (or better, its restriction to $]-\infty, b]$) belongs to $\mathscr{D}_{b,2m}(]-\infty, b]; \mathscr{H}(\Gamma))$, $j = 0, \ldots, m-1$.

Therefore for every $a < b$, there exist L_a and M_a such that

$$\varphi_j \in \mathscr{D}_{b,2m}([a,b], L_a; \mathscr{H}_{M_a}(\Gamma)).$$

It will suffice to consider a sequence a_n which decreases to $-\infty$ and to assume that the corresponding sequences L_{a_n} and M_{a_n} increase. For each n, we then make the construction of Theorem 4.2, in which we may assume that the neighborhood $\mathcal{O}_n$ of Γ contains the neighborhood $\mathcal{O}_{n+1}$, so that the function $w_{n+1}(x,t)$ coincides with $w_n(x,t)$ in $\mathcal{O}_{n+1} \times [a_n, b]$; and we may also assume that $\alpha_n(x,t)$ is the restriction to $\bar{\Omega} \times [a_n, b]$ of an infinitely differentiable function $\alpha(x,t)$ in $\bar{\Omega} \times]-\infty, b]$. ☐

4.2 The Existence of Solutions in the Space $\mathscr{D}'_+(\mathbf{R}; \mathscr{D}'(\Omega))$: The Space Y_+ and the Trace and Existence Theorems

Now, let us again consider Proposition (4.12); by transposition, we obtain:

$$(4.21) \qquad \begin{cases} \textit{for every continuous and antilinear form } v \to L(v) \\ \textit{on } X_-, \textit{ there exists a unique } u \textit{ in } \mathscr{D}'_+(\mathbf{R}; \mathscr{D}'(\Omega)) \\ \textit{such that } \langle u, \overline{P^* v} \rangle = L(v),\ v \in X_-. \end{cases}$$

Formally, we shall take $L(v)$ in the form:

$$L(v) = \langle f, \bar{v} \rangle + \sum_{j=0}^{m-1} \langle g_j, \overline{T_j v} \rangle$$

and seek the "best" spaces for f and g_j. We shall follow the general procedure of this text, in particular of Section 3. But, in order to avoid topological difficulties which might overburden the presentation, we shall make use of the weak dual topologies in the spaces intervening in the trace and existence theorems.

For the choice of f, we consider a topological vector space K_- such that

$$(4.22)\quad \begin{cases} X_- \subset K_- \subset L^2_{\text{loc}}(\mathbf{R}; L^2(\Omega)), \textit{ with continuous injection,} \\ \mathscr{D}_-(\mathbf{R}; \mathscr{D}(\Omega)) \textit{ is dense in } K_-. \end{cases}$$

Such spaces exist; for example $\mathscr{D}_-(\mathbf{R}; L^2(\Omega))$. In order to fix our ideas and in analogy with the choices made in the preceding Sections, we take for K_- the space

$$(4.23)\quad \mathscr{D}_-(\mathbf{R}; \Xi(\Omega)),$$

where $\Xi(\Omega)$ is the space defined in Section 3.2 of Chapter 8. Indeed, the space (4.23) satisfies conditions (4.22): in particular, we can easily see that $\mathscr{D}_-(\mathbf{R}; \mathscr{D}(\Omega))$ is dense in $\mathscr{D}_-(\mathbf{R}; \Xi(\Omega))$, using the fact that $\mathscr{D}(\Omega)$ is dense in $\Xi(\Omega)$ (see Proposition 3.3 of Chapter 8). ☐

Remark 4.2. One can even show that $\mathscr{D}_-(\mathbf{R}; \Xi(\Omega))$ is reflexive (see Lions-Magenes [2], Sections 4.2): but we shall not make use of this property here. ☐

We shall use the spaces

$$(4.24)\quad \mathscr{D}'_+(\mathbf{R}; \Xi'(\Omega)) = \text{dual of } \mathscr{D}_-(\mathbf{R}; \Xi(\Omega)),$$

$$(4.25)\quad \mathscr{D}'_+(\mathbf{R}; \mathscr{D}'(\Omega)) = \text{dual of } \mathscr{D}_-(\mathbf{R}; \mathscr{D}(\Omega)),$$

$$(4.26)\quad \mathscr{D}'_{+,2m}(\mathbf{R}; \mathscr{H}'(\Gamma)) = \text{dual of } \mathscr{D}_{-,2m}(\mathbf{R}; \mathscr{H}(\Gamma)),$$

all provided with the weak dual topology.

Note that, thanks to (4.22), $\mathscr{D}'_+(\mathbf{R}; \Xi'(\Omega))$ may be identified with a subspace of $\mathscr{D}'_+(\mathbf{R}; \mathscr{D}'(\Omega))$. ☐

We introduce the space

$$(4.27)\quad Y_+ = \{u \mid u \in \mathscr{D}'_+(\mathbf{R}; \mathscr{D}'(\Omega));\ Pu \in \mathscr{D}'_+(\mathbf{R}; \Xi'(\Omega))\},$$

provided with the coarsest locally convex topology which makes the mappings $u \to u$ and $u \to Pu$ of Y_+ into $\mathscr{D}'_+(\mathbf{R}; \mathscr{D}'(\Omega))$ (weak) and $\mathscr{D}'_+(\mathbf{R}; \Xi'(\Omega))$ (weak) respectively, continuous.

First we have

Theorem 4.3. *The space $\mathscr{D}(\mathbf{R}; \mathscr{D}(\bar{\Omega}))$ is dense in Y_+.*

Proof. Let $u \to M(u)$ be a continuous antilinear form on Y_+; it may be written, the spaces $\mathscr{D}'_+(\mathbf{R}; \mathscr{D}'(\Omega))$ and $\mathscr{D}'_+(\mathbf{R}; \Xi'(\Omega))$ being provided with the weak dual topologies:

(4.28) $$M(u) = \langle f, \bar{u} \rangle + \langle g, \overline{Pu} \rangle,$$

with $f \in \mathscr{D}_-(\mathbf{R}; \mathscr{D}(\Omega))$ and $g \in \mathscr{D}_-(\mathbf{R}; \Xi(\Omega))$.

Assume that we have $M(\varphi) = 0$, $\forall \varphi \in \mathscr{D}(\mathbf{R}; \mathscr{D}(\bar{\Omega}))$ and let us show that $M(u) = 0$, $\forall u \in Y_+$.

Thanks to the hypotheses on A, there exists a linear operator $\mathscr{A}(x, t; \mathrm{D}_x)$ defined in a suitable neighborhood $V(Q)$ of the cylinder Q, coinciding with A on Q with coefficients which are Gevrey functions of order $(1, 2m)$ in $V(Q)$ (i.e. analytic in x and of Gevrey class of order $2m$ in t), and such that the operator

$$\mathscr{A}(x, t_0; \mathrm{D}_x) + (-1)^m \, e^{i\theta} \mathrm{D}_y^{2m}$$

is properly elliptic in $V(Q) \cap \{(x, t), t = t_0\} \times \mathbf{R}_y$ for every $\theta \in [-\pi/2, \pi/2]$ and every $t_0 \in \mathbf{R}_t^1$ (see also the footnote to the proof of Theorem 2.2 of this Chapter).

Let $\tilde{f}$ and $\tilde{g}$ denote the extensions to $V(Q)$ of f and g by zero outside Q. Then if $\Phi \in \mathscr{D}(V(Q))$, we have

(4.29) $$\langle \tilde{f}, \bar{\Phi} \rangle + \langle \tilde{g}, \overline{\mathscr{A}\Phi} + \overline{\mathrm{D}_t \Phi} \rangle = 0,$$

the brackets being taken in the sense of distributions on $V(Q)$, for this expression is equivalent to $M(\varphi)$, $\varphi =$ restriction of Φ to Q (and therefore $\varphi \in \mathscr{D}(\mathbf{R}; \mathscr{D}(\bar{\Omega}))$. $\mathscr{A}^*$ denoting the formal adjoint of $\mathscr{A}$, it follows that

(4.30) $$\mathscr{A}^* \tilde{g} - \mathrm{D}_t \tilde{g} = -\tilde{f} \text{ in the sense of } \mathscr{D}'(V(Q)).$$

But then (Theorem 1.1), $\tilde{f}$ belonging to $\mathscr{D}(V(Q))$, it follows that $\tilde{g}$ is infinitely differentiable in $V(Q)$; furthermore, it follows from Theorem 1.2 that for each t_0, $\tilde{g}$ is analytic in x in

$$V(Q) \cap \{(x, t), t = t_0\} - K_t$$

(K_t support of $x \to f(x, t)$); now by definition $\tilde{g}$ vanishes in $V(Q) - Q$, therefore it also vanishes in a neighborhood of Σ and therefore $g \in \mathscr{D}_-(\mathbf{R}; \mathscr{D}(\Omega))$. Consequently (4.30) implies

(4.31) $$P^* g = -f \text{ in } Q$$

and therefore we have

(4.32) $$M(u) = -\langle P^* g, \bar{u} \rangle + \langle g, \overline{Pu} \rangle;$$

therefore $M(u) = 0$, $\forall u \in Y_+$, since $g \in \mathscr{D}_-(\mathbf{R}; \mathscr{D}(\Omega))$. ☐

Next, we have the trace theorem:

Theorem 4.4. *The mapping* $u \to Bu = \{B_0 u, \ldots, B_{m-1} u\}$ *of* $\mathscr{D}(\mathbf{R}; \mathscr{D}(\bar{\Omega}))$ *into* $[\mathscr{D}(\mathbf{R}; \mathscr{D}(\Gamma))]^m$ *extends by continuity to a continuous linear mapping, still denoted by* $u \to Bu$, *of* Y_+ *(weak) into* $[\mathscr{D}'_{+,2m}(\mathbf{R}; \mathscr{H}'(\Gamma)]^m$ *(weak); and we have Green's formula*

(4.33)

$$\langle Pu, \bar{v} \rangle - \langle u, \overline{P^* v} \rangle = - \sum_{j=0}^{m-1} \langle B_j u, \overline{T_j v} \rangle, \quad \forall u \in Y_+ \text{ and } v \in X_-,$$

the brackets denoting the duality between $\mathscr{D}'_+(\mathbf{R}; \Xi'(\Omega))$ *and* $\mathscr{D}_-(\mathbf{R}; \Xi(\Omega))$, $\mathscr{D}'_+(\mathbf{R}; \mathscr{D}'(\Omega))$ *and* $\mathscr{D}_-(\mathbf{R}; \mathscr{D}(\Omega))$, $\mathscr{D}'_{+,2m}(\mathbf{R}; \mathscr{H}'(\Gamma))$ *and* $\mathscr{D}_{-,2m}(\mathbf{R}; \mathscr{H}(\Gamma))$, *respectively.*

Proof. 1) Let fixed $u \in Y^+$; then u has support in t bounded on the left by t_0. Let $[a, b]$ be an interval with $a < t_0$ and let θ be fixed in $\mathscr{D}_{2m}(]a, b[)$. For given positive L and M, consider the mapping

$$\vec{\varphi} \to Z_\theta(\vec{\varphi}),$$

defined for $\vec{\varphi} = (\varphi_0, \ldots, \varphi_{m-1}) \in [\mathscr{D}_{b,2m}([a, b], L; \mathscr{H}_M(\Gamma))]^m$ by

$$Z_\theta(\vec{\varphi}) = \langle u, \overline{P^* v(\theta \vec{\varphi})} \rangle - \langle Pu, \overline{v(\theta \vec{\varphi})} \rangle, \tag{4.34}$$

where the first bracket denotes the duality between $\mathscr{D}'_+(\mathbf{R}; \mathscr{D}'(\Omega))$ and $\mathscr{D}_-(\mathbf{R}; \mathscr{D}(\Omega))$ and the second between $\mathscr{D}'_+(\mathbf{R}; \Xi'(\Omega))$ and $\mathscr{D}_-(\mathbf{R}; \Xi(\Omega))$, and $v(\theta\vec{\varphi})$ is the right-inverse of $\mathscr{T}$ defined by Theorem 4.2 and extended in $\bar{Q}$ so that $v(\theta\vec{\varphi}) = 0$ for $t \geq b$ and $v(\theta\vec{\varphi}) \in \mathscr{D}_-(\mathbf{R}; \mathscr{D}(\bar{\Omega}))$ with $P^* v \in \mathscr{D}_-(\mathbf{R}; \mathscr{D}(\Omega))$ (note that: 1) θ being fixed, $\theta\vec{\varphi}$ belongs to

$$[\mathscr{D}_{b,2m}([a, b], L'; \mathscr{H}_{M'}(\Gamma))]^m,$$

with suitable L' and M' depending on L and M, and 2) the construction given in the proof of Theorem 4.2 guarantees, see Remark 4.1, that $v(\theta\vec{\varphi})$ vanishes in neighborhoods of $t = a$ and $t = b$, since $\theta\vec{\varphi}$ vanishes in neighborhoods of $t = a$ and $t = b$; therefore, we can for example extend $v(\theta\vec{\varphi})$ by zero outside $[a, b]$, obtaining a function of $\mathscr{D}_-(\mathbf{R}; \mathscr{D}(\bar{\Omega}))$ with $P^* v \in \mathscr{D}_-(\mathbf{R}; \mathscr{D}(\Omega))$.

Let us verify that $Z_\theta(\vec{\varphi})$ does not depend on the right-inverse of $\mathscr{T}$ and on the extension into $\bar{Q}$ used for the function $v(\theta\vec{\varphi})$, but only on $\vec{\varphi}$ (θ being fixed). In fact, if v_1 and v_2 are two such functions $v(\theta\vec{\varphi})$, then $w = v_1 - v_2$ satisfies

$$\begin{cases} P^* w = \psi \text{ with } \psi \in \mathscr{D}_-(\mathbf{R}; \mathscr{D}(\Omega)), \\ C_j w = 0, \ T_j w = 0, \ j = 0, \ldots, m-1, \end{cases} \tag{4.35}$$

and we have

$$\langle u, \overline{P^*w}\rangle - \langle Pu, \overline{w}\rangle = 0,$$

whence

$$\langle u, \overline{P^*v_1}\rangle - \langle Pu, \bar{v}_1\rangle = \langle u, \overline{P^*v_2}\rangle - \langle Pu, \bar{v}_2\rangle.$$

This being set, it follows immediately from Theorem 4.2 and (4.34) that $\vec{\varphi} \to Z_\theta(\vec{\varphi})$ is a continuous antilinear form on $[\mathscr{D}_{b,2m}([a, b]; L, \mathscr{H}_M(\Gamma))]^m$.

2) We can now extend the definition of $Z_\theta(\vec{\varphi})$ ($[a, b]$ and θ being fixed) to the space $[\mathscr{D}_{b,2m}([a, b], \mathscr{H}(\Gamma))]^m$ by using the fact that (see Chapter 6, Section 4.4)

$$\text{(4.36)} \qquad \operatorname*{ind\,lim}_{L\to+\infty, M\to+\infty} \mathscr{D}_{b,2m}\big([a, b], L; \mathscr{H}_M(\Gamma)\big) = \mathscr{D}_{b,2m}\big([a, b]; \mathscr{H}(\Gamma)\big).$$

Indeed if

$$\vec{\varphi} \in [\mathscr{D}_{b,2m}\big([a, b], \mathscr{H}(\Gamma)\big)]^m,$$

then there exist L and M such that

$$\vec{\varphi} \in [\mathscr{D}_{b,2m}\big([a, b], L; \mathscr{H}_M(\Gamma)\big)]^m$$

and therefore we can define $Z_\theta(\vec{\varphi})$ by (4.34). Using (4.36) we then see that $Z_\theta(\vec{\varphi})$ is continuous on

$$[\mathscr{D}_{b,2m}\big([a, b]; \mathscr{H}(\Gamma)\big)]^m.$$

3) Now let $\vec{\varphi}$ be arbitrary in

$$[\mathscr{D}_{-,2m}\big(\mathbf{R}; \mathscr{H}(\Gamma)\big)]^m.$$

Then the support of $\vec{\varphi}$ in t is bounded on the right by a number t_1 and we may assume that $t_0 < t_1$. Consider a fixed interval $[a, b]$ with $a < t_0 < t_1 < b$; and let $\theta \in \mathscr{D}_{2m}(]a, b])$ with $\theta \geq 0$ and $\theta = 1$ in a neighborhood of $[t_0, t_1]$. Then $\theta\vec{\varphi}$ (or better, its restriction to $[a, b]$) belongs to

$$[\mathscr{D}_{b,2m}\big([a, b]; \mathscr{H}(\Gamma)\big)]^m$$

and furthermore $\theta\vec{\varphi} = 0$ in the neighborhood of a and b.

Therefore, we can define $\vec{\varphi} \to Z(\vec{\varphi})$ by

$$\text{(4.37)} \qquad Z(\vec{\varphi}) = Z_\theta(\vec{\varphi}),$$

where $Z_\theta(\vec{\varphi})$ is defined in 2).

We see that $Z(\vec{\varphi})$ is independent of θ. In fact if θ_1 and θ_2 are two such functions, we can find L and M such that, $\vec{\varphi}$ being fixed, $\theta_1\vec{\varphi}$ and $\theta_2\vec{\varphi}$ both belong to

$$[\mathscr{D}_{b,2m}([a, b], L; \mathscr{H}_M(\Gamma))]^m.$$

Then we can use the same right-inverse to define $Z_{\theta_1}(\vec{\varphi})$ and $Z_{\theta_2}(\vec{\varphi})$ by (4.34) and we shall therefore have

$$Z_{\theta_1}(\vec{\varphi}) - Z_{\theta_2}(\vec{\varphi}) = \langle u, \overline{P^* v((\theta_1 - \theta_2)\vec{\varphi})}\rangle - \langle Pu, \overline{v((\theta_1 - \theta_2)\vec{\varphi})}\rangle.$$

But $\theta_1 - \theta_2 = 0$ in a neighborhood of $[t_0, t_1]$, therefore $v((\theta_1 - \theta_2)\vec{\varphi}) = 0$ in this neighborhood (this follows from the same construction as in Theorem 4.2, see Remark 4.1).

Then, u (resp. $\vec{\varphi}$) having support bounded on the left by t_0 (resp. t_1), it follows that

$$Z_{\theta_1}(\vec{\varphi}) = Z_{\theta_2}(\vec{\varphi}).$$

Therefore $Z(\vec{\varphi})$, defined by (4.37), is independent of θ.

4) We easily verify that $Z(\vec{\varphi})$, defined in 3), is an antilinear form on

$$[\mathscr{D}_{-,2m}(\mathbf{R}; \mathscr{H}(\Gamma))]^m.$$

Let us now show that it is *continuous*. For this purpose, it is sufficient to show the continuity of Z on

$$[\mathscr{D}_{b,2m}(]-\infty, b]; \mathscr{H}(\Gamma))]^m,$$

for each fixed b. Consider a partition of unity in $]-\infty, b+\varepsilon]$ ($\varepsilon > 0$ fixed) by $\theta_i \in \mathscr{D}_{2m}(\mathbf{R})$, such that

$$\sum_{i=1}^{\infty} \theta_i(t) = 1 \quad \text{if} \quad t \leq b. \tag{4.38}$$

Note that

$$Z(\vec{\varphi}) = \sum_i Z(\theta_i \vec{\varphi}), \tag{4.39}$$

where the sum $\sum_i$ is taken over a finite number of indices i only, thanks to the properties of the supports of the θ_i's and of u.

Again recall that

$$\mathscr{D}_{b,2m}(]-\infty, b], \mathscr{H}(\Gamma)) = \operatorname*{proj\,lim}_{a_i \to -\infty} \mathscr{D}_{b,2m}([a_i, b]; \mathscr{H}(\Gamma)), \tag{4.40}$$

where, for a_i, we may take the left endpoint of the support of θ_i.

Consider the mapping

$$\vec{\varphi} \to \{\theta_i \vec{\varphi}\}_{i=1}^{\infty}$$

of

$$[\mathscr{D}_{b,2m}(]-\infty, b]; \mathscr{H}(\Gamma))]^m$$

into the topological product

$$\prod_{i=1}^{\infty} [\mathscr{D}_{b,2m}([a_i, b]; \mathscr{H}(\Gamma))]^m.$$

This mapping is one-to-one since $\sum_{i=1}^{\infty} \theta_i \vec{\varphi} = \vec{\varphi}$ in $]-\infty, b]$. Furthermore, thanks to (4.40),

$$[\mathscr{D}_{b,2m}(]-\infty, b]; \mathscr{H}(\Gamma))]^m$$

is provided with the topology of

$$\prod_{i=1}^{\infty} [\mathscr{D}_{b,2m}([a_i, b]; \mathscr{H}(\Gamma))]^m.$$

Thus, according to (4.39), it suffices to show that the mapping $\vec{\varphi} \to Z(\theta\vec{\varphi})$ is continuous on

$$[\mathscr{D}_{b,2m}([a, b]; \mathscr{H}(\Gamma))]^m$$

for fixed $[a, b]$ and fixed $\theta \in \mathscr{D}_{2m}([a, b])$, which we have seen in 2).

5) Therefore we can write $Z(\vec{\varphi})$ in the form

$$Z(\vec{\varphi}) = \sum_{j=0}^{m-1} \langle \tau_j u, \bar{\varphi}_j \rangle, \quad \forall \vec{\varphi} = (\varphi_0, \ldots, \varphi_{m-1}) \tag{4.41}$$

with $\tau_j u \in \mathscr{D}'_{+,2m}(\mathbf{R}; \mathscr{H}'(\Gamma))$, $u \to \tau u = (\tau_0 u, \ldots, \tau_{m-1} u)$ evidently being a linear mapping of Y_+ into

$$[\mathscr{D}'_{+,2m}(\mathbf{R}; \mathscr{H}'(\Gamma))]^m.$$

Let us show that $u \to \tau u$ is continuous in the *weak* topologies; to this end, we need to show that, for fixed $\vec{\varphi}$ in $[\mathscr{D}_{-,2m}(\mathbf{R}; \mathscr{H}(\Gamma))]^m$,

$$u \to \langle \tau u, \bar{\vec{\varphi}} \rangle = \sum_{j=0}^{m-1} \langle \tau_j u, \bar{\varphi}_j \rangle$$

is continuous on Y_+. But this follows from (4.41) and the definition of $Z(\vec{\varphi})$, by using the representation (4.34). of Z.

6) Finally, using Green's formula (4.11) and making an identification analogous to the one in Section 4.3, we verify that if $u \in \mathscr{D}(\mathbf{R}; \mathscr{D}(\bar{\Omega}))$, then we have $\tau_j u = B_j u$, $j = 0, \ldots, m-1$, and we have thus shown the existence of the traces Bu for $u \in Y_+$.

7) Whereas for Green's formula (4.33) it is sufficient to note that if $v \in X_-$, then

$$\varphi_j = T_j v \in \mathscr{D}_{-,2m}(\mathbf{R}; \mathscr{H}(\Gamma)),$$

$j = 0, \ldots, m-1$, and then the construction of $Z(\vec{\varphi})$ may be given directly by

$$Z(\vec{\varphi}) = \langle u, \overline{P^* v} \rangle - \langle u, \bar{v} \rangle. \quad \square$$

Now, choosing

$$f \in \mathscr{D}'_{+}(\mathbf{R}; \Xi'(\Omega))$$

and

$$g_j \in \mathscr{D}'_{+,2m}(\mathbf{R}; \mathscr{H}'(\Gamma))$$

and applying (4.21) and Theorem 4.4, no further obstacles remain to showing the following existence theorem:

Theorem 4.5. *Under the hypotheses* (4.1), ..., (4.7), *the boundary value problem*

$$(4.42)\quad \begin{cases} Pu = f \text{ in the sense of } \mathscr{D}'_{+}(\mathbf{R}; \mathscr{D}'(\Omega)), \\ B_j u = g_j,\ j = 0, \ldots, m-1, \text{ in the sense of Theorem 4.4,} \end{cases}$$

admits a unique solution in Y_+ *for every* $f \in \mathscr{D}'_{+}(\mathbf{R}; \Xi'(\Omega))$ *and* $g_j \in \mathscr{D}'_{+,2m}(\mathbf{R}; \mathscr{H}'(\Gamma))$; *furthermore,* $(f; g_0, \ldots, g_{m-1}) \to u$ *is a continuous mapping of*

$$\mathscr{D}'_{+}(\mathbf{R}; \Xi'(\Omega)) \times [\mathscr{D}'_{+,2m}(\mathbf{R}; \mathscr{H}'(\Gamma))]^m \text{ onto } Y_{+}. \quad \square$$

Remark 4.3. Thanks to (4.21), the mapping $(f; g_0, \ldots, g_{m-1}) \to u$ in Theorem 4.5 is continuous even if the intervening spaces are provided with the *strong dual* topologies. $\square$

Remark 4.4. For all the results obtained in this Section, one can take, instead of the space $\mathscr{D}'_{+}(\mathbf{R}; \Xi'(\Omega))$, a general space K'_{+}, (weak) dual of a space K_+ satisfying (4.22). $\square$

4.3 The Existence of Solutions in the Spaces $\mathscr{D}'_{+,s}(\mathbf{R}; \mathscr{D}'_r(\Omega))$, with $r > 1$, $s \geq 2m$

The methods and results of Sections 4.1 and 4.2 can be extended to the study of problem (4.8), (4.9) in spaces of ultra-distributions of Gevrey type $\mathscr{D}'_{+,s}(\mathbf{R}; \mathscr{D}'_r(\Omega))$ with $r > 1$ and $s \geq 2m$.

We shall restrict the present discussion to indicating the necessary changes to be made to Sections 4.1 and 4.2 to this effect.

The space X_- will be changed to the space

$$X^{s,r}_{-} = \{v \mid v \in \mathscr{D}_{-}(\mathbf{R}; \mathscr{D}(\overline{\Omega})),\ C_j v = 0,\ j = 0, \ldots, m-1,$$
$$P^* v \in \mathscr{D}_{-,s}(\mathbf{R}; \mathscr{D}_r(\Omega))\}$$

provided with the image topology of $\mathscr{D}_{-,s}(\mathbf{R}; \mathscr{D}_r(\Omega))$ under $(P^*)^{-1}$; we still have:

$$(4.43)\quad P^* \text{ is an isomorphism of } X^{s,r}_{-} \text{ onto } \mathscr{D}_{-,s}(\mathbf{R}; \mathscr{D}_r(\Omega)).$$

By a proof completely analogous to the one given for Theorem 4.1, we show that

$$
(4.44)\qquad \begin{cases} v \to \mathscr{C} v \text{ (defined in Section 4.1) is a continuous} \\ \text{linear mapping of} \\ X_-^{s,r} \text{ into } [\mathscr{D}_{-,2m}(\mathbf{R}; \mathscr{H}(\Gamma))]^m. \end{cases}
$$

It must only be pointed out that (4.14) must be replaced with

$$
(4.45)\qquad \mathscr{D}_{b,s}([a, b]; \mathscr{D}_r(\Omega)) = \operatorname*{ind\,lim}_{\substack{i\to\infty \\ L\to+\infty}} \mathscr{D}_{b,s}([a, b], L; \mathscr{D}_r(\mathscr{K}_i, \Omega)),
$$

which is valid (see (4.11) of Chapter 7); the spaces (4.15) must be replaced with

$$
(4.46)\qquad \begin{cases} X_-^{s,r}([a, b], L; \mathscr{K}_i) = \{v \mid v \in \mathscr{D}_b([a, b]; \mathscr{D}(\bar{\Omega})), \\ C_j v = 0 \text{ on } \Gamma \times [a, b], j = 0, \ldots, m-1, \\ P^* v \in \mathscr{D}_{b,s}([a, b], L; \mathscr{D}_r(\mathscr{K}_i, \Omega))\}, \\ X_-^{s,r}([a, b]) = \{v \mid v \in \mathscr{D}_b([a, b]; \mathscr{D}(\bar{\Omega})), \\ C_j v = 0 \text{ on } \Gamma \times [a, b], j = 0, \ldots, m-1, \\ P^* v \in \mathscr{D}_{b,s}([a, b]; \mathscr{D}_r(\Omega))\}, \\ X_-^{s,r}(b) = \{v \mid v \in \mathscr{D}_b(]-\infty, b]; \mathscr{D}(\bar{\Omega})), \\ C_j v = 0 \text{ on } \Gamma \times]-\infty, b], j = 0, \ldots, m-1, \\ P^* v \in \mathscr{D}_{b,s}(]-\infty, b]; \mathscr{D}_r(\Omega))\}, \end{cases}
$$

each of these spaces still being provided with the topology carried over by $(P^*)^{-1}$. □

Next, we show that $\mathscr{C}$ is surjective and more precisely that:

$$
(4.47)\qquad \begin{cases} \text{for every positive } L \text{ and } M \text{ and every fixed interval } [a, b], \\ \text{there exists a continuous linear mapping ("right-inverse"} \\ \text{of } \mathscr{C}) \ \vec{\varphi} \to v = R(\vec{\varphi}) \text{ of} \\ [\mathscr{D}_{b,2m}([a, b], L; \mathscr{H}_M(\Gamma))]^m \text{ into } X_-^{r,s}([a, b]; L', \mathscr{K}_i), \\ \text{where the number } L' \text{ and the compact set } \mathscr{K}_i \text{ depend} \\ \text{on } L, M \text{ and } [a, b], \text{ such that} \\ T_j v = \varphi_j, \ j = 0, \ldots, m-1. \end{cases}
$$

The proof is the same as for Theorem 4.2; we only have to point out that the function $\alpha(x,t)$ used in (4.20) must now belong to $\mathscr{D}_s([a,b];\mathscr{D}_r(\bar{\Omega}))$ (and not only to $\mathscr{D}(\bar{\Omega}\times[a,b])$), which is possible since r and s are greater than 1. (Also note that the hypothesis $s \geq 2m$ intervenes in order to guarantee that the function $v(x,t)$ given by (4.20) belongs to $\mathscr{D}_s([a,b];$ $\mathscr{D}_r(\bar{\Omega}))$ thanks to the fact that w, solution of (4.19), belongs to $\mathscr{D}_{2m}([a,b],$ $\mathscr{H}(\bar{\mathscr{O}})))$. $\square$

The space $Y_+^{s,r}$ (which replaces Y_+) is introduced in analogous fashion. The space K_- is replaced with a space $K_-^{s,r}$ such that

$$\text{(4.48)}\qquad \begin{cases} X_-^{s,r} \subset K_-^{s,r} \subset L^2_{\text{loc}}(\mathbf{R}; L^2(\Omega)) \textit{ with continuous injection,}\\ \mathscr{D}_{-,s}(\mathbf{R}; \mathscr{D}_r(\Omega)) \textit{ is dense in } K_-^{s,r}. \end{cases}$$

For $K_-^{s,r}$,we can choose the space

$$\text{(4.49)}\qquad \mathscr{D}_{-,s}(\mathbf{R}; \Xi_r(\Omega)),$$

where $\Xi_r(\Omega) = \Xi_{(k!)^r}(\Omega)$ is defined in Chapter 8, Section 4.2 (note that in this case the condition $X_-^{s,r} \subset K_-^{s,r}$ is a consequence of Theorem 1.2).

Then we have

$$\text{(4.50)}\qquad Y_+^{s,r} = \{u \mid u \in \mathscr{D}'_{+,s}(\mathbf{R}; \mathscr{D}'_r(\Omega)),\ Pu \in \mathscr{D}'_+(\mathbf{R}; \Xi'_s(\Omega))\},$$

provided with the coarsest locally convex topology which makes the mappings $u \to u$ and $u \to Pu$ of $Y_+^{s,r}$ into $\mathscr{D}'_{+,s}(\mathbf{R};\mathscr{D}'_r(\Omega))$ and $\mathscr{D}'_{+,s}(\mathbf{R};\Xi'_r(\Omega))$ respectively, continuous, these two spaces being provided with the *weak dual topologies*.

The density Theorem 4.3 extends without difficulty, by the same proof; it is sufficient to note that this time $g \in \mathscr{D}_{-,s}(\mathbf{R}; \mathscr{D}_r(\Omega))$, thanks to Theorem 1.2; then (4.32) may be taken in the sense of the duality between $\mathscr{D}'_{+,s}(\mathbf{R}; \mathscr{D}_r(\Omega))$ and $\mathscr{D}_{-,s}(\mathbf{R}; \mathscr{D}_r(\Omega))$; therefore

$$\text{(4.51)}\qquad \mathscr{D}(\mathbf{R}; \mathscr{D}(\bar{\Omega})) \textit{ is dense in } Y_+^{s,r}.$$

We also have the *trace theorem*:

$$\text{(4.52)}\qquad \begin{cases} \textit{the mapping } u \to Bu \textit{ extends by continuity to a}\\ \textit{mapping of } Y_+^{s,r} \textit{ (weak) into } [\mathscr{D}'_{+,2m}(\mathbf{R}; \mathscr{H}'(\Gamma))]^m\\ \textit{(weak) and we have Green's formula:}\\ \langle Pu, \bar{v}\rangle - \langle u, \overline{P^*v}\rangle = - \sum_{j=0}^{m-1} \langle B_j u, \overline{T_j v}\rangle,\\ \forall u \in Y_+^{s,r} \textit{ and } v \in X_-^{s,r}. \end{cases}$$

The reasoning follows the proof of Theorem 4.4, the brackets in (4.34) now denoting the duality between $\mathscr{D}'_{+,s}(\mathbf{R}; \mathscr{D}'_r(\Omega))$ and $\mathscr{D}_{-,s}(\mathbf{R}; \mathscr{D}_r(\Omega))$

and between $\mathscr{D}'_{+,s}(\mathbf{R}; \Xi'_r(\Omega))$ and $\mathscr{D}_{-,s}(\mathbf{R}; \Xi_r(\Omega))$, and using (4.47) instead of Theorem 4.2 and (4.51) instead of Theorem 4.3.

Finally, we obtain the following existence theorem:

Theorem 4.6. *Under the hypotheses* (4.1), ..., (4.7) *and if* $s \geq 2m$ *and* $r > 1$, *the boundary value problem*

$$(4.53) \qquad \begin{cases} Pu = f \text{ in the sense of } \mathscr{D}'_{+,s}(\mathbf{R}; \mathscr{D}'_r(\Omega)) \\ B_j u = g_j,\ j = 0, \ldots, m-1, \text{ in the sense of } (4.52), \end{cases}$$

admits a unique solution in $Y^{s,r}_+$ *for every* $f \in \mathscr{D}'_{+,s}(\mathbf{R}; \Xi'_r(\Omega))$ *and* $g_j \in \mathscr{D}'_{+,s}(\mathbf{R}; \mathscr{H}'(\Gamma))$, *and* $(f; g_0, \ldots, g_{m-1}) \to u$ *is a continuous mapping of*

$$\mathscr{D}'_{+,s}(\mathbf{R}; \Xi'_r(\Omega)) \times [\mathscr{D}'_{+,2m}(\mathbf{R}; \mathscr{H}'(\Gamma))]^m \text{ onto } Y^{s,r}_+$$

(*and even for the strong dual topologies*). □

Remark 4.5. As in Remark 4.4, instead of $\mathscr{D}'_{+,s}(\mathbf{R}; \Xi'_r(\Omega))$, one could take the dual of a space $K^{s,r}_-$ satisfying (4.48).

Remark 4.6. In Sections 4.1 and 4.2 we studied problem (4.8), (4.9) in the space $\mathscr{D}'_+(\mathbf{R}; \mathscr{D}'(\Omega))$ and in Section 4.3, in the space $\mathscr{D}'_{+,s}(\mathbf{R}; \mathscr{D}'_r(\Omega))$ with $s \geq m$ and $r > 1$. It is now evident that we could study problem (4.8), (4.9) by the same methods in the spaces

$$\mathscr{D}'_+(\mathbf{R}; \mathscr{D}'_r(\Omega)) \text{ and } \mathscr{D}'_{+,s}(\mathbf{R}; \mathscr{D}'(\Omega)),\ s \geq 2m,\ r > 1.$$

We do not specify the results. □

Remark 4.7. A corollary to Theorems 4.5 and 4.6, which is the analogue to Remark 3.6, concerns the regularity in the interior of ultra-distribution solutions of the equation $Pu = f$ in the cylinder Q; we easily see that: *under the hypotheses* (4.1), ..., (4.7), *each ultradistribution* $u \in \mathscr{D}'_{+,s}(\mathbf{R}; \mathscr{D}'_r(\Omega))$ *with* $s \geq 2m$ *and* $r > 1$, *solution in* Q *of* $Pu = f$ *with* (*for example*) $f \in \mathscr{D}'_+(\mathbf{R}; L^2(\Omega))$, *is a distribution of* $\mathscr{D}'_+(\mathbf{R}; \mathscr{D}'(\Omega))$, *and therefore* (see Chapter 7, Section 5.1) *is a scalar distribution on* Q, *i.e. belongs to* $\mathscr{D}'(Q)$.

4.4 Remarks on the Existence of Solutions and Trace Theorems in other Spaces of Ultra-Distributions

In Sections 4.1, 4.2 and 4.3 we have *characterized*, by the space $[\mathscr{D}'_{+,2m}(\mathbf{R}; \mathscr{H}'(\Gamma))^m]$, the "traces" on Σ (the $B_j u$'s) of solutions u of the equation $Pu = 0$, or more generally of $Pu = f$, with suitable f, which are *distributions or ultra-distributions whose supports in* t *are bounded on the left* (the spaces $\mathscr{D}'_+(\mathbf{R}; \mathscr{D}'(\Omega))$ and $\mathscr{D}'_{+,s}(\mathbf{R}; \mathscr{D}_r(\Omega))$).

Now, it seems natural to pose the more general problem of *characterizing the "traces" on* Σ *of solutions of* $Pu = f$ *which are distributions or ultra-distributions with arbitrary support in* t.

In fact, one could also pose the problem, tied to the preceding one, of studying, if possible, the boundary value problem (4.8), (4.9) *without the condition that the support in* t *of* f, g_j *and* u (or at least of g_j and u) *be bounded on the left*; the equation $Pu = f$ being a parabolic evolution equation, one can foresee the existence of "conditions" for $t = -\infty$ on the data f and g_j and the solution u.

The methods which we have studied can be adapted to these two problems, with, as a matter of fact, rather great technical difficulties. We shall restrict this discussion to giving an idea of the situation and to putting into evidence the main difficulties (see Problem 6.13). ☐

Let us, for example, study the problem in the space $\mathscr{D}'(\mathbf{R}; \mathscr{D}'(\Omega))$; analogous considerations can be developed in $\mathscr{D}'_s(\mathbf{R}; \mathscr{D}'_r(\Omega))$ with $s \geq 2m$, $r > 1$.

The starting point is the space

(4.54)

$$X = \{v \mid v \in \mathscr{D}_-(\mathbf{R}; \mathscr{D}(\bar{\Omega})),\ C_j v = 0,\ j = 0, \ldots, m-1,\ P^* v \in \mathscr{D}(\mathbf{R}; \mathscr{D}(\Omega))\}$$

and the adjoint problem (4.10), with $\varphi \in \mathscr{D}(\mathbf{R}; \mathscr{D}(\Omega))$; providing X with the image topology of $\mathscr{D}(\mathbf{R}; \mathscr{D}(\Omega))$ under $(P^*)^{-1}$, we see that P^* is an isomorphism of X onto $\mathscr{D}(\mathbf{R}; \mathscr{D}(\Omega))$

The essential point then (as always) is the *concrete characterization* of the image $\mathscr{T}(X)$ of X under the mapping $\mathscr{T}$ defined in Section 4.1 (for the space of "traces" which we seek will be the dual of this image). Since $X \subset X_-$ (where X_- is defined in Section 4.1), we have, thanks to Theorem 4.1:

$$\mathscr{T}(X) \subset [\mathscr{D}_{-,2m}(\mathbf{R}; \mathscr{H}(\Gamma))]^m. \tag{4.55}$$

But in this case $\mathscr{T}(X)$ is a *strict* subspace of $[\mathscr{D}_{-,2m}(\mathbf{R}; \mathscr{H}(\Gamma))]^m$. Let us just consider the case where $A(x, t, \mathrm{D}_x) = A(x, \mathrm{D}_x)$ and $B_j(x, t, \mathrm{D}_x) = B_j(x, \mathrm{D}_x)$ *do not depend on* t *and where furthermore* $-A$, *considered as an unbounded operator in* $L^2(\Omega)$ *with domain*

$$\mathrm{D}(A) = \{v \mid v \in H^{2m}(\Omega),\ B_j v = 0,\ j = 0, \ldots, m-1\},$$

is the infinitesimal generator of an analytic semi-group (this is the case, for example, of coercive variational problems; compare with hypotheses (3.13) and (3.14) of Section 3).

Then, applying Remark 7.11 of Chapter 9, we find that for $t < t_0$ the function v is *analytic in* t, *with values in* $\mathrm{D}(A^{*\infty}; k!)$ *and therefore* $T_j v$ *is, for* $t < t_0$, *analytic in* t *with values in the space* $T_j(\mathrm{D}(A^{*\infty}; k!))$ *of "traces" on* Γ *of the elements of* $\mathrm{D}(A^{*\infty}; k!)$, $j = 0, \ldots, m-1$.

Furthermore, we must study the behavior of v as $t \to -\infty$; *we obtain, for example, that* $\|v\|_{L^2(\Omega)}$ *tends to zero exponentially for* $t \to -\infty$, *which shows that there is a condition "for $t = -\infty$" on the elements of* $\mathscr{T}(X)$.

Thus, we see that the problem of the *characterization* of $\mathscr{T}(X)$ involves rather great difficulties; for a particular case (the classical heat equation in two variables on a strip) we refer the reader to Lions-Magenes [2], Section 8.

But we can show the existence of a "trace" on Σ without giving an explicit characterization of the space $\mathscr{T}(X)$ described by this "trace". Indeed, let us consider the kernel $N_{\mathscr{T}}$ of the mapping $\mathscr{T}$; it is a closed subspace of X. Let $\mathscr{T}^{\boldsymbol{\cdot}}$ be the quotient mapping of $\mathscr{T}$ by $N_{\mathscr{T}}$, which operates from $X^{\boldsymbol{\cdot}} = X/N_{\mathscr{T}}$ onto $\mathscr{T}(X)$ in one-to-one fashion; then we can provide $\mathscr{T}(X)$ with the topology such that $v^{\boldsymbol{\cdot}} \to \mathscr{T}^{\boldsymbol{\cdot}} v^{\boldsymbol{\cdot}}$ is a topological isomorphism of $X^{\boldsymbol{\cdot}}$ onto $\mathscr{T}(X)$. Denote by $\mathscr{G}$, the space $\mathscr{T}(X)$ provided with this topology.

This much being set, we can introduce the space Y (analogue of the space Y_+ of Section 4.2) in the following way:

$$Y = \{u \mid u \in \mathscr{D}'(\mathbf{R}; \mathscr{D}'(\Omega)),\ Pu \in \mathscr{D}'_+(\mathbf{R}; \Xi'(\Omega))\}, \tag{4.56}$$

provided with the coarsest locally convex topology which makes the mappings $u \to u$ and $u \to Pu$ of Y into $\mathscr{D}'(\mathbf{R}; \mathscr{D}'(\Omega))$ and $\mathscr{D}'_+(\mathbf{R}; \Xi'(\Omega))$ respectively, continuous, these two spaces being provided with the *weak* dual topology.

By the same proof as for Theorem 4.3, we show that:

$$\mathscr{D}(\mathbf{R}; \mathscr{D}(\overline{\Omega})) \text{ is dense in } Y. \tag{4.57}$$

Then we can state the trace theorem for Y:

(4.58) $\begin{cases} \textit{the mapping } u \to Bu = \{B_0 u, \ldots, B_{m-1} u\} \textit{ extends by} \\ \textit{continuity to a continuous linear mapping } u \to Bu \\ \textit{of } Y \textit{ into the space } \mathscr{G}', \textit{ weak dual of } \mathscr{G}. \end{cases}$

To prove (4.58), let u be given in Y; for $\vec{\varphi}$ in $\mathscr{G}$, $\vec{\varphi} = \{\varphi_0, \ldots, \varphi_{m-1}\}$, set

$$v^{\boldsymbol{\cdot}} = (\mathscr{T}^{\boldsymbol{\cdot}})^{-1} \vec{\varphi} \tag{4.59}$$

and for arbitrary v in $v^{\boldsymbol{\cdot}}$ $(v \in X)$ set

$$Z(\vec{\varphi}) = \langle u, \overline{P^* v} \rangle - \langle Pu, \bar{v} \rangle, \tag{4.60}$$

where the first bracket denotes the duality between $\mathscr{D}'(\mathbf{R}; \mathscr{D}'(\Omega))$ and $\mathscr{D}(\mathbf{R}; \mathscr{D}(\Omega))$ and the second between $\mathscr{D}'_+(\mathbf{R}; \Xi'(\Omega))$ and $\mathscr{D}_-(\mathbf{R}; \Xi(\Omega))$. As for Theorem 4.4, using the uniqueness of the Cauchy problem, we verify that $Z(\vec{\varphi})$ depends only on $\vec{\varphi}$.

Thus, we have defined an antilinear form on $\mathscr{G}$. It is continuous; indeed the form $v \to \Phi(v)$ defined by

$$\Phi(v) = \langle u, \overline{P^* v}\rangle - \langle Pu, \bar{u}\rangle$$

is continuous on X and null on $N_{\mathscr{C}}$; therefore the form $v^{\cdot} \to \Phi^{\cdot}(v^{\cdot}) = \Phi(v)$, $v \in v^{\cdot}$, is continuous on $X^{\cdot}$ and therefore

$$\vec{\varphi} \to Z(\vec{\varphi}) = \Phi^{\cdot}\big((\mathscr{C}^{\cdot})^{-1}\vec{\varphi}\big)$$

is continuous on $\mathscr{G}$.

Consequently

$$Z(\vec{\varphi}) = \langle \tau u, \overline{\vec{\varphi}}\rangle, \quad \tau u \in \mathscr{G}' = \text{dual of } \mathscr{G}, \tag{4.61}$$

and we have in this manner defined a linear mapping $u \to \tau u$ of Y into $\mathscr{G}'$. It is also continuous for the topologies of Y and $\mathscr{G}'$ (weak); indeed, thanks to (4.61) and (4.60),

$$\langle \tau u, \overline{\vec{\varphi}}\rangle = \langle u, \overline{P^* v}\rangle - \langle Pu, \bar{v}\rangle, \quad \forall v \in X,$$

and if $u \to 0$ in Y, then $u \to 0$ in $\mathscr{D}'(\mathbf{R}; \mathscr{D}'(\Omega))$ (weak) and $Pu \to 0$ in $\mathscr{D}'_+(\mathbf{R}; \Xi'(\Omega))$ (weak), whence the result.

Finally, applying Green's formula, we see that $\tau u = Bu$ if $u \in \mathscr{D}(\mathbf{R}; \mathscr{D}(\bar{\Omega}))$. And therefore (4.58) is proved. ☐

There are no difficulties in showing the existence theorem:

$$\left\{\begin{array}{l} \textit{for every } f \in \mathscr{D}'_+\big(\mathbf{R}; \Xi'(\Omega)\big) \textit{ and } g \in \mathscr{G}', \textit{ there exists} \\ \textit{a unique } u \textit{ in } Y, \textit{ solution of} \\ \left\{\begin{array}{l} Pu = f \textit{ in the sense of } \mathscr{D}'\big(\mathbf{R}; \mathscr{D}'(\Omega)\big), \\ Bu = g \textit{ in the sense of } (4.58), \end{array}\right. \\ \textit{and } u \textit{ depends continuously on } \{f, g\}. \end{array}\right. \tag{4.62}$$

Thus we can say that *every solution in* $\mathscr{D}'(\mathbf{R}; \mathscr{D}'(\Omega))$ *of equation* $Pu = f$ with $f = 0$, or more generally $f \in \mathscr{D}'_+(\mathbf{R}; \Xi'(\Omega))$, admits "traces" Bu on Σ in the space $\mathscr{G}'$.

And we have also solved problem (4.8), (4.9) by imposing *only on* f (*and not on* g_j) *the condition of having the support in t bounded on the left.* ☐

Remark 4.8. According to Chapter 7, Section 5, the space $\mathscr{D}'(\mathbf{R}; \mathscr{D}'(\Omega))$ may be identified with a subspace of $\mathscr{D}'(Q)$, but this subspace does not coincide with $\mathscr{D}'(Q)$. Thus, we can ask the questions of Section 4.4, taking, instead of $\mathscr{D}'(\mathbf{R}; \mathscr{D}'(\Omega))$, the space $\mathscr{D}'(\Omega)$ of scalar distributions on Q (and we could also take $\mathscr{D}'_{s,r}(Q)$ instead of $\mathscr{D}'_s(\mathbf{R}; \mathscr{D}'_r(\Omega))$, with $s \geq 2m$, $r > 1$).

The methods of this Chapter still apply. Indeed, it is sufficient to take as a starting point, instead of the space X defined by (4.54), the space

$$X_* = \{v \mid v \in \mathscr{D}_-(\mathbf{R}; \mathscr{D}(\bar{\Omega})),\ C_j v = 0,\ j = 0, \ldots, m-1,\ P^* v \in \mathscr{D}(Q)\}$$

and the theory can be developed as in Section 4.4, with the same type of difficulties; we find, in particular, that every solution of equation $Pu = 0$ in the sense of scalar distributions on Q, admits traces Bu on Σ, in a suitable space.

But we shall no longer insist on this point or on other possible generalizations to different spaces of distributions or ultra-distributions (see for example Lions-Magenes [2] and Problem 6.13 below).

5. Comments

The hypoellipticity of parabolic operators in the sense of Petrowski [2] has been proven by Mizohata [1], Eidelman [3] and Browder [1] (see also Friedman [6]); Theorem 1.1 is a particular case of these results, but the type of proof given here, which makes use of boundary value problems in the spaces $H^{2mr,r}(Q)$, seems to be new.

The *regularity in the interior in Gevrey type spaces* is due to Friedman [4, 5] (see also Petrowski [2] and Eidelman [2, 3], for the analyticity in the space variables). Our proof of Theorem 1.2 follows Cavallucci [1]. The result for the heat and second order equations is classical: Holmgren [1], Levi [1], Gevrey [1].

For the theorem of Whitney [1] in the form used in Section 2.1 (compatibility relations (2.8)), see Malgrange [1, 2].

We have already cited Cavallucci [1] and Matsuzawa [1, 2] in connection with Theorem 2.3 on the *regularity at the boundary*; in these works, the authors more generally, study the regularity of solutions of quasi-elliptic equations.

We must again note here the regularity results for hypoelliptic operators already cited in the Comments to Chapter 8: Hörmander [1, 2], Friberg [1], Pini [1, 2], Cavallucci [2], Volevich [1], Shilov [1], Friedman [1], ...

Concerning the boundary value problem (2.9), we must also recall the results on the regularity of the solution from the point of view of the analyticity with respect to the variable t, which we have mentioned in Chapter 9. For the analyticity results with respect to all the variables, see Tanabe [2]; for the case of solutions of a second order parabolic equation, the following is a method due to K. Masuda [1]: let u be a solution of

$$\frac{\partial u}{\partial t} + A(t)\, u = 0. \tag{5.1}$$

We already know the *analyticity* in t; therefore $\partial u/\partial \bar{t} = 0$, so that (5.1) implies

$$\frac{\partial u}{\partial t} + A(t)u - \frac{\partial^2 u}{\partial t\, \partial \bar{t}} = 0, \tag{5.2}$$

an *elliptic* equation, from which the result will follow.

If $A(t)$ is of order $2m$, consider, instead of (5.2), the equation

$$\frac{\partial u}{\partial t} + A(t)\, u + (-1)^m \frac{\partial^{2m} u}{\partial t^m\, \partial \bar{t}^m} = 0. \tag{5.3}$$

The results of Section 3 (finite cylinder case), announced in Magenes [4], are proven here "in extenso" for the first time; less precise results had been given in Lions-Magenes [2].

The regularity *in the interior* of ultra-distribution solutions (Remark 3.6) in the case of hypoelliptic equations with constant coefficients has been studied by different methods by Björck [1].

The results of Section 4 (infinite cylinder case) extend and specify the results of the authors [3, 5].

6. Problems

6.1. General study of the differentiation operator in the spaces $H_{r,s}(Q)$ (see Lemma 1.1 of this Chapter and Proposition 2.3 of Chapter 4).

6.2. What are the "optimal" hypotheses on the sequences M_k and N_h and on the coefficients of P and B_j for Theorems 1.2 and 2.3 still to be valid in the spaces $\mathscr{E}_{M_k,N_h}(Q)$ and $\mathscr{D}_{M_k,N_h}(\bar{Q})$ respectively? (see Remark 2.3 and Friedman [4, 5]).

6.3. To find the suitable extensions of the elliptic iterates theorem (Chapter 8) to parabolic, or more generally quasi-elliptic operators in the spaces $\mathscr{D}_{M_k,N_h}(\bar{Q})$ (see Matsuzawa [1, 2] and Nelson [1], Goodman [1]).

6.4. Regularity theorems in Beurling type spaces (see Björck [1]).

6.5. Examples of spaces of type $K(Q)$, $K_{r,s}(Q)$, K_-, $K_-^{s,r}$, different from the ones given in the text; see Problem 6.4, Chapter 8.

6.6. Study of the traces $u(x, 0)$ and Bu, *separately*, for the u's in Y (see Remark 3.2).

6.7. Structure of the elements of the space $\mathscr{V}'$.

6.8. Non-homogeneous boundary value problems in unbounded open sets (see Cavallucci [2], Pini [2]).

6.9. Non-homogeneous boundary value problems in other spaces of ultra-distributions, for example the hyperfunctions of Satô.

6.10. Generalization of the theory to parabolic systems.

6.11. Is Theorem 4.4 valid if the spaces are provided with the strong dual topology? (it is possible for Theorem 4.3 by using Remark 4.3).

6.12. Do there exist counter-examples to the validity of the regularity in the interior of ultra-distribution solutions (in $\mathscr{D}'_{s,r}(Q)$) of equation $Pu = 0$, in the case $1 < s < 2m$, $1 < r$ (see Remark 3.6 and Bjorck [1]).

6.13. The Remarks of Section 4.4 pose several problems in relation with the various spaces of distributions or ultra-distributions which can be chosen (see Lions-Magenes [2], for an example).

6.14. Non-homogeneous boundary value problems for quasi-elliptic operators.

6.15. Non-homogeneous boundary value problems in non-cylindrical open sets.

6.16. Problem of the compatibility relations in Gevrey classes; see Remark 2.1.

6.17. To see to what extend the theory of this Chapter can be extended to the settings of Problems 17.6, 17.7, 17.9, 17.11, 17.12, 17.13 of Chapter 4.

Chapter 11

Evolution Equations of the Second Order in t and of Schroedinger Type

This Chapter extends, in so far as possible, the problems studied in Chapter 5 in the Hilbert space setting, to spaces of distributions or ultra-distributions.

1. Equations of the Second Order in t; Regularity of the Solutions of Boundary Value Problems

1.1 The Regularity in the Space $\mathscr{D}(\bar{Q})$

We shall reconsider, in the setting of spaces of distributions and of Gevrey functionals, the boundary value problems for the equations of the second order in t, studied in Chapter 5, with the same notations and hypotheses (Section 1.1, Chapter 5).

Thus, let Ω be a bounded open set in $\mathbf{R}^n$, with:

$$(1.1)\quad \begin{cases} \textit{the boundary } \Gamma \textit{ of } \Omega \textit{ is an}(n-1) \textit{ dimensional,} \\ \textit{infinitely differentiable variety, } \Omega \textit{ being locally} \\ \textit{on only one side of } \Gamma. \end{cases}$$

In $\mathbf{R}^{n+1} = \mathbf{R}^n_x \times \mathbf{R}^1_t$, we consider the cylinder

$$Q = \Omega \times]0, T[, \text{ with } \Sigma = \Gamma \times]0, T[, \ T < +\infty.$$

We propose to study the problem:

$$(1.2)\quad \begin{cases} Pu = A\left(x, t, \dfrac{\partial}{\partial x}\right) u + \dfrac{\partial^2 u}{\partial t^2} = f \text{ in } Q, \\ B_j u = g_j.\ j = 0, \ldots, m-1, \text{ on } \Sigma \\ u(x, 0) = u_0(x), \ \dfrac{\partial u(x, 0)}{\partial t} = u_1(x) \text{ in } \Omega, \end{cases}$$

where

$$Au = \sum_{|p|,|q| \leq m} (-1)^{|p|} D_x^p (a_{pq}(x, t) D_x^q u) \tag{1.3}$$

with

$$a_{pq} \in \mathscr{D}(\overline{Q}), \tag{1.4}$$

$$A \textit{ symmetric} \text{ i.e. } A = A^* \text{ or } a_{pq} = \bar{a}_{qp}, \tag{1.5}$$

and where

$$B_j u = \sum_{|h| \leq mj} b_{jh}(x, t) D_x^h, \tag{1.6}$$

with

$$b_{jh} \in \mathscr{D}(\overline{\Sigma}), \tag{1.7}$$

$$\begin{cases} 0 \leq m_j < 2m \textit{ and the system } \{B_j\}, \textit{ for every } t \in [0 \quad T], \\ \textit{is normal on } \Gamma. \end{cases} \tag{1.8}$$

Furthermore we assume that

$$\begin{cases} \textit{the boundary conditions } B_j u = g_j, \ j = 0, \ldots \ m-1, \\ \textit{correspond, according to } \text{Chapter 2, Section 9.4}, \textit{ to a form} \\ a(t; u, v) = \sum_{|p|,|q| \leq m} \int_\Omega a_{pq}(x, t) D_x^q u \overline{D_x^p v} \, dx \\ \textit{and a closed vector subspace } V \textit{ of } H^m(\Omega), \textit{ with} \\ H_0^m(\Omega) \subset V \subset H^m(\Omega), \\ \textit{the form } a(t; u, v) \textit{ being } V\textit{-coercive, uniformly} \\ \textit{with respect to } t, \text{ i.e. } \textit{there exists } \alpha > 0 \textit{ independent of } t \textit{ such that} \\ a(t; v, v) \geq \alpha \|v\|^2_{H^m(\Omega)}, \forall v \in V, \forall t \in [0, T]. \end{cases} \tag{1.9}$$

We are given f, g_j, u_0, u_1 with

$$f \in \mathscr{D}(\overline{Q}), \ g_j \in \mathscr{D}(\overline{\Sigma}), \ u_0 \text{ and } u_1 \in \mathscr{D}(\overline{\Omega}) \tag{1.10}$$

and with the *compatibility relations* (see Whitney [1], Malgrange [1] implying the existence of $w \in \mathscr{D}(\overline{Q})$ with

$$\begin{cases} B_j w = g_j \text{ on } \Sigma, \ j = 0, \ldots, m-1, \\ w(x, 0) = u_0(x), \ \dfrac{\partial w}{\partial t}(x, 0) = u_1(x) \text{ on } \Omega, \\ D_t^k P w(x, 0) = D_t^k f(x, 0) \text{ on } \Omega, \ k = 0, 1, \ldots. \end{cases} \tag{1.11}$$

We seek u in $\mathscr{D}(\overline{Q})$, solution of (1.2).

To this end, it suffices to solve the problem

$$
(1.12) \qquad \begin{cases} Pv = f - Pw \text{ in } Q, \\ B_j v = 0 \text{ on } \Sigma, j = 0, \ldots, m-1, \\ v(x, 0) = \dfrac{\partial v}{\partial t}(x, 0) = 0 \text{ in } \Omega; \end{cases}
$$

using Theorem 7.1 of Chapter 5, we obtain $v \in \mathscr{D}(\overline{Q})$, solution of (1.12), and then $u = v + w$ solves (1.2) and $u \in \mathscr{D}(\overline{Q})$. Furthermore u is unique, thanks to the results of Chapter 3 (Theorem 8.1 applied to the present situation).

Therefore, we have

Theorem 1.1. *Under hypotheses* (1.1), (1.2), ..., (1.11), *there exists a unique solution of* (1.2) *in the space* $\mathscr{D}(\overline{Q})$. ☐

Remark 1.1. As we have already said in Chapter 5 (see Remark 1.2), instead of (1.5) and the V-coerciveness of $a(t; u, v)$, it would be sufficient that *the principal part of A be symmetric and that there exist $\alpha > 0$ and λ such that*:

$$
\textit{principal part of } a(t; v, v) \geq \alpha \|v\|^2_{H^m(\Omega)} - \lambda \|v\|^2_{L^2(\Omega)}, \quad \forall v \in V, t \in [0, T].
$$

1.2 The Regularity in Gevrey Spaces

Let us now study the regularity of the solution of problem (1.2) when the data $(\Omega, f, g_j, u_0, u, a_{pq}, b_{jh})$ are in Gevrey classes. Here, the situation is more complicated than for the parabolic problems studied in Chapter 10, for the operator P is not hypoelliptic and therefore there is no Gevrey class "associated" to the operator; in particular we do not have *local* regularity results of the type of Theorems 1.2 and 2.3 of Chapter 10. Nevertheless, it is possible for the boundary value problem (1.2) to prove *global* regularity results in certain Gevrey spaces, using the regularity in t studied in the abstract case in Chapter 9, Section 2 and the Theorem on "elliptic iterates" of Chapter 8, according to an idea previously developed for the parabolic case in Remarks 2.2 and 2.3 of Chapter 10.

We seek regularity results in the spaces $\mathscr{D}_{ms,s}(\overline{Q})$, with $s \geq 1$ (for a generalization, see Remark 1.4 below). Therefore, we shall first make the hypothesis:

$$
(1.13) \qquad \Omega \textit{ is of class } \{M_k\} \textit{ with } M_k = (k!)^s, \textit{ real } s \geq 1.
$$

We shall also assume that the coefficients of the operators A and B_j, given by (1.3) and (1.6), are independent of t and of Gevrey class of order s, i.e.

$$a_{pq}(x, t) = a_{pq}(x) \in \mathscr{D}_s(\overline{\Omega}), \tag{1.14}$$

$$b_{jh}(x, t) = b_{jh}(x) \in \mathscr{D}_s(\Gamma). \tag{1.15}$$

Then we have

Theorem 1.2. *Let P and B_j, $j = 0, \ldots, m-1$, be given by* (1.2), (1.3), (1.6) *with* (1.13), (1.14), (1.15), (1.5), (1.8), (1.9) *and*

$$s \geq 1, \; sm > 1; \tag{1.16}$$

let f, g_j, u_0, u_1, $j = 0, \ldots, m-1$, be given with

$$f \in \mathscr{D}_{ms,s}(\overline{Q}), \quad g_j \in \mathscr{D}_{ms,s}(\overline{\Sigma}), \quad u_0 \in \mathscr{D}_s(\overline{\Omega}), \quad u_1 \in \mathscr{D}_s(\overline{\Omega}), \tag{1.17}$$

and the compatibility relations which guarantee the existence of w such that

$$\begin{cases} w \in \mathscr{D}_{ms,s}(\overline{Q}), \\ w(x, 0) = u_0(x) \text{ and } \dfrac{\partial w}{\partial t}(x, 0) = u_1(x) \text{ on } \Omega, \\ B_j w = g_j, \; j = 0, \ldots, m-1, \text{ on } \Sigma, \\ \mathrm{D}_t^k P w(x, 0) = \mathrm{D}_t^k f(x, 0) \quad \text{on } \Omega, \; \forall k. \end{cases} \tag{1.18}$$

Then there exists one and only one solution of problem (1.2) *belonging to the space $\mathscr{D}_{ms,s}(\overline{Q})$.*

Proof. As for Theorem 1.1, we reduce the problem to (1.12). Then it is possible, noting that $\mathrm{D}_t^k(f - Pw)|_{t=0} = 0$, $\forall k$, and by an obvious extension into the cylinder $\Omega \times]-\infty, +\infty[$, to apply Theorem 2.2 of Chapter 9; therefore the solution v of (1.12) belongs to $\mathscr{D}_{sm}([0, T]; L^2(\Omega))$ (and more precisely, the extension of v belongs to $\mathscr{D}_{+;sm}(\mathbf{R}; L^2(\Omega))$).

Differentiating the equation $Pv = \varphi$ ($\varphi = f - Pw$) with respect to t and applying the operator A, it follows that

$$A^i v = (-1)^i \mathrm{D}_t^{2i} v + \sum_{h=0}^{i-1} (-1)^{i+h+1} A^h(\mathrm{D}_t^{2(i-h-1)} \varphi), \quad i = 1, 2, \ldots \tag{1.19}$$

and

$$B_j(A^i v) = \sum_{h=0}^{i-1} (-1)^{i+h+1} B_j\big(A^h(\mathrm{D}_t^{2(i-h-1)} \varphi)\big), \quad i = 1, 2, \ldots; \; j = 0, \ldots, m-1. \tag{1.20}$$

Thanks to the fact that $v \in \mathscr{D}_{sm}([0, T]; L^2(\Omega))$ and that $\varphi \in \mathscr{D}_{sm,s}(\overline{Q})$, it follows that there exist c_0 and L_0 such that, for $0 \leq t \leq T$,

$$\|A^i v(x\ t)\|_{L^2(\Omega)} \leq c_0 L_0^i ((2mi)!)^s, \quad i = 0, 1, 2, \ldots, \tag{1.21}$$

$$\sum_{j=0}^{m-1} \| B_j(A^i v)\|_{H^{2m+2mk-m_j-1/2}(\Gamma)} \leq c_0 L_0^{k+i+1}((2m(i+k+1))!)^s, \quad i, k = 0, 1, \ldots \tag{1.22}$$

Therefore, applying the theorem on *elliptic iterates* (Theorem 1.2, Chapter 8), we obtain that $v(x, t)$ belongs, for each $t \in [0, T]$, to a bounded set of $\mathscr{D}_s(\overline{\Omega})$.

By the same reasoning, we also find that $D_t v(x, t)$ belongs, for each $t \in [0, T]$, to a bounded set of $\mathscr{D}_s(\overline{\Omega})$; indeed we only need to note that $D_t v = z$ is a solution of the problem:

$$\begin{cases} Pz = D_t \varphi \text{ in } Q, \\ B_j z = 0 \text{ on } \Sigma,\ j = 0, \ldots, m-1, \\ z(x, 0) = D_t z(x, 0) = 0 \text{ on } \Omega \end{cases}$$

and that $D_t\varphi$ again belongs to $\mathscr{D}_{sm}([0, T]; L^2(\Omega))$, with $D_t^k(D_t\varphi)|_{t=0} = 0$, $\forall k$.

Then, from (1.19) and the analogous equality for $D_t v$:

$$A^i(D_t v) = (-1)^i D_t^{2i+1} v + \sum_{h=0}^{i-1} (-1)^{i+h+1} A^h(D_t^{2(i-h-1)+1}\varphi), \quad i = 0,1, \ldots,$$

we deduce the existence of c_* and L_* such that

$$|D_x^p D_t^l D v(x, t)| \leq c_* L_*^{|p|+l} (p!)^s (l!)^{ms}, \quad (x, t) \in \overline{Q}$$

and therefore

$$v \in \mathscr{D}_{sm,s}(\overline{Q}). \quad \square$$

Remark 1.1. Condition (1.16) *implies that for the hyperbolic case* ($m = 1$) *we must have* $s > 1$ *and therefore we do not prove the analyticity in* x. □

Remark 1.2. As we have already stated in Remark 2.1 of Chapter 10 for the parabolic case, we do not know whether the "natural" explicit compatibility conditions on f, g_j, u_0, u_1 are sufficient for the validity of (1.18). □

Remark 1.3. The hypothesis that the operators A and B_j do *not depend on* t is used only in point (1.19) and (1.20). It would be of interest to generalize the result to the case of time-dependent coefficients, and more precisely to the case

(1.23) $$a_{pq} \in \mathscr{D}_{sm,s}(\overline{Q}), \quad b_{jh} \in \mathscr{D}_{sm,s}(\overline{\Sigma}). \ \square$$

Remark 1.4. The method used for Theorem 1.2 can be generalized to spaces $\mathscr{D}_{M_k,N_h}(\overline{Q})$ in the same way as for the parabolic case; Remark 2.3 of Chapter 10. Instead of the condition (2.26) of Chapter 10, for the sequences $\{N_h\}$ and $\{M_k\}$, we have the relation

$$N_{2h} \leq M_{2hm}, \quad \forall h,$$

therefore we can take $N_h = (h!)^s$, $M_k = (k!)^r$ with $r \geq 1$ and $1 < s \leq mr$ (and therefore $u \in \mathscr{D}_{s,r}(\overline{Q})$). $\square$

2. Equations of the Second Order in t; Application of Transposition and Existence of Solutions in Spaces of Distributions

2.1 Generalities

The notation and the hypotheses on Ω, P and $\{B_j\}_{j=0}^{m-1}$ are as in Section 1.1 (hypotheses (1.1), ..., (1.9)).

We recall Green's formula (see Chapter 5, Section 1.1):

(2.1) $$\int_Q (Pu)\,\overline{v}\, dx\, dt - \int_Q u\overline{P^*v}\, dx\, dt = \sum_{j=0}^{m-1} \int_\Sigma S_j u \overline{C_j v}\, d\sigma -$$

$$- \sum_{j=0}^{m-1} \int_\Sigma B_j u \overline{T_j v}\, d\sigma + \int_\Omega \frac{\partial u(x,T)}{\partial t}\, \overline{v(x,T)}\, dx - \int_\Omega \frac{\partial u(x,0)}{\partial t}\, \overline{v(x,0)}\, dx -$$

$$- \int_\Omega u(x,T)\, \overline{\frac{\partial v(x,T)}{\partial t}}\, dx + \int_\Omega u(x,0)\, \overline{\frac{\partial v(x,0)}{\partial t}}\, dx,$$

where P^* is the *formal adjoint* of P and therefore, under hypothesis (1.5), coincides with P (but we shall keep the notation P^*, because, as we have already stated in Remark 1.1, we could, for the sequel, assume that only the *principal part* of A is symmetric).

We propose to study problem (1.2), i.e.

(2.2) $$Pu = f \text{ in } Q,$$

(2.3) $$B_j u = g_j \text{ on } \Sigma,\ j = 0, \ldots, m-1,$$

(2.4) $$u(x,0) = u_0(x), \quad \frac{\partial u}{\partial t}(x,0) = u_1(x) \text{ on } \Omega$$

in certain spaces of distributions on Q.

Again, the starting point is the *adjoint problem*:

$$(2.5)\qquad \begin{cases} P^*v = \varphi \text{ in } Q, \\ C_j v = 0 \text{ on } \Sigma, j = 0, \ldots, m-1, \\ v(x, T) = \dfrac{\partial v(x, T)}{\partial t} = 0 \text{ on } \Omega. \end{cases}$$

The "most general" situation for problem (2.2), (2.3), (2.4) (i.e. to find u in $\mathscr{D}'(Q)$), would lead us to study problem (2.5) for

$$(2.6)\qquad \varphi \in \mathscr{D}(Q).$$

Thanks to Theorem 1.1, we would obtain the existence and uniqueness of the solution v of (2.5) belonging to the space

$$(2.7)\qquad \tilde{X} = \{v \mid v \in \mathscr{D}(\overline{Q}),\, v(x, T) = \frac{\partial v(x, T)}{\partial t} = 0,\, C_j v = 0,\quad j = 0, \ldots, m-1,\, P^*v \in \mathscr{D}(Q)\}$$

and we would then have to study the space described by $\mathscr{C}v$ as v describes $\tilde{X}$, where

$$(2.8)\qquad \mathscr{C}v = \left\{ v(x, 0), \frac{\partial v(x, 0)}{\partial t}; T_0 v, \ldots, T_{m-1} v \right\}$$

(see for example Section 3.2 of Chapter 10). But here we meet great technical difficulties which lead to the elimination of the case (2.6): for example (compare with (3.25) of Chapter 10) we would have to study the Cauchy problem:

$$(2.9)\qquad \begin{cases} Pv = 0 \text{ in a neighborhood of } \Sigma, \\ C_j v = 0,\; T_j v = \varphi_j \text{ on } \Sigma,\; j = 0, \ldots, m-1, \end{cases}$$

where φ_j is infinitely differentiable on Σ; but P^* is not hyperbolic with respect to Σ (except for the very particular case corresponding to the wave equation in *one* space variable); and therefore (see Hörmander [1], page 130, Courant-Hilbert [1], page 759) the space of φ_j's (i.e. the space described by $\{T_0 v, \ldots, T_{m-1} v\}$ as v describes $\tilde{X}$) depends on the operator P^* is an essential way. Furthermore, there are difficulties due to the "linking" of $v(x, 0), \dfrac{\partial v}{\partial t}(x, 0)$ and $T_j v|_{t=0}$ on the variety Γ.

Therefore, in the following Sections, we shall seek another possibility, which will lead to the study of problem (2.2), (2.3), (2.4) in a still very general *subspace* of $\mathscr{D}'(Q)$.

2.2 The Space $\mathscr{D}_{-,\gamma}([0, T]; \mathscr{D}_\gamma(\bar{\Omega}))$ and its Dual

The space $\mathscr{D}_\gamma(\bar{\Omega})$.

We define

$$\mathscr{D}_\gamma(\bar{\Omega}) = \{\varphi \mid \varphi \in \mathscr{D}(\bar{\Omega}),\ \gamma_j\varphi = 0 \text{ on } \Gamma,\ \forall j = 0, 1, \ldots\}^{((1))}. \tag{2.10}$$

This is a closed subspace of $\mathscr{D}(\bar{\Omega})$. If we note that, thanks to Corollary 9.2 of Chapter 1, $\mathscr{D}(\bar{\Omega})$ can be considered as a Frechet space with respect to the family of semi-norms:

$$\|D^p\varphi\|_{L^2(\Omega)},\ \forall p, \tag{2.11}$$

then we easily see that

$$\begin{cases} \mathscr{D}_\gamma(\bar{\Omega}), \text{ provided with the family of semi-norms (2.11),} \\ \text{is a Frechet space.} \end{cases} \tag{2.12}$$

Also note that

$$\mathscr{D}_\gamma(\bar{\Omega}) = \textit{closure of } \mathscr{D}(\Omega) \textit{ in } \mathscr{D}(\bar{\Omega}). \quad \square \tag{2.13}$$

The space $\mathscr{D}'_\gamma(\bar{\Omega})$.

We denote by $\mathscr{D}'_\gamma(\bar{\Omega})$, the (strong) dual of $\mathscr{D}_\gamma(\bar{\Omega})$. Thanks to (2.13),

$$\mathscr{D}'_\gamma(\bar{\Omega}) \textit{ is a space of distributions on } \Omega. \tag{2.14}$$

By applying the Hahn-Banach theorem, we also obtain a representation fo the elements f of $\mathscr{D}'_\gamma(\bar{\Omega})$:

$$\begin{cases} \textit{every element } f \textit{ of } \mathscr{D}'_\gamma(\bar{\Omega}) \textit{ can be written in the form} \\ \qquad f = \sum_{\text{finite}} D^p f_p,\ f_p \in L^2(\Omega). \end{cases} \tag{2.15}$$

As in Proposition 3.5 of Chapter 8, it follows that

$$\textit{the space } \mathscr{D}_\gamma(\bar{\Omega}) \textit{ is reflexive.} \tag{2.16}$$

Finally, comparing $\mathscr{D}_\gamma(\bar{\Omega})$ with the space $\Xi(\Omega)$ introduced in Section 3.2 of Chapter 8, we have the following inclusions:

$$\mathscr{D}(\Omega) \subset \mathscr{D}_\gamma(\bar{\Omega}) \subset \mathscr{D}(\bar{\Omega}) \subset \Xi(\Omega) \tag{2.17}$$

and

$$\Xi'(\Omega) \subset \mathscr{D}'_\gamma(\bar{\Omega}) \subset \mathscr{D}'(\Omega). \quad \square \tag{2.18}$$

((1)) These spaces should not be confused with the spaces of Gevrey functions.

The space $\mathscr{D}_{-,\gamma}([0, T]; \mathscr{D}_\gamma(\bar{\Omega}))$.

First, for fixed a such that $0 < a < T$, we define

$$\mathscr{D}^a_{-,\gamma}([0, T]; \mathscr{D}_\gamma(\bar{\Omega})) = \{\varphi \mid \varphi \in \mathscr{D}([0, T]; \mathscr{D}_\gamma(\bar{\Omega})),\ \varphi^{(j)}(0) = 0,\ \forall j \text{ and } \varphi(t) = 0 \text{ for } t \geq T - a\}, \tag{2.19}$$

where $\mathscr{D}([0, T]; \mathscr{D}_\gamma(\bar{\Omega}))$ is the space of infinitely differentiable functions $t \to \varphi(t)$ on $[0, T]$, with values in $\mathscr{D}_\gamma(\bar{\Omega})$, provided wth the natural Frechet space topology. The space $\mathscr{D}^a_{-,\gamma}([0, T]; \mathscr{D}_\gamma(\bar{\Omega}))$ provided with the topology of $\mathscr{D}([0, T]; \mathscr{D}_\gamma(\bar{\Omega}))$ is a Frechet space.

Then we define

$$\mathscr{D}_{-,\gamma}([0, T]; \mathscr{D}_\gamma(\bar{\Omega})) = \operatorname{ind}\lim_{a \to 0} \mathscr{D}^a_{-,\gamma}([0, T]; \mathscr{D}_\gamma(\bar{\Omega})). \tag{2.20}$$

We thus obtain a strict $(\mathscr{L}\mathscr{F})$-space; and we easily see that

$$\mathscr{D}(Q) \text{ is dense in } \mathscr{D}_{-,\gamma}([0, T]; \mathscr{D}_\gamma(\bar{\Omega})). \tag{2.21}$$

The space $\mathscr{D}'_{+,\gamma}([0, T]; \mathscr{D}'_\gamma(\bar{\Omega}))$.

By definition

$$\mathscr{D}'_{+,\gamma}([0, T]; \mathscr{D}'_\gamma(\bar{\Omega})) \text{ is the dual of } \mathscr{D}_{-,\gamma}([0, T]; \mathscr{D}_\gamma(\bar{\Omega})). \tag{2.22}$$

In order to avoid any topological difficulties, we provide $\mathscr{D}'_{+,\gamma}([0, T]; \mathscr{D}'_\gamma(\bar{\Omega}))$ with the *weak dual* topology.

Note that, thanks to (2.21), $\mathscr{D}'_{+,\gamma}([0, T]; \mathscr{D}'_\gamma(\bar{\Omega}))$ can be identified *algebraically* with a subspace of distributions on Q. ▯

2.3 The Spaces X and Y

Let us now come back to the *adjoint problem* (2.5) and take, instead of (2.6),

$$\varphi \in \mathscr{D}_{-,\gamma}([0, T]; \mathscr{D}_\gamma(\bar{\Omega})). \tag{2.23}$$

Thanks to Theorem 1.1 (applied to the adjoint problem), we see that there exists a unique solution of (2.5) belonging to $\mathscr{D}(\bar{\Omega})$ for every

$$\varphi \in \mathscr{D}_{-,\gamma}([0, T]; \mathscr{D}_\gamma(\bar{\Omega})).$$

Let us introduce the space

$$X = \{v \mid v \in \mathscr{D}(\bar{Q}),\ v(x, T) = \frac{\partial v(x, T)}{\partial t} = 0,\ C_j v = 0,\ j = 0, \dots, m-1,\ P^* v \in \mathscr{D}_{-,\gamma}([0, T]; \mathscr{D}_\gamma(\bar{\Omega}))\}. \tag{2.24}$$

The operator P^* defines a one-to-one mapping of X onto $\mathscr{D}_{-,\gamma}([0, T]; \mathscr{D}_\gamma(\bar{\Omega}))$; therefore, we can define on X the image topology of $\mathscr{D}_{-,\gamma}([0, T]; \mathscr{D}_\gamma(\bar{\Omega}))$ under $(P^*)^{-1}$; and we have:

$$(2.25)\quad \begin{cases} P^* \textit{ is an (algebraic and topological) isomorphism} \\ \textit{of } X \textit{ onto } \mathscr{D}_{-,\gamma}\big([0, T]; \mathscr{D}_\gamma(\bar{\Omega})\big). \end{cases} \square$$

Note that, as $\mathscr{D}_{-,\gamma}([0, T]; \mathscr{D}_\gamma(\bar{\Omega}))$, X is a strict $(\mathscr{LF})$-space; more precisely

$$X = \operatorname{ind}\lim_{a \to 0} X^a,$$

where

$$X^a = \Big\{ v \mid v \in \mathscr{D}(\bar{Q}),\ v(x, T) = \frac{\partial v(x, T)}{\partial t} = 0,\ C_j v = 0,$$

$$j = 0, \ldots, m-1,\ P^* v \in \mathscr{D}^a_{-,\gamma}\big([0, T]; \mathscr{D}_\gamma(\bar{\Omega})\big) \Big\},$$

provided with the image topology of $\mathscr{D}^a_{-,\gamma}([0, T]; \mathscr{D}_\gamma(\bar{\Omega}))$ under $(P^*)^{-1}$. $\square$

By transposition of (2.25), we obtain

$$(2.26)\quad \begin{cases} \textit{for every continuous antilinear form } v \to L(v) \textit{ on } X, \textit{ there exists} \\ \textit{a unique } u \textit{ in } \mathscr{D}'_{+,\gamma}\big([0, T]; \mathscr{D}'_\gamma(\bar{\Omega})\big) \textit{ such that} \\ \qquad \langle u, \overline{P^* v} \rangle = L(v),\ \forall v \in X. \end{cases}$$

Formally, we choose $L(v)$ in the form

$$(2.27)\quad L(v) = \langle f, \bar{v} \rangle + \langle u_1, \overline{v(x, 0)} \rangle - \langle u_0, \overline{\frac{\partial v(x, 0)}{\partial t}} \rangle + \sum_{j=0}^{m} \langle g_j, \overline{T_j v} \rangle$$

the brackets being taken in the sense of spaces which we shall specify. $\square$

For the choice of f, we can consider a topological vector space K such that:

$$(2.28)\quad \begin{cases} X \subset K \subset L^2(Q), \quad \textit{with continuous injections,} \\ \mathscr{D}_{-,\gamma}\big([0, T]; \mathscr{D}_\gamma(\bar{\Omega})\big) \textit{ is dense in } K. \end{cases}$$

Such spaces exist; for example $K = L^2(Q)$. In order to fix our ideas and in analogy with the choices made in this text for similar situations, we shall take for K the spaces

$$(2.29)\quad \Xi_-\big(0, T; \Xi(\Omega)\big),$$

defined in the following way. Let $t \to \mathrm{d}(t)$ be an infinitely differentiable function on $[0, T]$ such that

$$(2.30)\quad \begin{cases} \mathrm{d}(t) = 1 \text{ in the neighborhood of } 0, \\ \mathrm{d}(t) > 0 \text{ for } t > 0. \end{cases}$$

For integer $k > 0$ and real a with $0 < a < T$, we define the (Hilbert) space

$$\Xi(k, a) = \{v \mid d^j v^{(j)} \in L^2(0, T; \Xi^{k-j}(\Omega)),\ 0 \leq j \leq k, \quad v(t) = 0 \text{ for } t \geq T - a\}, \tag{2.31}$$

provided with the norm

$$\left(\sum_{j=0}^{k} \| d^j v^{(j)} \|^2_{L^2(0,T;\Xi^{k-j}(\Omega))} \right)^{1/2} \tag{2.32}$$

(see Chapter 2, Section 6.3, for the definition of $\Xi^{k-j}(\Omega)$).

Next, let

$$\Xi(a) = \bigcap_{k=0}^{\infty} \Xi(k, a), \tag{2.33}$$

a Frechet space for the family of norms (2.32), $k = 0, 1, \ldots$ Finally, let

$$\Xi_-(0, T; \Xi(\Omega)) = \operatorname*{ind\,lim}_{a \to 0} \Xi(a). \tag{2.34}$$

Since the functions v of X vanish in the neighborhood of $t = T$, we see that

$$X \subset \Xi_-(0, T; \Xi(\Omega)) \subset L^2(Q).$$

And, by the same type of proof as for Proposition 9.1 of Chapter 4 and Proposition 3.3 of Chapter 8, we see that $\mathscr{D}(Q)$ (and therefore also $\mathscr{D}_{-,\gamma}([0, T]; \mathscr{D}_\gamma(\bar{\Omega}))$ is dense in $\Xi_-(0, T; \Xi(\Omega))$.

Therefore $\Xi_-(0, T; \Xi(\Omega))$ satisfies (2.28). ☐

Now we denote by

$$\Xi'_+(0, T; \Xi'(\Omega)) \text{ the dual of } \Xi_-(0, T; \Xi(\Omega)), \tag{2.35}$$

provided with the *weak dual* topology. ☐

Having set this, let us introduce the space

$$Y = \{u \mid u \in \mathscr{D}'_{+,\gamma}([0, T]; \mathscr{D}'_\gamma(\bar{\Omega})), Pu \in \Xi'_+(0, T; \Xi'(\Omega))\}, \tag{2.36}$$

provided with the coarsest locally convex topology which makes the mappings $u \to u$ and $u \to Pu$ of Y into $\mathscr{D}'_{+,\gamma}([0, T]; \mathscr{D}'_\gamma(\bar{\Omega}))$ (weak) and $\Xi'_+(0, T; \Xi'(\Omega))$ (weak) respectively, continuous.

First, we have

Theorem 2.1. *The space $\mathscr{D}(\bar{Q})$ is dense in Y.*

Proof. Let $u \to M(u)$ be a continuous antilinear form on Y; all the spaces being provided with the weak dual topologies, it may be written

$$M(u) = \langle f, \bar{u} \rangle + \langle g, \overline{Pu} \rangle, \tag{2.37}$$

with $f \in \mathscr{D}_{-,\gamma}([0, T]; \mathscr{D}_\gamma(\bar{\Omega}))$ and $g \in \Xi_-(0, T; \Xi(\Omega))$.

Assume that we have

$$M(\varphi) = 0, \forall \varphi \in \mathscr{D}(\overline{Q}). \tag{2.38}$$

Introduce the open set $\mathcal{O} = \mathbf{R}^n \times]-\infty, T[$ and denote by $\tilde{f}, \tilde{g}$ the extensions of f and g to $\mathcal{O}$ by zero outside Q and by $\tilde{P}$ the extension of P to $\mathcal{O}$ having "the same properties" as P (actually, it is sufficient to extend A into $\mathbf{R}^n$ in such a way that the extension $\tilde{A}$ has infinitely differentiable coefficients in $\mathbf{R}^n$ and that the operator $\tilde{P}^* = \tilde{A}^* + \partial^2/\partial t^2$ satisfies the conditions for the "regularity" of the Cauchy problem in the sense of (2.39) and (2.40) below; this is always possible).

Then, by (2.38), we have

$$\langle \tilde{f}, \overline{\Phi} \rangle + \langle \tilde{g}, \overline{\tilde{P}\Phi} \rangle = 0, \forall \Phi \in \mathscr{D}(\mathcal{O})$$

and therefore

$$\tilde{P}^* \tilde{g} + \tilde{f} = 0 \text{ in } \mathcal{O}, \tag{2.39}$$

with the "initial" conditions for $t = T$:

$$\tilde{g}(x, T) = 0, \ \frac{\partial \tilde{g}}{\partial t}(x, T) = 0, \ \forall x \in \mathbf{R}^n. \tag{2.40}$$

Then, according to the regularity of the solution of the Cauchy problem (see Hörmander [1]), we have

$$\tilde{g} \text{ is infinitely differentiable in } \mathcal{O}. \tag{2.41}$$

But since $\tilde{g}$ has support in $\overline{Q}$, (2.41) implies

$$\gamma_j \tilde{g} = 0 \text{ on } \Sigma \text{ for every } j,$$

$$\frac{\partial^k \tilde{g}(x, 0)}{\partial t^k} = 0, \ x \in \Omega, \text{ for every } k.$$

Therefore

$$g \in \mathscr{D}_{-,\gamma}([0, T]; \mathscr{D}_\gamma(\overline{\Omega}))$$

and consequently, for $u \in Y$, we have

$$\langle g, \overline{Pu} \rangle = \langle P^* g, \bar{u} \rangle,$$

so that

$$M(u) = \langle f + P^* g, \bar{u} \rangle$$

and therefore, according to (2.39), $M(u) = 0$, $\forall u \in Y$. ☐

2.4 Study of the Operator $\mathscr{C}$

We now have to choose the "boundary data" in (2.27); and, as always, the essential point is the study of the space described by (2.8), i.e. by

$$\mathscr{C}v = \left\{v(x, 0), \frac{\partial v(x, 0)}{\partial t}; T_0 v, \ldots, T_{m-1} v\right\}$$

as v describes X.

If $v \in X$, then v vanishes in a neighborhood of $t = T$, therefore

$$\mathscr{C}v \in \mathscr{D}(\bar{\Omega}) \times \mathscr{D}(\bar{\Omega}) \times [\mathscr{D}_-([0, T]; \mathscr{D}(\Gamma))]^m, \tag{2.42}$$

where $\mathscr{D}_-([0, T]; \mathscr{D}(\Gamma))$ is the space of infinitely differentiable functions on $[0, T]$ with values in $\mathscr{D}(\Gamma)$ and vanishing in a neighborhood of T, provided with the usual strict $(\mathscr{L}\mathscr{F})$-space topology:

$$\mathscr{D}_-([0, T]; \mathscr{D}(\Gamma)) = \operatorname{ind}\lim_{a \to 0} \mathscr{D}_a([0, T]; \mathscr{D}(\Gamma)),$$

$\mathscr{D}_a([0, T]; \mathscr{D}(\Gamma))$ being the closed subspace of $\mathscr{D}([0, T]; \mathscr{D}(\Gamma))$ of functions vanishing for $T - a < t < T$, $a < T$.

Of course, the mapping $\mathscr{C}$ is not surjective from X onto

$$\mathscr{D}(\bar{\Omega}) \times \mathscr{D}(\bar{\Omega}) \times [\mathscr{D}_-([0, T]; \mathscr{D}(\Gamma))]^m,$$

for the images $v(x, 0)$, $\partial v(x, 0)/\partial t$, $T_0 v, \ldots, T_{m-1} v$ must satisfy the *compatibility relations* (which we denote by $\mathscr{R} \cdot \mathscr{C}$), i.e. linking conditions on the variety Γ, for $t = 0$; these are the *pointwise differential* conditions which can be "formally" specified by taking into account the fact that, for every $v \in X$, we must have

$$\begin{cases} v \in \mathscr{D}(\bar{Q}),\ C_j v = 0 \text{ on } \Sigma, & j = 0, \ldots, m-1, \\ \gamma_k(P^* v) = 0 \text{ on } \Sigma, \quad \forall k, & \left.\dfrac{\partial^k (P^* v)}{\partial t^k}\right|_{t=0} = 0 \text{ on } \Omega, \forall k. \end{cases} \tag{2.43}$$

For $0 < a < T$, we set

$$\mathscr{V}_a = \{(\chi_0, \chi_1; \varphi_0, \ldots, \varphi_{m-1}) / (\chi_0, \chi_1; \varphi_0, \ldots, \varphi_{m-1}) \in \mathscr{D}(\bar{\Omega}) \times \times \mathscr{D}(\bar{\Omega}) \times [\mathscr{D}_a([0, T]; \mathscr{D}(\Gamma))]^m, \text{ satisfying the } \mathscr{R} \cdot \mathscr{C}\}; \tag{2.44}$$

we provide $\mathscr{V}_a$ with the topology induced by the topology of

$$\mathscr{D}(\bar{\Omega}) \times \mathscr{D}(\bar{\Omega}) \times [\mathscr{D}_a[0, T]; \mathscr{D}(\Gamma)]^m;$$

and then

(2.45) $$\mathscr{V} = \operatorname*{ind\,lim}_{a \to 0} \mathscr{V}_a.$$

Thus $\mathscr{V}$ is a strict $(\mathscr{L}\mathscr{F})$-space. Thanks to the closed graph theorem, we have

(2.46) *$v \to \mathscr{C}v$ is a continuous linear mapping of X into $\mathscr{V}$.* □

We shall now show that $\mathscr{C}$ is surjective:

Theorem 2.2. *The mapping $v \to \mathscr{C}v$ is surjective from X onto $\mathscr{V}$.*

Proof. Let $\{\chi_0, \chi_1, ;\varphi_0 \ldots, \varphi_{m-1}\}$ be given in $\mathscr{V}$. We seek $v \in \mathscr{D}(\overline{Q})$ satisfying

(2.47) $$P^*v \in \mathscr{D}_{-,\gamma}([0, T]; \mathscr{D}_\gamma(\overline{\Omega})),$$

(2.48) $$C_j v = 0 \text{ on } \Sigma, j = 0, \ldots, m-1,$$

(2.49) $$T_j v = \varphi_j \text{ on } \Sigma, j = 0, \ldots, m-1,$$

(2.50) $$v(x, T) = 0, \frac{\partial v}{\partial t}(x, T) = 0, x \in \Omega,$$

(2.51) $$v(x, 0) = \chi_0(x), \frac{\partial v(x, 0)}{\partial t} = \chi_1(x), \quad x \in \Omega.$$

First, we note that, the system $\{C_j, T_j\}_{j=0}$ being of Dirichlet of order $2m$, the conditions (2.48), (2.49) are *equivalent* to (see Chapter 2, Lemma 2.1):

(2.52) $$\gamma_j v = \psi_j \text{ on } \Sigma, j = 0, \ldots, 2m-1, \psi_j \text{ given in } \mathscr{D}_-([0, T]; \mathscr{D}(\Gamma)).$$

Since the ψ_j's vanish for $T - a \leq t \leq T$, we shall take (also taking into account (2.50) and (2.47)) $v = 0$ for $T - a \leq t \leq T$.

Furthermore condition (2.47) decomposes as follows:

(2.53) $$\begin{cases} \gamma_k(P^*v) = 0 \text{ on } \Sigma, \forall k, \\ \left.\dfrac{\partial^k(P^*v)}{\partial t^k}\right|_{t=0} = 0 \text{ on } \Omega, \forall k. \end{cases}$$

Since $P^* = A^* + \partial^2/\partial t^2$ and since A^* is elliptic in $\overline{\Omega}$ (this is a consequence of hypothesis (1.9)) and we have (2.51) and (2.52), we easily see by induction (see Theorem 10.2, Chapter 4) that (2.53) is equivalent to

(2.54) $$\begin{cases} \gamma_j v = \psi_j \text{ on } \Sigma, j = 2m, 2m+1, \ldots, \psi_j \text{ determined} \\ \text{uniquely in } \mathscr{D}_-([0, T]; \mathscr{D}(\Gamma)) \text{ and zero for } t \geq T - a, \end{cases}$$

and

$$\frac{\partial^k v(x, 0)}{\partial t^k} = \chi_k, \ k = 2, 3, \ldots, \chi_k \text{ uniquely determined in } \mathscr{D}(\bar{\Omega}). \tag{2.55}$$

Furthermore, according to the $\mathscr{R}.\mathscr{C}$ (this can in fact be the *definition* of the $\mathscr{R}.\mathscr{C}$), we have

$$\gamma_j \chi_k = \left. \frac{\partial^k \psi_j}{\partial t^k} \right|_{t=0}, \quad \forall j, k. \tag{2.56}$$

Conversely, if $v \in \mathscr{D}(\bar{Q})$ satisfies (2.51), (2.52), (2.54), (2.55) with (2.56) and if furthermore

$$\frac{\partial^k v}{\partial t^k}(x, T - a) = 0, \ \forall k \text{ and } v(x, t) = 0, \ \text{for } t > T - a, \tag{2.57}$$

then v is the element we seek.

Now, by Whitney's theorem (see Malgrange [2]), such a v exists, and this proves the theorem. ☐

By passage to the quotient with respect to the kernel $\text{Ker}(\mathscr{T})$ of $\mathscr{T}$, which is a closed subspace of X, denoting by $\mathscr{T}^{\cdot}$ the *quotient mapping* and recalling that $X/\text{Ker}(\mathscr{T})$ is an $(\mathscr{L}\mathscr{F})$-space (see Grothendieck [1]) and therefore that the closed graph theorem applies, we have:

$$\begin{cases} \mathscr{T}^{\cdot} \textit{ is an algebraic and topological isomorphism} \\ \textit{of } X/\text{Ker}(\mathscr{T}) \textit{ onto } \mathscr{V}. \ \square \end{cases} \tag{2.58}$$

2.5 Trace and Existence Theorems in the Space Y

Let $\mathscr{V}'$ denote the dual of $\mathscr{V}$, provided with the *weak dual* topology. We are now ready to prove the trace theorem:

Theorem 2.3. *The mapping*

$$u \to \pi u = \left\{ \frac{\partial u(x, 0)}{\partial t}, -u(x, 0); B_0 u, \ldots, B_{m-1} u \right\} \tag{2.59}$$

of $\mathscr{D}(\bar{Q})$ into $\mathscr{D}(\bar{\Omega}) \times \mathscr{D}(\bar{\Omega}) \times [\mathscr{D}(\bar{\Sigma})]^m$ extends by continuity to continuous linear mapping, still denoted by $u = \pi u$, of Y into $\mathscr{V}'$; and we have Green's formula:

$$\langle u, \overline{P^* v} \rangle - \langle Pu, \bar{v} \rangle = \langle \pi u, \overline{\mathscr{T} v} \rangle, \ \forall u \in Y, \ v \in X, \tag{2.60}$$

the brackets denoting the duality between $\mathscr{D}'_{+,\gamma}([0, T]; \mathscr{D}'_\gamma(\bar{\Omega}))$ and $\mathscr{D}_{-,\gamma}([0, T]; \mathscr{D}_\gamma(\bar{\Omega}))$, $\Xi'_+(0, T; \Xi'(\Omega))$ and $\Xi_-(0, T; \Xi(\Omega))$, and $\mathscr{V}'$ and $\mathscr{V}$, respectively.

Proof. 1) As in Section 3.3 for the parabolic case, we first note that we can *identify* (which we shall do) an arbitrary element $(u_1, -u_0; g_0, \ldots, g_{m-1})$ of $\mathscr{D}(\overline{\Omega}) \times \mathscr{D}(\overline{\Omega}) \times [\mathscr{D}(\overline{\Sigma})]^m$ satisfying the compatibility relations (1.11) of Section 1.1, with an element of $\mathscr{V}'$ (by the same type of arguments as in Section 3.3, using Green's formula (2.1)).

2) Let u be given in Y. For every $\vec{\psi} = \{\chi_1, \chi_0; \varphi_0, \ldots, \varphi_{m-1}\} \in \mathscr{V}$, we set

$$v^{\boldsymbol{\cdot}} = (\mathscr{T}^{\boldsymbol{\cdot}})^{-1}\vec{\psi}, \quad v^{\boldsymbol{\cdot}} \in X/\mathrm{Ker}(\mathscr{T}), \tag{2.61}$$

and for arbitrary v in $v^{\boldsymbol{\cdot}}$ $(v \in X)$, we set

$$Z(\vec{\psi}) = \langle u, \overline{P^* v}\rangle - \langle Pu, \bar{v}\rangle, \tag{2.62}$$

the brackets denoting the duality between $\mathscr{D}'_{+,\gamma}([0, T]; \mathscr{D}'_\gamma(\overline{\Omega}))$ and $\mathscr{D}_{-,\gamma}([0, T]; \mathscr{D}_\gamma(\overline{\Omega}))$ and between $\Xi'_+(0, T; \Xi'(\Omega))$ and $\Xi_-(0, T; \Xi(\Omega))$, respectively.

Let us verify that $Z(\vec{\psi})$ depends only on $\vec{\psi}$; let w be another element of the class $v^{\boldsymbol{\cdot}}$; then $w \in X$ and satisfies

$$w(x, T) = \frac{\partial w(x, T)}{\partial t} = 0, \quad C_j w = 0, \quad T_j w = \varphi_j,$$

$$j = 0, \ldots, m-1, \; w(x, 0) = \chi_0(x), \frac{\partial w}{\partial t}(x, 0) = \chi_1(x)$$

and therefore we obtain

$$\langle u, \overline{P^*(v-w)}\rangle - \langle Pu, \overline{v-w}\rangle = 0 \tag{2.63}$$

(using Green's formula (2.1), we first prove (2.63) for $u \in \mathscr{D}(\overline{Q})$ and then by passage to the limit, since $\mathscr{D}(\overline{Q})$ is dense in Y).

3) The mapping $\vec{\psi} \to Z(\vec{\psi})$ is continuous antilinear on $\mathscr{V}$; indeed, if we set

$$\mathscr{L}(v) = \langle u, \overline{P^* v}\rangle - \langle Pu, \bar{v}\rangle,$$

we obtain a continuous form on X, vanishing on $\mathrm{Ker}(\mathscr{T})$ (according to 2)); and therefore the form

$$v^{\boldsymbol{\cdot}} \to \mathscr{L}^{\boldsymbol{\cdot}}(v^{\boldsymbol{\cdot}}) = \mathscr{L}(v) \quad (v \in v^{\boldsymbol{\cdot}})$$

is continuous on $X/\mathrm{Ker}(\mathscr{T})$. Now

$$Z(\vec{\psi}) = \mathscr{L}^{\boldsymbol{\cdot}}((\mathscr{T}^{\boldsymbol{\cdot}})^{-1}\vec{\psi}),$$

whence the result.

4) Therefore

$$Z(\vec{\psi}) = \langle \tilde{\pi} u, \overline{\vec{\psi}} \rangle, \quad \tilde{\pi} u \in \mathscr{V}'$$

and we have in this way defined a linear mapping of Y into $\mathscr{V}'$.

Let us show the continuity of π for the *weak* topologies; we need to show that, for every $\vec{\psi} \in \mathscr{V}$, $u \to \langle \tilde{\pi} u, \overline{\vec{\psi}} \rangle$ is continuous on Y. Now

$$\langle \tilde{\pi} u, \overline{\vec{\psi}} \rangle = Z(\vec{\psi}) = \langle u, \overline{P^* v} \rangle - \langle Pu, \bar{v} \rangle, \quad v \text{ fixed in } X,$$

whence the result.

5) Finally, applying Green's formula (2.1) and the *identification* made in 1), we see that, if $u \in \mathscr{D}(\overline{Q})$, we have

$$\tilde{\pi} u = \pi u .$$

From which we have the first part of the Theorem, according to the density Theorem.

6) Finally, Green's formula (2.60) follows from the above, for if $v \in X$, then $\mathscr{C} v \in \mathscr{V}$ and we can take precisely this function v in (2.62). ▯

Finally, we have the existence theorem:

Theorem 2.4. *Under hypotheses* (1.1), ..., (1.9), *the problem*:

$$(2.64) \quad \begin{cases} Pu = f \text{ in the sense of } \mathscr{D}'_{+,\gamma}([0, T]; \mathscr{D}'_\gamma(\overline{\Omega})) \text{ (and therefore also of } \mathscr{D}'(Q)), \\ \pi u = g^* \text{ in the sense of Theorem } 2.3, \end{cases}$$

admits a unique solution in Y for every $f \in \Xi'_+(0, T; \Xi'(\Omega))$ and $g^* \in \mathscr{V}'$, u depending continuously on f and g^*.

Proof. Given $f \in \Xi'_+(0, T; \Xi'(\Omega))$ and $g^* \in \mathscr{V}'$, take

$$L(v) = \langle f, \bar{v} \rangle + \langle g^*, \overline{\mathscr{C} v} \rangle$$

in (2.26).

It follows that the solution u of (2.26) satisfies

$$Pu = f \text{ in the sense of } \mathscr{D}'_{+,\gamma}([0, T]; \mathscr{D}'_\gamma(\overline{\Omega}))$$

and therefore $u \in Y$; but then, thanks to Green's formula (2.60) and Theorem 2.3, we have

$$\langle \pi u - g^*, \overline{\mathscr{C} v)} \rangle = 0, \quad \forall v \in X,$$

whence $\pi u = g^*$; and the Theorem follows. ▯

Remark 2.1. It is very likely that the results of Theorems 2.1, 2.3 and 2.4 remain valid if the spaces $\mathscr{D}'_{+,\gamma}([0, T]; \mathscr{D}'_\gamma(\bar{\Omega}))$, $\Xi_+(0, T; \Xi'(\Omega))$ and $\mathscr{V}'$ are provided with the *strong* dual topologies (for an analogous situation, see Lions-Magenes [2], N°. 4). ☐

Remark 2.2. Particular elements of $\mathscr{V}'$ can be constructed in the following way. Define the space $\Xi_-(0, T; \mathscr{D}(\Gamma))$ in analogous fashion to the space $\Xi_-(0, T; \Xi(\Omega))$: first introduce the Hilbert space

$$\tilde{\Xi}(k, a) = \{v \mid \mathrm{d}^j v^{(j)} \in L^2(0, T; H^{k-j}(\Gamma)),\ 0 \leq j \leq k,\ v(t) = 0 \quad \text{if } t \geq T - a\}$$

for integer $k > 0$ and $0 < a < T$. Next, set

$$\tilde{\Xi}(a) = \bigcap_{k=0}^{\infty} \tilde{\Xi}(k, a) \quad \text{(Frechet space)}$$

and finally define

$$\Xi_-(0, T; \mathscr{D}(\Gamma)) = \operatorname*{ind\,lim}_{a \to 0} \tilde{\Xi}(a).$$

We see that $\mathscr{D}(\Sigma)$ is dense in $\Xi_-(0, T; \mathscr{D}(\Gamma))$; therefore the dual of $\Xi_-(0, T; \mathscr{D}(\Gamma))$, which we denote by $\Xi'_+(0, T; \mathscr{D}'(\Gamma))$ and provide with the weak dual topology, is an algebraic subspace of $\mathscr{D}'(\Sigma)$.

If

$$g_j \in \Xi'_+(0, T; \mathscr{D}'(\Gamma)),\ u_0 \text{ and } u_1 \in \Xi'(\Omega),$$

the form

$$\langle u_1, \chi_0 \rangle - \langle u_0, \chi_1 \rangle + \sum_{j=0}^{m-1} \langle g_j, \varphi_j \rangle$$

is continuous linear on $\mathscr{V}$, according to (2.46) and the inclusions:

$$\mathscr{D}(\bar{\Omega}) \subset \Xi(\Omega), \quad \mathscr{D}_-([0, T]; \mathscr{D}(\Gamma)) \subset \Xi_-(0, T; \mathscr{D}(\Gamma)).$$

Remark 2.3. Results which are more general in the time variable and somewhat less general in the space variables can be obtained for this type of problem by application of Chapter 9, Section 10.6. ☐

2.6 Complements on the Trace Theorems

The (trace) Theorem 2.3 is of a type analogous to Theorem 3.4 of Chapter 10 for parabolic equations (see also Section 12.3 of Chapter 4 and the same question in Chapter 5): the boundary conditions, $B_j u = g_j$, $u(x, 0) = u_0(x)$, $\dfrac{\partial u}{\partial t}(x, 0) = u_1(x)$, are "contained" in the condition

"$\pi u = g^*$". Now the problem is to "separate", if possible, the operators $B_j u$ on Σ and $u(x, 0)$ and $\frac{\partial u}{\partial t}(x, 0)$ on Ω. For the parabolic case the matter is delicate (see Remark 3.2 of Chapter 10). In the present case, the situation is somewhat simpler[1]: it is rather similar to the one encountered in Volume 2, Chapters 4 and 5, in the case of the spaces $D_P^{-(r-1)}(Q)$ and $D_{A+D_t^2}^{-(2r-1)}(Q)$.

Indeed, we can give meaning *separately* to the operators $B_j u$, $u(x, 0)$ and $\frac{\partial u}{\partial t}(x, 0)$ in spaces which, in a certain sense, are probably *larger than the "optimal" spaces*. Let us first try to define $u(x, 0)$ and $\frac{\partial u}{\partial t}(x, 0)$ for $u \in Y$. To this end, we introduce the space

$$X_1 = \{v \mid v \in X,\ T_j v = 0,\ j = 0, \ldots, m-1\}, \tag{2.65}$$

which is a closed subspace of X.

The mapping

$$v \to \left\{v(x, 0), \frac{\partial v}{\partial t}(x, 0)\right\} \tag{2.66}$$

then is a continuous, linear and surjective mapping of X_1 onto $\mathcal{D}_\gamma(\bar{\Omega}) \times \mathcal{D}_\gamma(\bar{\Omega})$.

In order to prove this, it is sufficient to go through the proofs of Section 2.4 again, adapted to the new situation (the mapping $\mathcal{T}$ is replaced by (2.66) and X by X_1): in particular we note that in (2.52) and (2.54) we now have $\psi_j = 0$ for every j and therefore the $\mathcal{R}.\mathcal{C}.$ (2.56) reduce precisely to

$$\gamma_j \chi_k = 0, \ \forall j, k.$$

By a proof analogous to the one for Theorem 2.3, still replacing $\mathcal{T}$ with (2.66) and X with X_1, we obtain:

$$\begin{cases} \textit{the mapping } u \to \left\{\frac{\partial u}{\partial t}(x, 0), -u(x, 0)\right\} \textit{ of } \mathcal{D}(\bar{Q}) \textit{ into} \\ \mathcal{D}(\bar{\Omega}) \times \mathcal{D}(\bar{\Omega}) \textit{ extends by continuity to a continuous} \\ \textit{linear mapping of } Y \textit{ into } \mathcal{D}'_\gamma(\bar{\Omega}) \times \mathcal{D}'_\gamma(\bar{\Omega}). \quad \square \end{cases} \tag{2.67}$$

[1] We have the same simplification in the parabolic case if we consider a functional setting analogous to the one chosen for this Chapter; but the setting of Chapter 10 leads to more general results. In fact, see Section 4.3 below.

Concerning the operators $B_j u$, we follow an analogous procedure by introducing the closed subspace of X:

$$X_2 = \left\{ v \mid v \in X, \;\; v(x, 0) = \frac{\partial v}{\partial t}(x, 0) = 0 \right\} \tag{2.68}$$

and the mapping

$$v \to \{T_0 v, \ldots, T_{m-1} v\}. \tag{2.69}$$

We see that (2.69) is *continuous and surjective from* X_2 onto $[\mathscr{D}_{-,\gamma}([0, T]; \mathscr{D}(\Gamma))]^m$, where $\mathscr{D}_{-,\gamma}([0, T]; \mathscr{D}(\Gamma))$ is defined in completely analogous fashion to $\mathscr{D}_{-,\gamma}([0, T]; \mathscr{D}_\gamma(\overline{\Omega}))$ (we just replace the space $\mathscr{D}_\gamma(\overline{\Omega})$ with $\mathscr{D}(\Gamma)$ in the definition given in Section 2.2).

It follows that

(2.70) $\begin{cases} \textit{the mapping } u \to Bu = \{B_0 u, \ldots, B_{m-1} u\} \textit{ of } \mathscr{D}(\overline{Q}) \\ \textit{into } [\mathscr{D}(\overline{\Sigma})]^m \textit{ extends by continuity to a continuous} \\ \textit{linear mapping of } Y \textit{ into } [\mathscr{D}'_{+,\gamma}([0, T]; \mathscr{D}'(\Gamma))]^m = \textit{weak} \\ \textit{dual of } [\mathscr{D}_{-,\gamma}([0, T]; \mathscr{D}(\Gamma))]^m. \;\; \square \end{cases}$

Of course, these results do not completely solve the problem of "separating" $B_j u$, $u(x, 0)$ and $\frac{\partial u}{\partial t}(x, 0)$ in the mapping πu; the question still remains of whether the space $\mathscr{V}'$ can be interpreted as a product of spaces of distributions on Ω, Σ and Γ in analogous fashion to what can be done in the parabolic case relative to the spaces $\mathrm{D}_P^{-(r-1)}(Q)$ (see Baiocchi [1] and Section 12.3 of Chapter 4). $\square$

2.7 The Infinite Cylinder Case

We now consider the cylinder $Q = \Omega \times \mathbf{R}_t$, with boundary $\Sigma = \Gamma \times \mathbf{R}_t$, and the operator

$$P = A + \frac{\partial^2 u}{\partial t^2},$$

where A is given by (1.3) with (1.5) and

$$a_{pq} \in \mathscr{E}(\mathbf{R}_t; \mathscr{D}(\overline{\Omega})). \tag{1.4'}$$

The operators $B_j u$ are again given by (1.6) with (1.8) (where $\mathbf{R}_t$ replaces $[0, T]$) and

$$b_{jh} \in \mathscr{D}(\bar{\Sigma}). \tag{1.7'}$$

We again make the hypothesis (1.9), still replacing $[0, T]$ with $\mathbf{R}_t$.

Then, "formally", the problem we wish to study is

$$\begin{cases} Pu = f \text{ in } Q, \\ B_j u = g_j \text{ on } \Sigma, \ j = 0, \ldots, m-1. \end{cases} \tag{1.2'}$$

Now the methods of Sections 2.1, ..., 2.6, completed by some of the methods of Section 4, Chapter 10, are also applicable to the study of problem (1.2'). We shall restrict the present discussion to giving a sketch of the main points.

We introduce the space $\mathscr{D}_-(\mathbf{R}; \mathscr{D}_\gamma(\bar{\Omega}))$ (note that it is a strict $(\mathscr{L}\mathscr{F})$-space) and its dual $\mathscr{D}'_+(\mathbf{R}; \mathscr{D}_\gamma(\bar{\Omega}))$ according to the definitions of Chapter 7. We provide $\mathscr{D}'_+(\mathbf{R}; \mathscr{D}'_\gamma(\bar{\Omega}))$ with the weak dual topology. Note that, algebraically,

$$\mathscr{D}'_+(\mathbf{R}; \mathscr{D}'_\gamma(\bar{\Omega})) \subset \mathscr{D}'_+(\mathbf{R}; \mathscr{D}'(\Omega)).$$

We consider the *adjoint problem*:

$$\begin{cases} P^*v = \varphi \ \text{ in } Q, \text{with } \varphi \in \mathscr{D}_-(\mathbf{R}; \mathscr{D}_\gamma(\bar{\Omega})), \\ C_j v \ = 0 \ \text{ on } \Sigma, \ j = 0, \ldots, m-1 \end{cases} \tag{2.71}$$

and the space

$$X_- = \{v \mid v \in \mathscr{D}_-(\mathbf{R}; \mathscr{D}(\bar{\Omega})), \ C_j v = 0, \ j = 0, \ldots, m-1, \\ P^*v \in \mathscr{D}_-(\mathbf{R}; \mathscr{D}_\gamma(\bar{\Omega}))\}, \tag{2.72}$$

provided with the image topology of $\mathscr{D}_-(\mathbf{R}; \mathscr{D}(_\gamma\bar{\Omega}))$ under $(P^*)^{-1}$.

Then we are required to study the mapping $\mathscr{T}$ defined (compare with (2.28)) by

$$\mathscr{T}v = \{T_0 v, \ldots, T_{m-1} v\}. \tag{2.73}$$

Applying the closed graph theorem (X_- and $\mathscr{D}_-(\mathbf{R}; \mathscr{D}(\Gamma))$ being strict $(\mathscr{L}\mathscr{F})$-spaces) we show that

(2.74) *$v \to \mathscr{T}v$ is a continuous linear mapping of*

X_- into $[\mathscr{D}_-(\mathbf{R}; \mathscr{D}(\Gamma))]^m$

and then using Whitney's theorem in analogous fashion to Theorem 2.2 (but now we do not have the compatibility relations on Γ for $t = 0$!) we show that

$$(2.75)\qquad v \to \mathscr{T} v \text{ is a surjective mapping of } X_- \text{ onto } [\mathscr{D}^-(\mathbf{R}; \mathscr{D}(\Gamma))]^m.$$

Taking, instead of the space $\Xi'_+(0, T; \Xi'(\Omega))$, the space $\mathscr{D}'_+(\mathbf{R}; \Xi'(\Omega))$ (dual of $\mathscr{D}_-(\mathbf{R}; \Xi(\Omega))$ provided with the weak dual topology as in Section 4 of Chapter 10) we can introduce the space

$$(2.76)\qquad Y_+ = \{u \mid u \in \mathscr{D}'_+(\mathbf{R}; \mathscr{D}'_\gamma(\Omega)),\ \ Pu \in \mathscr{D}'_+(\mathbf{R}; \Xi'(\Omega))\},$$

provided with the coarsest locally convex topology which makes the mappings $u \to u$ and $u \to Pu$ of Y_+ into $\mathscr{D}'_+(\mathbf{R}; \mathscr{D}'_\gamma(\bar{\Omega}))$ and $\mathscr{D}'_+(\mathbf{R}; \Xi'(\Omega))$ respectively, continuous.

Then, by the same method as for Theorem 2.1, we show that

$$(2.77)\qquad \mathscr{D}(\mathbf{R}; \mathscr{D}(\bar{\Omega})) \text{ is dense in } Y_+.$$

Finally, it is easy to adapt the arguments used for Theorems 2.3 and 2.4 to obtain the trace and existence theorems:

$$(2.78)\qquad \begin{cases} \text{the mapping } u \to Bu = \{B_0 u, \ldots, B_{m-1} u\} \text{ of } \mathscr{D}(\mathbf{R}; \mathscr{D}(\bar{\Omega})) \\ \text{into } [\mathscr{D}(\mathbf{R}; \mathscr{D}(\Gamma))]^m \text{ extends by contuinuity to a} \\ \text{continuous linear mapping of } Y_+ \text{ into } [\mathscr{D}'_+(\mathbf{R}; \mathscr{D}'(\Gamma))]^m; \\ \text{and we have Green's formula} \\ \langle u, \overline{P^* v}\rangle - \langle Pu, \bar{v}\rangle = \sum_{j=0}^{m-1} \langle B_j u, \overline{T_j v}\rangle,\ \ \forall u \in Y_+,\ v \in X_-, \end{cases}$$

and

$$(2.79)\qquad \begin{cases} \text{the boundary value problem} \\ Pu = f \text{ in the sense of } \mathscr{D}'_+(\mathbf{R}; \mathscr{D}'_\gamma(\bar{\Omega})), \\ B_j u = g_j,\ j = 0, \ldots, m-1, \text{ in the sense of (2.78)}, \\ \text{admits a unique solution in } Y_+ \text{ for every} \\ f \in \mathscr{D}'_+(\mathbf{R}; \Xi'(\Omega)) \text{ and } g_j \in \mathscr{D}'_+(\mathbf{R}; \mathscr{D}'(\Gamma)), \\ u \text{ depending continuously on } f \text{ and } g_j. \end{cases}$$

Remark 2.4. The preceding arguments are simpler than those of Chapter 10, Section 4.1; this is due to the fact that here we know that we can use the closed graph theorem, whereas the analogous question in the setting of Chapter 10, Section 4.1 seems to be open.

3. Equations of the Second Order in t; Application of Transposition and Existence of Solutions in Spaces of Ultra-Distributions

3.1 The Difficulties in the Finite Cylinder Case

We shall now attempt to apply the regularity results of Section 1.2 to the study of non-homogeneous boundary value problems in spaces of ultra-distributions, for equations of the second order in t.

We have the regularity hypotheses of Section 1.2 on the data, and furthermore the following hypotheses:

(3.1) Ω *satisfies* (1.1) *with* Γ *an analytic variety*,

(3.2) P *and* B_j *satisfy the hypotheses* (1.2), ..., (1.9) *in the cylinder* $Q = \Omega \times]0, T[$ *and furthermore*

$$a_{pq}(x, t) = a_{pq}(x) \in \mathscr{H}(\bar{\Omega}),$$

$$b_{jh}(x, t) = b_{jh}(x) \in \mathscr{H}(\Gamma).$$

The starting point is still the adjoint problem (2.5). The regularity results of Section 1.2 lead us to take in (2.5):

$$\varphi \in \mathscr{D}_{sm,s}(Q), \quad s > 1, \tag{3.3}$$

and to introduce the space

$$X_s = \{v \mid v \in \mathscr{D}_{ms,s}(\bar{Q}),\ v(x, T) = \frac{\partial v(x, T)}{\partial t} = 0,\ C_j v = 0, \quad j = 0, \ldots, m-1,\ P^* v \in \mathscr{D}_{ms,s}(Q)\}. \tag{3.4}$$

But by considerations analogous to those of Section 2.1 on the Cauchy problem (2.9), we are led to eliminate the case (3.3), since the results on the Cauchy problem in Gevrey classes of order sm with respect to t and s with respect to x are not valid for $s > 1$ (see Talenti [2]). □

Thus we are led to study the adjoint problem (2.5) in spaces analogous to the space $\mathscr{D}_{-,\gamma}([0, T]; \mathscr{D}_\gamma(\bar{\Omega}))$, in order to develop a theory of non-homogeneous problems in spaces of Gevrey ultra-distributions in parallel with the theory of Sections 2.2, ..., 2.6. We can introduce the space

$$\mathscr{D}_{s,\gamma}(\bar{\Omega}) = \{\varphi \mid \varphi \in \mathscr{D}_s(\bar{\Omega}),\ \gamma_j \varphi = 0 \text{ on } \Gamma,\ j = 0, \ldots, m-1\}$$

and then the space

$$\mathscr{D}_{-,ms,\gamma}([0, T]; \mathscr{D}_{s,\gamma}(\bar{\Omega})) = \{\varphi \mid \varphi \in \mathscr{D}_{ms}([0, T]; \mathscr{D}_{s,\gamma}(\bar{\Omega})), \quad \varphi^{(j)}(0) = 0, j = 0, \ldots, m-1,\ \varphi(t) = 0 \text{ in a neighbourhood of } T\} \tag{3.5}$$

and, in problem (2.5), take φ belonging to

$$\mathscr{D}_{-,ms,\gamma}\big([0, T]; \mathscr{D}_{s,\gamma}(\overline{\Omega})\big).$$

Then the essential difficulty is the analogue to Whitney's theorem in the Gevrey spaces $\mathscr{D}_{ms,s}(\overline{Q})$ (see Section 2.4, Theorem 2.2), which we do not know to be valid. Now the use of this theorem is *essential* here, in order to know the space analogous to $\mathscr{V}$.

Thus we cannot at this time develop the analogue of the theory of Sections 2.2, ..., 2.6 in Gevrey spaces. ☐

Finally, let us point out another possibility: in the adjoint problem (2.5), take φ belonging to

$$\mathscr{D}_{-,m}\big(]0, T[; \mathscr{D}(\Omega)\big)$$

with $m > 1$, which leads to the study of homogeneous problems in ultra-distributions of

$$\mathscr{D}'_{+,m}\big(]0, T[; \mathscr{D}'(\Omega)\big).$$

In this case, we can avoid the difficulties tied to the Cauchy problem (2.9) (we can apply Talenti [2, 3]). But we shall have the difficulties due to the linking of $v(x, 0)$, $\dfrac{\partial v}{\partial t}(x, 0)$, $T_j v|_{t=0}$ on the variety Γ. We shall develop this idea for the infinite cylinder case in the next Section.

3.2 The Infinite Cylinder Case for $m > 1$

Under hypothesis (3.1) on Ω, we consider the infinite cylinder

$$Q = \Omega \times \mathbf{R}_t \tag{3.6}$$

and again make hypothesis (3.2), replacing $[0, T]$ with $\mathbf{R}_t$, and furthermore assume that

$$m > 1. \tag{3.7}$$

We shall study the problem

$$\begin{cases} Pu = f \text{ in } Q, \\ B_j u = g_j \text{ on } \Sigma,\ j = 0, \ldots, m-1, \end{cases} \tag{3.8}$$

in the space of ultra-distributions $\mathscr{D}'_{+,m}(\mathbf{R}; \mathscr{D}'(\Omega))$ (the Remarks of the preceding Section lead us to eliminate the spaces $\mathscr{D}'_{+}(\mathbf{R}; \mathscr{D}'(\Omega))$ and $\mathscr{D}'_{+,sm}(\mathbf{R}; \mathscr{D}'_s(\Omega))$ and the case $m = 1$).

We start from the adjoint problem

$$(3.9)\qquad \begin{cases} P^*v = \varphi \text{ with } \varphi \in \mathcal{D}_{-,m}(\mathbf{R}; \mathcal{D}(\Omega)), \\ C_j v = 0, \; j = 0, \ldots, m-1, \end{cases}$$

and introduce the space

$$(3.10)\qquad X_- = \{v \mid v \in \mathcal{D}_-(\mathbf{R}; \mathcal{D}(\overline{\Omega})), \; C_j v = 0, \; j = 0, \ldots, m-1, \\ P^* v \in \mathcal{D}_{-,m}(\mathbf{R}; \mathcal{D}(\Omega))\}.$$

Thanks to Theorem 1.1, we easily see that P^* defines a one-to-one mapping of X_- onto $\mathcal{D}_{-,m}(\mathbf{R}; \mathcal{D}(\Omega))$; therefore, on X_-, we can define the image topology of $\mathcal{D}_{-,m}(\mathbf{R}; \mathcal{D}(\Omega))$ under $(P^*)^{-1}$; so that we have

$$(3.11)\qquad \begin{cases} P^* \textit{ is an (algebraic and topological) isomorphism} \\ \textit{of } X_- \textit{ onto } \mathcal{D}_{-,m}(\mathbf{R}; \mathcal{D}(\Omega)). \quad \square \end{cases}$$

Let us study the space described by $\mathcal{T}v$ as v describes X_-, where

$$(3.12)\qquad \mathcal{T}v = \{T_0 v, \ldots, T_{m-1} v\}.$$

Applying Theorem 2.2 of Chapter 9 (see also Remark 2.3 of Chapter 9) we see that (for example):

$$(3.13)\qquad v \in \mathcal{D}_{-,m}(\mathbf{R}; L^2(\Omega)).$$

But we have

$$(3.14)\qquad P^*v = A^*v + \frac{\partial^2 v}{\partial t^2} = 0 \text{ in a neighborhoodof } \Sigma,$$

$$(3.15)\qquad C_j v = 0, \text{ on } \Sigma, \quad j = 0, \ldots, m-1.$$

Then, by the same type of arguments as in the proof of Theorem 1.2, it follows from (3.13), (3.14), (3.15) and the Theorem on "elliptic iterates" of Chapter 8, that, for each t, $v(x, t)$ is analytic in x in a neighborhood of Γ and finally that

$$(3.16)\qquad \mathcal{T}v \in [\mathcal{D}_{-,m}(\mathbf{R}; \mathcal{H}(\Gamma))]^m. \quad \square$$

Next, by a proof completely analogous to the proof of Theorem 4.1 of Chapter 10, we obtain that

$$(3.17)\qquad v \to \mathcal{T}v \textit{ is a continuous linear mapping of } X_- \textit{ into} \\ [\mathcal{D}_{-,m}(\mathbf{R}; \mathcal{H}(\Gamma))]^m.$$

Still following Chapter 10 (see Theorem 4.2 and Corollary) we find that

(3.18) $v \to \mathscr{C}v$ *is a surjective mapping of* X_- *onto*

$$[\mathscr{D}_{-,m}(\mathbf{R};\mathscr{H}(\Gamma))]^m.$$

In this regard, we note that the application of the Theorem of Talenti [3] to the Cauchy problem analogous to (4.19) is possible even for the operator of the second order in t, when the data are in $\mathscr{D}_{-,m}(\mathbf{R};\mathscr{H}(\Gamma))$. ☐

Now we introduce the space

$$(3.19)\qquad Y_+ = \{u \mid u \in \mathscr{D}'_{+,m}(\mathbf{R};\mathscr{D}'(\Omega)),\ Pu \in \mathscr{D}'_{+,m}(\mathbf{R};\Xi'(\Omega))\},$$

provided with the coarsest locally convex topology which makes the mappings $u \to u$ and $u \to Pu$ of Y_+ into $\mathscr{D}'_{+,m}(\mathbf{R};\mathscr{D}'(\Omega))$ and $\mathscr{D}'_{+,m}(\mathbf{R};\Xi'(\Omega))$ respectively, continuous, these two spaces being provided with the *weak* dual topologies (resp. of $\mathscr{D}_{-,m}(\mathbf{R};\mathscr{D}(\Omega))$ and $\mathscr{D}_{-,m}(\mathbf{R};\Xi(\Omega))$.

Then we show that

(3.20) *the space* $\mathscr{D}(\mathbf{R};\mathscr{D}(\overline{\Omega}))$ *is dense in* Y_+.

The proof of (3.20) is slightly different from the proof of Theorem 4.3 of Chapter 10, since now the operator P is not parabolic (and not even hypoelliptic). Let $u \to M(u)$ be a continuous linear form on Y_+; it may be written

$$M(u) = \langle f, \bar{u}\rangle + \langle g, \overline{Pu}\rangle$$

with $f \in \mathscr{D}_{-,m}(\mathbf{R};\mathscr{D}(\Omega))$ and $g \in \mathscr{D}_{-,m}(\mathbf{R};\Xi(\Omega))$. Assume that we have $M(\varphi) = 0\ \ \forall \varphi \in \mathscr{D}(\mathbf{R};\mathscr{D}(\overline{\Omega}))$ and let $\mathscr{A}$ be an extension of A into a suitable neighborhood $\mathscr{O}$ of $\overline{\Omega}$, with analytic coefficients in $\mathscr{O}$ and elliptic in $\mathscr{O}$ (which is always, possible since A is elliptic in $\overline{\Omega}$). Then let $\tilde{f}$ and $\tilde{g}$ be the extensions to $\mathscr{O} \times \mathbf{R}_t$ of f and g by zero outside Q.

Then if $\Phi \in \mathscr{D}(\mathscr{O} \times \mathbf{R}_t)$, we have

$$\langle \tilde{f}, \bar{\Phi}\rangle + \langle g, \overline{\mathscr{A}\Phi + D_t^2\Phi}\rangle = 0,$$

the brackets being taken in the sense of distributions on $\mathscr{O} \times \mathbf{R}_t$.

Therefore we have

$$(3.21)\qquad \mathscr{A}^*\tilde{g} + D_t^2\tilde{g} = \tilde{f} \text{ in } \mathscr{O} \times \mathbf{R}_t,$$

with, in particular, $\tilde{g} \in \mathscr{D}_{-,m}(\mathbf{R};L^2(\mathscr{O}))$ and $\tilde{f} \in \mathscr{D}_{-,m}(\mathbf{R};\mathscr{D}(\mathscr{O}))$. By an obvious application of Theorem 1.1, it follows that $\tilde{g} \in \mathscr{D}_-(\mathbf{R};\mathscr{D}(\overline{\mathscr{O}}))$. Let us now consider a fixed interval $[a, b]$ of $\mathbf{R}_t$; then $\tilde{f}$ vanishes in $\mathscr{U} \times [a, b]$

($\mathscr{U}$ a suitable neighborhood of Γ) since $f \in \mathscr{D}_{-,m}(\mathbf{R}; \mathscr{D}(\Omega))$ and therefore in $\mathscr{U} \times [a, b]$ it follows from (3.21) that

$$\mathscr{A}^*\tilde{g} + D_t^2\tilde{g} = 0.$$

Differentiating and applying the operator $\mathscr{A}^*$ as in the proof of Theorem 1.2, we obtain

$$\mathscr{A}^{*i}\tilde{g} = (-1)^i D_t^{2i}\tilde{g}$$

and therefore, since $\tilde{g} \in \mathscr{D}_{-,m}(\mathbf{R}; L^2(\mathscr{O}))$, we have

$$\| \mathscr{A}^{*i}\tilde{g} \|_{L^2(\mathscr{U})} \leq c_0 L_0^i (2mi)!,\ i = 0, 1, \ldots,$$

and therefore (see Theorem 2.4 of Chapter 8)

$$\tilde{g} \text{ is analytic in } x \text{ in } \mathscr{U} \text{ for every } t \in [a, b].$$

But $\tilde{g}$ vanishes outside Q and therefore g vanishes in $\mathscr{U} \times]a, b[$. Also, thanks to the fact that $g \in \mathscr{D}_{-,m}(\mathbf{R}; \Xi'(\Omega))$, it follows that $g \in \mathscr{D}_{-;m}(\mathbf{R}; \mathscr{D}(\Omega))$ and then (3.21) implies

$$A^*g + D_t^2 g = -f \text{ in } Q$$

and therefore $M(u) = -\langle P^*g, \bar{u}\rangle + \langle g, \overline{Pu}\rangle$ and therefore $M(u) = 0$, $\forall u \in Y_+$. ☐

Finally, by the same methods as for Theorems 4.4 and 4.5 of Chapter 10, we have no difficulty in showing the following trace and existence theorems:

(3.22) *the mapping* $u \to Bu = \{B_0 u, \ldots, B_{m-1}\}$ *of* $\mathscr{D}(\mathbf{R}; \mathscr{D}(\bar{\Omega}))$ *into* $[\mathscr{D}(\mathbf{R}; \mathscr{D}(\Gamma))]^m$ *extends by contuinuity to a continuous linear mapping of* Y_+ *into* $[\mathscr{D}'_{+,m}(\mathbf{R}; \mathscr{H}'(\Gamma))]^m$; *and we have Green's formula*

$$\langle u, \overline{P^*v}\rangle - \langle Pu, \bar{v}\rangle = \sum_{j=0}^{m-1} \langle B_j u, \overline{T_j v}\rangle,\ \forall u \in Y_+,\ v \in X_-,$$

and

(3.23) *the boundary value problem*

$$Pu = f \text{ in the sense of } \mathscr{D}'_{+,m}(\mathbf{R}; \mathscr{D}'(\Omega)),$$

$$B_j u = g_j,\ j = 0, \ldots, m-1, \text{ in the sense of } (3.22),$$

admits a unique solution u *in* Y_+ *for every* $f \in \mathscr{D}'_{+,m}(\mathbf{R}; \Xi'(\Omega))$ *and* $g_j \in \mathscr{D}'_{+,m}(\mathbf{R}; \mathscr{H}'(\Gamma))$, u *depending continuously on* f *and* g_j. ☐

4. Schroedinger Equations; Complements for Parabolic Equations

4.1 Regularity Results for the Schroedinger Equation

Results comparable to those obtained in the preceding Sections for equations of the second order in t, can also be obtained for Schroedinger's equation, by the same type of proofs. We shall therefore restrict this discussion to the brief statement of results, and refer the reader to the preceding Sections for the proofs. □

Concerning the regularity results, we have either regularity in the space $\mathscr{D}(\bar{Q})$, or regularity in Gevrey spaces ($Q = \Omega \times]0, T[$).

On the open set Ω and the operators A and B_j, let us make the hypotheses (1.1), (1.3), (1.5), (1.6), (1.8) *and* (1.9) *and let us replace* (1.4) *with*

$$a_{pq}(x, t) = a_{pq}(x) \in \mathscr{D}(\bar{\Omega}) \tag{4.1}$$

and (1.7) *with*

$$b_{jh}(x, t) = b_{jh}(x) \in \mathscr{D}(\Gamma), \tag{4.2}$$

so that the form $a(t; u, v)$ in (1.9) does not depend on t.

We are given f, g_j, u_0 with

$$f \in \mathscr{D}(\bar{Q}), \quad g_j \in \mathscr{D}(\bar{\Sigma}), \quad u_0 \in \mathscr{D}(\bar{\Omega}), \ j = 0, \ldots, m-1, \tag{4.3}$$

and with *the compatibility relations* which guarantee the existence of $w \in \mathscr{D}(\bar{Q})$ with

$$\begin{cases} B_j w = g_j \text{ on } \Sigma, \ j = 0, \ldots, m-1, \ w(x, 0) = u_0(x) \text{ on } \Omega, \\ D_t^k P w(x, 0) = D_t^k f(x, 0) \text{ on } \Omega, \ \forall k, \end{cases} \tag{4.4}$$

where

$$P = iA + \frac{\partial}{\partial t}. \tag{4.5}$$

Applying Theorem 12.2 of Chapter 5, we obtain in analogy to Theorem 1.1:

Theorem 4.1. *Under the preceding hypotheses, there exists a unique u in $\mathscr{D}(\bar{Q})$, solution of*

$$\begin{cases} Pu = f \ \text{in } Q, \\ B_j u = g_j \ \text{on } \Sigma, \ j = 0, \ldots, m-1, \\ u(x, 0) = u_0(x) \ \text{on } \Omega. \ \square \end{cases} \tag{4.6}$$

Now assume that Ω *satisfies* (1.13) *and that* (4.1) *and* (4.2) *are replaced with* (1.14) *and* (1.15), *with*

$$s > 1. \tag{4.7}$$

Let f, g_j, u_0 be given with

$$f \in \mathscr{D}_{2ms,s}(\overline{Q}), \quad g_j \in \mathscr{D}_{2ms,s}(\overline{\Sigma}), \quad u_0 \in \mathscr{D}_s(\overline{\Omega}) \tag{4.8}$$

and the compatibility relations which guarantee the existence of $w \in \mathscr{D}_{2ms,s}(\overline{Q})$ satisfying (4.4). Then

Theorem 4.2. *Under the above hypotheses, there exists a unique u in $\mathscr{D}_{2ms,s}(\overline{Q})$, solution of* (4.6).

After reduction of the problem to the case $g_j = 0$ and $u_0 = 0$, the proof is exactly the same as in Remark 2.2 of Chapter 10 (parabolic case); and by the same considerations as in Remark 2.3 of Chapter 10, we can replace in the Theorem the spaces $\mathscr{D}_{2ms,s}(\overline{Q}), \ldots$ with $\mathscr{D}_{N_h,M_k}(\overline{Q}), \ldots$ under the same hypotheses on the sequences $\{N_h\}$ and $\{M_k\}$ as in Remark 2.3 of Chapter 10.

4.2 The Non-Homogeneous Boundary Value Problems for the Schroedinger Equation

The analogy with the case of equations of the second order in t is complete for the application of the transposition method. It will be sufficient to state the main results.

Let Ω satisfy (1.1), *Q be the cylinder $Q = \Omega \times]0, T[$, A and B_j the operators satisfying* (1.3), (1.5), (1.6), (1.8), (1.9), (4.1) *and* (4.3).

We introduce the following spaces:

$$X = \{v \mid v \in \mathscr{D}(\overline{Q}),\ v(x, T) = 0,\ C_j v = 0,\ j = 0, \ldots, m-1,$$

$$P^* v \left(= -\mathrm{i} A v - \frac{\partial v}{\partial t}\right) \in \mathscr{D}_{-,\gamma}([0, T]; \mathscr{D}_\gamma(\overline{\Omega}))\},$$

$$Y = \{u \mid u \in \mathscr{D}'_{+,\gamma}([0, T]; \mathscr{D}'_\gamma(\overline{\Omega})),\ Pu \in \Xi'_+(0, T; \Xi'(\Omega))\}$$

with topologies analogous to the topologies of X and Y in Section 2.

We consider the mapping

$$v \to \mathscr{T} v = \{v(x, 0); T_0 v, \ldots, T_{m-1} v\}$$

and show that it is a continuous, linear and surjective mapping of X onto the subspace $\mathscr{V}$ of $\mathscr{D}(\overline{\Omega}) \times [\mathscr{D}_-([0, T]; \mathscr{D}(\Gamma))]^m$ characterized by the *compatibility relations on* Γ, for $t = 0$:

$$\begin{cases} v \in \mathscr{D}(\bar{Q}), \ C_j v = 0 \text{ on } \Sigma, \ j = 0, \ldots, m-1, \\ \gamma_k(P^* v) = 0 \text{ on } \Sigma, \ \forall k, \ \left.\dfrac{\partial^k (P^* v)}{\partial t^k}\right|_{t=0} = 0 \text{ on } \Omega, \ \forall k. \end{cases}$$

$\mathscr{V}$ can be provided with a strict $(\mathscr{L}\mathscr{F})$-space topology in analogous fashion as in Section 2.4. Then *$\mathscr{V}'$ is the (weak) dual of $\mathscr{V}$.*

This being set, we show, following the methods of Sections 2.1, ..., 2.5:

Theorem 4.3. *The space $\mathscr{D}(\bar{Q})$ is dense in Y and the mapping*

$$u \to \sigma u = \{u(x, 0);\, B_0 u, \ldots, B_{m-1} u\}$$

of $\mathscr{D}(\bar{Q})$ into $\mathscr{D}(\bar{\Omega}) \times [\mathscr{D}(\bar{\Sigma})]^m$ extends by continuity to a continuous linear mapping of Y into $\mathscr{V}'$; furthermore, for every $f \in \Xi'_+(0, T; \Xi'(\Omega))$ and $g^* \in \mathscr{V}'$, the problem

$$\begin{cases} Pu = f \text{ in the sense of } \mathscr{D}'_{+,\gamma}\big([0, T]; \mathscr{D}'_\gamma(\bar{\Omega})\big) \\ \qquad \big(\text{and therefore also of } \mathscr{D}'(Q)\big), \\ \qquad \sigma u = g^* \text{ in the sense defined above}, \end{cases}$$

admits a unique solution u in Y; and u depends continuously on f and g^*. □

Remark 4.1. We also have complements analogous to those of Section 2.6: *$u \to u(x, 0)$ is a continuous linear mapping of Y into $\mathscr{D}'_\gamma(\bar{\Omega})$ and $u \to Bu = \{B_0 u, \ldots, B_{m-1} u\}$ is a continuous linear mapping of Y into*

$$[\mathscr{D}'_{+,\gamma}\big([0, T]; \mathscr{D}'(\Gamma)\big)]^m. \quad □$$

There are no difficulties in extending the results of Section 2.7 for the infinite cylinder case to Schroedinger's equation. We do not specify the results. □

Remark 4.2. As in Remark 2.3, we see that the application of the results of Chapter 9, Section 10.6, leads to problems which are more general in t and less general in the space variables. □

Finally, we can also adapt to Schroedinger's equation all that we have seen in Section 3 for the equation of the second order in t; we have the same technical difficulties and the same type of results. Let us only note the fact that now ms must be replaced with $2ms$ and therefore $2ms > 1$ if $m \geq 1$ and $s \geq 1$. Therefore, in particular, *the results of Section* 3.2 *are valid, for every integer $m \geq 1$, in the space $\mathscr{D}'_{+,2m}(\mathbf{R}; \mathscr{D}'(\Omega))$*; more precisely, once having introduced the space

$$Y_+ = \{u \mid u \in \mathscr{D}'_{+,2m}\big(\mathbf{R}; \mathscr{D}'(\Omega)\big),\ Pu \in \mathscr{D}'_{+,2m}\big(\mathbf{R}; \Xi'(\Omega)\big)\},$$

we have:
the mapping $u \to Bu = \{B_0 u, \ldots, B_{m-1} u\}$ *extends by continuity to a continuous linear mapping of* Y_+ *into* $[\mathscr{D}'_{+,2m}(\mathbf{R}; \mathscr{H}'(\Gamma))]^m$ *and the problem*

$$(4.9) \qquad \begin{cases} Pu = f, \\ B_j u = g_j, \; j = 0, \ldots, m-1, \end{cases}$$

admits a unique solution in Y_+ *for every*

$$f \in \mathscr{D}'_{+,2m}(\mathbf{R}; \Xi'(\Omega))$$

and

$$g_j \in \mathscr{D}'_{+,2m}(\mathbf{R}; \mathscr{H}'(\Gamma)),$$

u depending continuously on f and g_j.

4.3 Remarks on Parabolic Equations

The idea introduced in Section 2 for equations of the second order in t (and in Section 4.2 for Schroedinger's equation) to use the spaces $\mathscr{D}_{-,\gamma}([0, T]; \mathscr{D}_\gamma(\bar{\Omega}))$ and $\mathscr{D}'_{+,\gamma}([0, T]; \mathscr{D}'_\gamma(\bar{\Omega}))$, may also be useful for parabolic equations.

One obtains results in less general spaces for the solution u than those obtained in Chapter 10, but under more general hypotheses for the operators $P = A + \partial/\partial t$ and $\{B_j\}_{j=0}^{m-1}$.

It is not difficult to see that by the methods of Sections 2.2, ..., 2.5 and applying the regularity theorems of Chapter 10, Sections 1 and 2, we arrive at the following results.

The notation and the hypotheses on Ω, $P = A + \partial/\partial t$, $\{B_j\}_{j=0}^{m-1}$ are those of Section 2.1 of Chapter 10; note that *we shall not need hypotheses* (3.13) *and* (3.14) of Chapter 10. We introduce the spaces (different from X and Y of Chapter 10):

$$X_\gamma = \{v \mid v \in \mathscr{D}(\bar{Q}), v(x, T) = 0, \; C_j v = 0, \; j = 0, \ldots, m-1,$$

$$P^* v \left(= A^* v - \frac{\partial v}{\partial t}\right) \in \mathscr{D}_{-,\gamma}([0, T]; \mathscr{D}_\gamma(\bar{\Omega}))\},$$

$$Y_\gamma = \{u \mid u \in \mathscr{D}'_{+,\gamma}([0, T]; \mathscr{D}'_\gamma(\bar{\Omega})), Pu \in \Xi'_+(0, T; \Xi'(\Omega))\},$$

with topologies analogous to the topologies of the spaces X and Y of Section 2.

We consider the mapping

$$v \to \mathscr{T} v = \{v(x, 0); T_0 v, \ldots, T_{m-1}\}$$

and show that it is a continuous, linear and surjective mapping of X_γ onto the subspaces $\mathscr{V}_\gamma$ of $\mathscr{D}(\bar{\Omega}) \times [\mathscr{D}_-([0, T]; \mathscr{D}(\Gamma))]^m$ of elements satisfying the *compatibility relations* on Γ and for $t = 0$ (obtained "formally" from the definition of X_γ). Next, we set $\mathscr{V}'_\gamma =$ (weak) dual of $\mathscr{V}_\gamma$.

Then we can prove

Theorem 4.4. *The space $\mathscr{D}(\overline{Q})$ is dense in Y_γ and the mapping*

$$u \to \sigma u = \{u(x, 0); B_0 u, \ldots, B_{m-1} u\}$$

extends by continuity to a continuous linear mapping of Y_γ into $\mathscr{V}'_\gamma$. The problem:

$$Pu = f,$$

$$\sigma u = g^*,$$

admits a unique solution u in Y_γ for every $f \in \Xi'_+(0, T; \Xi'(\Omega))$ and $g^* \in \mathscr{V}'_\gamma$; u depends continuously on f and g^*. ☐

We can also, in opposition to what we have done for the space Y in Chapter 10 (see Remark 3.2), define $u(x, 0)$ and Bu *separately* for the elements u of Y_γ, by following the method of Section 2.6; we obtain:

1) *$u \to u(x, 0)$ is a continuous linear mapping of Y_γ into $\mathscr{D}'_\gamma(\overline{\Omega})$;*

2) *$u \to Bu$ is a continuous linear mapping of Y_γ into $[\mathscr{D}'_{+,\gamma}([0, T]; \mathscr{D}'(\Gamma))]^m$.* ☐

Finally, the results of Section 2.7 for the infinite cylinder case can be extended to parabolic equations; but in this case the hypotheses on A and B_j will be the same as in Section 4 of Chapter 10 and the spaces will be less general ($\mathscr{D}'_+(\mathbf{R}; \mathscr{D}_\gamma(\overline{\Omega}))$ being contained in $\mathscr{D}'_+(\mathbf{R}; \mathscr{D}'(\Omega))$ and $\mathscr{D}'_+(\mathbf{R}; \mathscr{D}'(\Gamma))$ in $\mathscr{D}'_{+,2m}(\mathbf{R}; \mathscr{H}'(\Gamma))$.)

5. Comments

The regularity in Gevrey spaces of solutions of boundary value problems for equations of the second order in t and Schroedinger equations (see Sections 1.2 and 4.1) has been given in Lions-Magenes [4]. In this same work, we study non-homogeneous problems for the case of the infinite cylinder and the spaces $\mathscr{D}'_{+,m}(\mathbf{R}; \mathscr{D}'(\Omega))$ (see Section 3.2 and the end of Section 4.2).

The other results of this Chapter are given here for the first time. For the space $\mathscr{D}_\gamma(\overline{\Omega})$ of Section 2.2, see also Triebel [1].

6. Problems

6.1. We recall the problem (already noted in Remark 1.2 and in Section 3.1) of specifying the compatibility conditions in Gevrey spaces.

6.2. Is Theorem 1.2 valid under hypothesis (1.23) (case of coefficients depending on t, see Remark 1.3)? Same problem for Schroedinger's equation.

6.3. If $m = 1$ (*hyperbolic* case), is Theorem 1.2 still valid for $s = 1$?

6.4. Reflexivity of the spaces $\mathscr{D}_{-,\gamma}([0, T]; \mathscr{D}_\gamma(\bar{\Omega}))$ and $\Xi_-(0, T; \Xi(\Omega))$ (for a similar case, see Lions-Magenes [2], Section 4.2).

6.5. Structure of the various spaces $\mathscr{V}'$ (see Remark 2.2, Sections 4.2, 4.3).

6.6. Do Theorems 2.1, 2.3, ... remain valid if the spaces $\mathscr{D}'_{+,\gamma}([0, T]; \mathscr{D}'_\gamma(\bar{\Omega}))$, $\Xi'_+(0, T; \Xi'(\Omega))$, ... are provided with the *strong* dual topologies? (See Remark 2.1 and Problem 6.4).

6.7. "Separation" of the components $u(x, 0)$, $\partial u(x, 0)/\partial t$, $T_0 v, \ldots, T_{m-1} v$ in the operator πu of Section 2.5.

6.8. Same problem as above for the operator σu which comes up for Schroedinger's equation (Section 4.2) and parabolic equations (Section 4.3).

6.9. In opposition to the case of elliptic and parabolic equations, we do not have "optimal" regularity results for the case of equations of the second order in t or of Schroedinger. Thus, there are *other* possible choices than the spaces $\mathscr{D}'_{+,\gamma}([0, T]; \mathscr{D}'_\gamma(\bar{\Omega}))$; it may be of interest to study them in a more systematic way.

6.10. In particular, note the case of the operator

$$Pu = \frac{\partial^2 u}{\partial t^2} - \frac{\partial^2 u}{\partial x^2}$$

(wave operator in two variables) which is hyperbolic either with respect to t or with respect to x; this should allow for a study of non-homogeneous boundary value problems in spaces other than $\mathscr{D}'_{+,\gamma}([0, T]; \mathscr{D}'_\gamma(\bar{\Omega}))$; see Section 2.1.

6.11. It is very likely that a theory analogous to the one given here is valid for first order hyperbolic systems (see M. S. Agranovich [1], K. O. Friedrichs and P. Lax [1], P. Lax and R. S. Phillips [1], C. Bardos [1]); but this remains to be made explicit.

It would be of great interest to study the analogous problems for higher order hyperbolic equations; see S. Agmon [1], S. Miyatake [1], S. Mizohata [2], T. Shirota and K. Asano [1].

6.12. In Chapter 9, Section 5, we have seen how it was possible to "approximate" hyperbolic systems or systems of the second order in t with parabolic problems; then it is very likely that the solutions obtained in this Chapter can be "approximated" with solutions of parabolic problems (solved in the preceding Chapter).

Same remark concerning the approximation by systems of Cauchy-Kowalevski type (see Chapter 9, Section 5).

6.13. Extension of the theory of this Chapter to the settings of Problems 14.8, 14.9, 14.10, 14.12, 14.13, 14.14, 14.17 of Chapter 5.

Appendix

Calculus of Variations in Gevrey-Type Spaces

1. Generalities

1.1 Duality Operators in Gevrey-Type Spaces

Let Ω be a bounded open set in $\mathbf{R}^n$. We are given

$$s > 1 \tag{1.1}$$

and a number $L > 0$.

For φ given in $\mathscr{D}(\bar{\Omega})$, we set

$$\|\varphi\|_E = \left(\sum_\alpha \frac{1}{(|\alpha|!)^{2s} L^{2|\alpha|}} \|D^\alpha \varphi\|^2_{L^2(\Omega)} \right)^{1/2} \tag{1.2}$$

which in general is infinite, and define the space E by[1]

$$E = \{\varphi \mid \varphi \in \mathscr{D}(\bar{\Omega}),\ \|\varphi\|_E < \infty\}. \tag{1.3}$$

Provided with the norm $\|\varphi\|_E$, E is a *Hilbert space*.

Next, we introduce

$$E_0 = \{\varphi \mid \varphi \in E,\ D^\alpha \varphi = 0 \text{ on } \Gamma,\ \forall \alpha\} \tag{1.4}$$

(where, as usual, Γ denotes the boundary of Ω).

We verify that:

$$\mathscr{D}_s(\Omega)\ E \text{ is dense in } E_0, \tag{1.5}$$

$$\begin{cases} \mathscr{D}_{s_1}(\Omega) \subset E_0 \text{ if } 1 < s_1 < s, \\ \mathscr{D}_{s_1}(\Omega) \text{ is dense in } E_0. \end{cases} \tag{1.6}$$

---

[1] The space E (which depends on L) coincides with the space $\mathscr{D}_{(k!)^s}(\bar{\Omega}; L)$, in the notation of Chapter 7, Section 1.3.

Then the dual E_0' of E_0 can be identified with a subspace of $\mathscr{D}'_{s_1}(\Omega)$, arbitrary s_1 with $1 < s_1 < s$.

Since E_0 is a Hilbert space, there exists a *canonical isomorphism* Λ_0 *of* E_0 *onto* E_0', *given by*

$$\Lambda_0 \varphi = \sum_\alpha \frac{(-1)^{|\alpha|}}{(|\alpha|!)^{2s} L^{2|\alpha|}} \mathrm{D}^{2\alpha} \varphi. \tag{1.7}$$

The operator Λ_0 is the "duality operator" of E_0 onto E_0'.

Every element f of E_0' can be represented, non-uniquely, by

$$\begin{cases} f = \sum_\alpha \mathrm{D}^\alpha f_\alpha, \ f_\alpha \in L^2(\Omega), \\ \sum_\alpha (|\alpha|!)^{2s} L^{2|\alpha|} \|f_\alpha\|^2_{L^2(\Omega)} < \infty. \end{cases} \tag{1.8}$$

Then, *for* f *given in the form* (1.8), *there exists a unique* u *in* E_0, *solution of*

$$\Lambda_0(u) = f. \tag{1.9}$$

This is *a* (*homogeneous*) *Dirichlet problem of infinite order.* ☐

Remark 1.1. The duality operator Λ of E onto E' solves an *infinite order Neumann problem.* ☐

Remark 1.2. It is possible to construct spaces analogous to E and E_0, replacing L^2 with L^p, $1 < p < \infty$, $p \neq 2$. Then, the duality operators Λ (and Λ_0) of E onto E' (and E_0 onto E_0') are *nonlinear* operators of infinite order. ☐

1.2 Orientation

The remarks made in Section 1.1 show to what *type* of problems we are led when we consider calculus of variations problems[1] in Gevrey spaces.

We shall examine more closely what one obtains in *optimal control* theory in the setting of Gevrey spaces or of analytic functions. The following results are taken from notes by the authors (see Lions-Magenes [6]).

We shall distinguish two cases:

(i) elliptic systems;

(ii) evolution systems.

In either case, one obtains boundary value problems of infinite order.

---

[1] Indeed (analogue of the Dirichlet principle) problem (1.9) is equivalent to the search for the minimum of $\|v\|^2_E - 2(f, v)$ on E_0, (f, v) = scalar product between $f \in E_0'$ and $v \in E_0$.

2. Elliptic Systems

2.1 Notation

Let A be the second order operator

$$Au = -\sum_{i,j=1}^{n} \frac{\partial}{\partial x_i}\left(a_{ij}(x)\frac{\partial}{\partial x_j}\right)$$

and assume that Ω is of class $\{k!\}$ (see Chapter 8, Section 1.2), that the coefficients a_{ij} are real analytic in $\bar{\Omega}$ and that A is elliptic in $\bar{\Omega}$ ((1)).

Suppose that the state of a physical system is given by

$$y = y(x; v), \; x \in \Omega, \; v = \text{control},$$

solution of

$$Ay(x; v) = 0 \text{ in } \Omega, \tag{2.1}$$

$$y(x; v) = v \text{ on } \Gamma. \tag{2.2}$$

(we shall also denote the function $x \to y(x; v)$ by $y(v)$).

We shall assume that v belongs to a space of analytic functions on Γ. More precisely (see Chapter 7, Section 3.2) we introduce:

$$\Delta_\Gamma = \text{Laplace-Beltrami operator on } \Gamma,$$

$$\begin{cases} \mathscr{H}_L(\Gamma) = \{\text{space of functions } \varphi \text{ such that} \\ \displaystyle\sum_{k=0}^{\infty} \frac{1}{L^{2k}((2k)!)^2} \|\Delta_\Gamma^k \varphi\|_{L^2(\Gamma)}^2 = \|\varphi\|_{\mathscr{H}_L(\Gamma)}^2 < \infty\}^{((2))}. \end{cases} \tag{2.3}$$

We know (see Chapter 8) that $\mathscr{H}_L(\Gamma)$ is a space of real analytic functions on Γ; provided with the norm $\|\varphi\|_{\mathscr{H}_L(\Gamma)}$, $\mathscr{H}_L(\Gamma)$ is a Hilbert space. □

In the sequel, we shall assume that

$$v \in \mathscr{U} = \mathscr{H}_L(\Gamma) \quad (L > 0 \text{ fixed}). \tag{2.4}$$

Then

$$\frac{\partial y}{\partial \nu_A} \in \mathscr{H}_{L_1}(\Gamma) \quad (\nu_A \text{ normal to } \Gamma \text{ with respect to } A), \tag{2.5}$$

where L_1 depends on L, on the coefficients of A and on Γ.

((1)) We have taken A to be a second order operator (and of a particular type) and we consider the Dirichlet problem relative to A, but only in order to simplify the presentation; what is to follow can be extended to regular problems for elliptic operators of arbitrary order with analytic coefficients and, more precisely, under the hypotheses of Chapter 8, Section 3.1.

((2)) For the sake of simplicity, all the functions are taken to be real-valued.

Therefore, for every $v \in \mathcal{U}$, we can define the function:

$$J(v) = \left\| \frac{\partial y(v)}{\partial \nu_A} - z_d \right\|^2_{\mathscr{H}_{L_1}(\Gamma)} + \nu \, \|v\|^2_{\mathscr{H}_L(\Gamma)}, \tag{2.6}$$

where:

z_d is given in $\mathscr{H}_{L_1}(\Gamma)$,

ν is a given positive number.

2.2 Control Problem

Now we consider

$$\mathcal{U}_{\text{ad}} = \text{closed convex set} \subset \mathcal{U}, \tag{2.7}$$

and we seek

$$\operatorname*{Inf.}_{v \in \mathcal{U}_{\text{ad}}} J(v)^{((1))} \tag{2.8}$$

It is easy to see that there exists a unique element $u \in \mathcal{U}_{\text{ad}}$ such that

$$J(u) \leq J(v), \ \forall v \in \mathcal{U}_{\text{ad}}.$$

u is called the optimal control.

We aim to give the system of equations or inequalities which characterize the optimal control.

2.3 Necessary and Sufficient Conditions for Optimality

2.3.1 First Characterization

If $J'(v)$ denotes the derivative of J (the existence of which is easily verified), then the optimal control u is characterized by

$$J'(u)\,(v - u) \geq 0, \ \ \forall v \in \mathcal{U}_{\text{ad}}, \tag{2.9}$$

which is equivalent to

$$\left(\frac{\partial y(u)}{\partial \nu_A} - z_d, \ \frac{\partial y(v)}{\partial \nu_A} - \frac{\partial y(u)}{\partial \nu_A} \right)_{\mathscr{H}_{L_1}(\Gamma)} + \nu(u, v - u)_{\mathscr{H}_L(\Gamma)} \geq 0 \ \forall v \in \mathcal{U}_{\text{ad}}. \tag{2.10}$$

((1)) Thus we consider problems analogous to those studied in Chapter **6** (Vol. 2), but the Sobolev spaces are replaced with *Hilbert spaces of Gevrey type or of analytic functions.*

2.3.2 Adjoint State

We know (see Chapter 7, Section 3.2) that the series

$$\sum_{k=0}^{\infty} \frac{1}{L_1^{2k}((2k)!)^2} \Delta_\Gamma^{2k} \left(\frac{\partial y(u)}{\partial \nu_A} - z_d \right)$$

represents a continuous linear form on $\mathscr{H}(\Gamma)$, i.e. an element of the space $\mathscr{H}'(\Gamma)$ of analytic functionals on Γ. And then, according to the results of Chapter 8, Section 3.6, the Dirichlet problem

$$A^* p(u) = 0, \tag{2.11}$$

$$p(u) = \sum_{k=0}^{\infty} \frac{1}{L_1^{2k}((2k)!)^2} \Delta_\Gamma^{2k} \left(\frac{\partial y(u)}{\partial \nu_A} - z_d \right), \tag{2.12}$$

where A^* is the formal adjoint of A, admits a unique solution $p(u) \in \mathscr{D}'(\Omega)$, condition (2.12) being taken in $\mathscr{H}'(\Gamma)$. The distribution $p(u)$ is called the adjoint state of the control problem.

2.3.3 Transformation of (2.10)

Applying Green's formula (see Chapter 8, Section 3.5), we can then write the following relation:

$$-\left\langle \frac{\partial p(u)}{\partial \nu_{A^*}}, y(v) - y(u) \right\rangle + \left\langle p(u), \frac{\partial}{\partial \nu_A} (y(v) - y(u)) \right\rangle = 0 \tag{2.13}$$

where the brackets denote the duality between $\mathscr{H}'(\Gamma)$ and $\mathscr{H}(\Gamma)$. It follows from (2.13) and (2.12) that

$$\left(\frac{\partial y(u)}{\partial \nu_A} - z_d, \frac{\partial y(v)}{\partial \nu_A} - \frac{\partial y(u)}{\partial \nu_A} \right)_{\mathscr{H}_{L_1}(\Gamma)} = \left\langle \frac{\partial p(u)}{\partial \nu_{A^*}}, v - u \right\rangle. \tag{2.14}$$

Set:

$$\Lambda^L = \Sigma \frac{1}{L^{2k}((2k)!)^2} \Delta_\Gamma^{2k}. \tag{2.15}$$

Then

$$(u, v - u)_{\mathscr{H}_L(\Gamma)} = \langle \Lambda^L u, v - u \rangle, \tag{2.16}$$

so that, with (2.14) and (2.16), condition (2.10) is equivalent to

$$\left\langle \frac{\partial p(u)}{\partial \nu_{A^*}} + \nu \Lambda^L u, v - u \right\rangle \geq 0 \quad \forall v \in \mathscr{U}_{\text{ad}}. \tag{2.17}$$

2.4 Conclusion

We summarize the results in

Theorem 2.1. *The optimal control u is given by the resolution of the system:*

$$\begin{cases} Ay(u) = 0 \text{ in } \Omega, \\ A^*p(u) = 0 \text{ in } \Omega, \end{cases} \tag{2.18}$$

$$\begin{cases} y(u) = u \text{ on } \Gamma, \\ p(u) = \Lambda^{L_1}\left(\dfrac{\partial y(u)}{\partial \nu_A} - z_d\right) \text{ on } \Gamma, \end{cases} \tag{2.19}$$

$$\left\langle \frac{\partial p(u)}{\partial \nu_{A^*}} + \nu \Lambda^L u, v - u \right\rangle \geq 0, \quad \forall v \in \mathscr{U}_{ad}{}^{((1))}. \tag{2.20}$$

2.5 Applications (I). Unconstrained Case

The problem is said to be unconstrained when $\mathscr{U}_{ad} = \mathscr{U}$. Then (2.20) reduces to (since $\mathscr{H}_L(\Gamma)$ is dense in $\mathscr{H}(\Gamma)$):

$$\frac{\partial p}{\partial \nu_{A^*}}(u) + \nu \Lambda^L u = 0. \tag{2.21}$$

u can be eliminated from (2.18), (2.19), (2.20); we obtain:

$$\begin{cases} Ay = 0, \\ A^*p = 0, \end{cases} \text{ in } \Omega, \tag{2.22}$$

$$\begin{cases} \dfrac{\partial p}{\partial \nu_{A^*}} + \nu \Lambda^L y = 0 \text{ on } \Gamma, \\ p = \Lambda^{L_1}\left(\dfrac{\partial y}{\partial \nu_A} - z_d\right) \text{ on } \Gamma. \end{cases} \tag{2.23}$$

Therefore the optimal control is obtained by:

(i) solving system (2.22), (2.23) [2];

(ii) setting $u = y|_\Gamma$.

((1)) (2.18), (2.19), (2.20) is a "unilateral, *infinite order*" problem.

[2] The system (2.22), (2.23) is a coupled elliptic system of infinite order, with the particularity that the component y of the solution (y, p) is "very regular", whereas the component p is "very irregular". The components y and p are "in duality".

2.6 Applications (II). Constrained Case

Now let

$$\mathcal{U}_{\mathrm{ad}} = \{v \mid v \in \mathcal{U} = \mathcal{H}_L(\Gamma),\, v \geq 0 \text{ on } \Gamma\}. \tag{2.24}$$

Then (2.20) is equivalent to (since $\mathcal{U}_{\mathrm{ad}}$ is a cone with apex $\{0\}$):

$$\left\langle \frac{\partial}{\partial \nu_{A^*}} p(u) + \nu \Lambda^L u,\, v \right\rangle \geq 0 \;\forall v \in \mathcal{U}_{\mathrm{ad}}, \tag{2.25}$$

$$\left\langle \frac{\partial}{\partial \nu_{A^*}} p(u) + \nu \Lambda^L u,\, u \right\rangle = 0. \tag{2.26}$$

But if $v \in \mathcal{H}(\Gamma)$, $v \geq 0$ on Γ, then there exists $v_j \in \mathcal{H}_L(\Gamma)$, $v_j \geq 0$, $v_j \to v$ in $\mathcal{H}(\Gamma)$, and therefore (2.25) implies (and is equivalent to) the same inequality $\forall v \in \mathcal{H}(\Gamma)$, $v \geq 0$ on Γ. Therefore

$$\frac{\partial p}{\partial \nu_{A^*}}(u) + \nu \Lambda^L u \text{ is an analytic functional} \geq 0. \tag{2.27}$$

But, by arguments analogous to those used by Schwartz [6] in order to show that every distribution ≥ 0 is a measure ≥ 0, we see that (2.27) is equivalent to:

$$\frac{\partial}{\partial \nu_{A^*}} p(u) + \nu \Lambda^L u \text{ is a measure} \geq 0 \text{ on } \Gamma. \tag{2.28}$$

Then condition (2.26) is equivalent to

$$u\left(\frac{\partial}{\partial \nu_{A^*}} p(u) + \nu \Lambda^L u\right) = 0 \text{ on } \Gamma. \tag{2.29}$$

Finally we obtain: the optimal control is provided by the resolution of the system:

$$\begin{cases} Ay = 0, \\ A^* p = 0, \end{cases} \quad \text{in } \Omega, \tag{2.30}$$

$$\begin{cases} p = \Lambda^{L_1}\left(\dfrac{\partial y}{\partial \nu_A} - z_d\right) \text{ on } \Gamma, \\ y \geq 0 \text{ on } \Gamma, \\ \dfrac{\partial p}{\partial \nu_{A^*}} + \nu \Lambda^L y \geq 0 \text{ on } \Gamma, \\ y\left(\dfrac{\partial p}{\partial \nu_{A^*}} + \nu \Lambda^L y\right) = 0 \text{ on } \Gamma. \end{cases} \tag{2.31}$$

Then

$$u = y|_\Gamma.$$

3. Evolution Systems

3.1 Generalities

It is possible, in the setting of Gevrey and analytic function spaces, to study optimal control problems for systems governed by evolution operators like those considered in Chapter 4, 5 and 9, 10, 11 of this text.

In order to fix our ideas, we consider the setting of Chapter 9, Section 7.

We take the notation of Chapter 9, Section 7, but with *E a Hilbert space on* **R**.

Let A be an operator such that *$-A$ is the infinitesimal generator of a semi-group $G(t)$ in E.*

We introduce (there is a slight modification with respect to the definition (7.27), Chapter 9, and this so as to be able to benefit from the *Hilbert* structure of E):

$$(3.1)\quad \begin{cases} \mathrm{D}^L(A^\infty; M_k) = \{e \mid e \in \mathrm{D}(A^\infty),\ \|e\|_{D^L(A^\infty; M_k)} = \\ = \left(\sum_{k=0}^{\infty} \frac{1}{L^{2k} M_k^2} \|A^k e\|_E^2\right)^{1/2} < \infty\}; \end{cases}$$

provided with the norm $\|e\|_{D^L(A^\infty; M_k)}$, $\mathrm{D}^L(A^\infty; M_k)$ is a Hilbert space.

Note that

$$(3.2)\quad G(t) \in \mathscr{L}(\mathrm{D}^L(A^\infty; M_k); \mathrm{D}^L(A^\infty; M_k)),$$

and therefore that we have: let v be given in $\mathrm{D}^L(A^\infty; M_k)$; there exists a unique solution $y(t, v) = y(v)$, with values in $\mathrm{D}^L(A^\infty; M_k)$, of

$$(3.3)\quad \frac{\mathrm{d}}{\mathrm{d}t} y(t; v) + A y(t; v) = 0,\ t > 0,$$

$$(3.4)\quad y(0; v) = v.$$

We shall consider v as the "control"[1] and $y(v)$ as the state of the system[2].

3.2 The Optimal Control Problem

Setting

$$(3.5)\quad \mathscr{U} = \mathrm{D}^L(A^\infty; M_k),$$

we consider the cost function

$$(3.6)\quad J(v) = \|y(T; v) - z_d\|_{\mathscr{U}}^2 + \nu \|v\|_{\mathscr{U}}^2,$$

[1] Actually this is more like a "filtering" problem.

[2] One could also consider systems governed by equations of the second order in t (and therefore, in particular, hyperbolic equations).

where

$$T > 0 \text{ is fixed,}$$

$$z_d \text{ is given in } \mathrm{D}^L(A^\infty; M_k),$$

$$\nu > 0 \text{ is fixed.}$$

Next, we are given

$$\mathscr{U}_{\mathrm{ad}} = \text{closed convex set in } \mathscr{U}, \tag{3.7}$$

and we seek to "characterize" the unique element u of $\mathscr{U}_{\mathrm{ad}}$ (the optimal control) such that

$$\operatorname*{Inf}_{v \in \mathscr{U}_{\mathrm{ad}}} J(v) = J(u). \tag{3.8}$$

3.3 Necessary and Sufficient Conditions for Optimality

3.3.1 First Characterization

The control u is optimal if and only if

$$\begin{cases} \big(y(T; u) - z_d, y(T; v) - y(T; u)\big)_{D^L(A^\infty; M_k)} + \\ + \nu(u, v - u)_{D^L(A^\infty; M_k)} \geq 0 \ \forall v \in \mathscr{U}_{\mathrm{ad}}. \end{cases} \tag{3.9}$$

Let us introduce the operator

$$\nabla^L = \sum_{k=0}^{\infty} \frac{1}{L^{2k} M_k^2} A^{*k} A^k. \tag{3.10}$$

We easily see that

$$\nabla^L \in \mathscr{L}\big(\mathrm{D}^L(A^\infty; M_k); (\mathrm{D}^L(A^\infty; M_k))'\big). \tag{3.11}$$

Then (3.9) is equivalent to

$$\begin{cases} \langle \nabla^L(y(T; u) - z_d), y(T; v) - y(T; u)\rangle + \nu\langle \nabla^L u, v - u\rangle \geq 0 \\ \forall v \in \mathscr{U}_{\mathrm{ad}}, \end{cases} \tag{3.12}$$

where the brackets denote the duality between $\mathrm{D}^L(A^\infty; M_k)$ and its dual.

3.3.2 Adjoint State

The adjoint state $p(u)$ is defined by the solution of (apply Chapter 9, Section 7.4):

$$-\frac{\mathrm{d}}{\mathrm{d}t} p(u) + A^* p(u) = 0 \quad \text{in} \quad]0, T[, \tag{3.13}$$

$$p(T; u) = \nabla^L(y(T; u) - z_d) \quad , \tag{3.14}$$

$$p(t; u) \in \big(\mathrm{D}^L(A^\infty; M_k)\big)'. \tag{3.15}$$

The solution is given by

$$p(t; u) = G^*(T - t) \, \nabla^L(y(T; u) - z_d). \tag{3.16}$$

Then we see that

$$\langle \nabla^L(y(T; u) - z_d), y(T; v) - y(T; u)\rangle$$
$$= \langle p(0; u), y(0; v) - y(0; u)\rangle = \langle p(0; u), v - u\rangle$$

and therefore (3.12) is equivalent to

$$\langle p(0; u) + \nu \nabla^L u, v - u\rangle \geq 0 \quad \forall v \in \mathscr{U}_{\mathrm{ad}}. \tag{3.17}$$

3.4 Conclusion

In summary, we have:

Theorem 3.1. *The optimal control u is given by the resolution of the system*:

$$\begin{cases} \dfrac{\mathrm{d}}{\mathrm{d}t} y(u) + A y(u) = 0, \\ -\dfrac{\mathrm{d}}{\mathrm{d}t} p(u) + A^* p(u) = 0, \end{cases} \quad \text{in }]0, T[\tag{3.18}$$

$$\begin{cases} y(0; u) = u, \\ p(T; u) = \nabla^L(y(T; u) - z_d), \end{cases} \tag{3.19}$$

$$\langle p(0; u) + \nu \nabla^L u, v - u\rangle \geq 0, \ \forall v \in \mathscr{U}_{\mathrm{ad}}. \tag{3.20}$$

3.5 Application

We only give one application. We take Ω and A as in Section 2 and $E = L^2(\Omega)$. Then we have

$$\mathrm{D}(A) = \{\psi \mid \psi \in H^2(\Omega), \gamma_0 \psi = 0 \text{ on } \Gamma\}.$$

We are within the conditions of applicability of the preceding theory. If we take

$$M_k = ((2k)!)^s \ (s \geq 1), \tag{3.21}$$

then, thanks to Theorem 1.2 of Chapter 8:

$$\mathrm{D}^L(A^\infty; M_k) \subset \mathscr{D}_s(\bar{\Omega}) = \{\text{Gevrey functions of order } s \text{ in } \bar{\Omega}\}. \tag{3.22}$$

We take

$$\mathscr{U}_{\mathrm{ad}} = \{v \mid v \in \mathrm{D}^L(A^\infty; M_k), v \geq 0 \text{ in } \Omega\}. \tag{3.23}$$

(Applying Theorem 3.1) we end up with the following problem:

$$\left\{\begin{aligned}&\frac{\partial}{\partial t} y + Ay = 0,\\ &-\frac{\partial p}{\partial t} + A^* p = 0,\end{aligned}\right. \qquad \text{in } Q = \Omega \times]0, T[\tag{3.24}$$

$$y = 0, p = 0 \text{ on } \Sigma, \tag{3.25}$$

$$\left\{\begin{aligned}&p(x, T) = \nabla^L(y(x, T) - z_d(x)),\\ &y(x, 0) \geq 0,\\ &p(x, 0) + \nu \nabla^L y(x, 0) \geq 0 \text{ in } \Omega^{((1))},\\ &y(x, 0)\,[p(x, 0) + \nu \nabla^L y(x, 0)] = 0.\end{aligned}\right. \tag{3.26}$$

This is a nonlinear problem (of "unilateral" type) containing *differential operators of infinite order*, and admitting a unique solution.

((1)) We write the functional $p(T) \in (\mathrm{D}^L(A^\infty; M_k))'$ like a function. This is a symbolic notation.

Bibliography

Agmon, S.

1. Problèmes mixtes pour les équations hyperboliques d'ordre supérieur. Colloque Intern. C. N. R. S., N° 117 sur Les équations aux dérivées partielles, Paris, 1962, 13—18.

Agmon, S., Nirenberg, L.

1. Properties of solutions of ordinary differential equations in Banach spaces. Comm. Pure Appl. Math. **16**, 121—239 (1963).

Agranovich, M. S.

1. Sur la théorie des problèmes aux limites pour les systèmes symétriques du premier ordre. Mat. Sbornik **73**, 161—197 (1967).

Amerio, L., Prouse, G.

1. Abstract almost periodic functions and functional analysis. New York: van Nostrand 1970.

Aronszajn, N.

1. Sur un théorème de la théorie des fonctions de plusieurs variables complexes. C. R. Acad. Sc. Paris **205**, 16—18 (1937).
2. Polyharmonic functions. (Book to be published).

Artola, M.

1. Sur les perturbations des équations d'évolution. Application à des problémes de retard. Annales Ec. Norm. Sup., t. 2, 137—253 (1969).

Asano, K.

Cf. T. Shirota et K. Asano.

Aubin, J. P.

1. Behavior of the error of the approximate solutions of boundary value problems for linear elliptic operators by Galerkin's and finite difference methods. Annali Scuola Norm. Sup. Pisa **21**, 599—637 (1967).
2. Approximation of non-homogenous Neumann problems. M. R. C., Univ. of Wisconsin (1968).
3. Approximation of elliptic boundary value problems. New York: Academic Press 1972.

Baiocchi, C.

1. Sui problemi ai limiti per le equazioni paraboliche del tipo del calore. Boll. U. M. I. **3**, 19, 407—422 (1964).
2. Teoremi di esistenza e regolarità per certe classi di equazioni differenziali astratte. Ann. Mat. Pura Appl. **4**, 72, 365—418 (1966).
3. Sulle equazioni differenziali astratte lineari del primo e del secondo ordine negli spazi di Hilbert. Ann. Mat. pura appl. **4**, 76, 233—304 (1967).

Baouendi, M. S.

1. Sur une classe d'opérateurs elliptiques dégénérés. Bull. Soc. Math. France **95**, 45—87 (1967).

Baouendi, M. S., Geymonat, G.
1. Quelques résultats de dualité dans les problèmes aux limites linéaires elliptiques. C. R. A. S: Paris, J. Diff. Equations **11**, 207—220 (1972).

Baouendi, M. S., Goulaouic, C.
1. Étude de la régularité et du spectre d'une classe d'opérateurs elliptiques dégénérés. C. R. A. S. Paris, 3. A **266**, 336—339 (1968); and Arch. for Rat. Mech. and Anal. **34**, 361—379 (1969).

Bardos, G.
1. Problèmes aux limites pour les équations aux dérivées partielles du premier ordre à coefficients réels; théorèmes d'approximation; application à l'équation de transport. Thesis, Paris, 1969; Ann. Ec. Norm. Sup. **3**, 185—233 (1970).

Beltrami, E. J., Wohlers, M. R.
1. Distributional boundary values of functions holomorphic in a half plane. J. of Math. and Mech. **15**, 137—146 (1966).

Bengel, G.
1. Sur une extension de la théorie des hyperfunctions. C. R. Acad. Sc. Paris **262**, serie A, 499—501 (1966).
2. Régularité des solutions hyperfunctions d'une équation elliptique. C. R. Acad. Sc. Paris **262**, série A, 569—570 (1966).
3. Das Weylsche Lemma in der Theorie der Hyperfunktionen. Math. Zeitschr. **96**, 373—392 (1967).

Beurling, A.
1. Quasi-analyticity and general distributions, Lectures 4 and 5, Summer Institute, Stanford 1961.

Björck, G.
1. Linear partial differential operators and generalized distributions. Ark Mat. **6**, 21, 351—407 (1966).

Boman, J.
1. On the intersection of classes of infinitely differentiable functions. Ark. Mat. **5**, 301—309 (1964).

Bossavit, A.
1. Régularisation d'équations variationnelles et applications. Thèse. Paris, Juin 1970.

Bourbaki, N.
1. Espaces vectoriels topologiques, Chap. I and II. Paris: Hermann 1966 (second edition).
2. Espaces vectoriels topologiques, Chap. III and IV. Paris: Hermann 1955.
3. Intégration, Chap. 1, 2, 3, 4. Paris: Hermann 1965.

Boutet de Monvel, L.
1. Opérateurs pseudo-différentiels analytiques et problèmes aux limites elliptiques. Ann. Inst. Fourier XIX, 169—268 (1970).
2. Opérateurs pseudo-différentiels analytiques. Problèmes aux limites pseudo-différentiels. C. I. M. E., Stresa 1968, Roma: Cremonese 1969, 37—83.

Boutet de Monvel, L., Krée, P.
1. Pseudo-differential operators and Gevrey classes. C. R. Acad. Sc. Paris **263**, (1966). Série A, Ann. Inst. Fourier **17**, 295—323 (1967).

Browder, F. E.
1. Parabolic systems of differential equations with time dependent coefficients. Proc. Nat. Acad. Sci. **42**, 914—917 (1956).

Carleson, L.
1. On universal moment problems. Math. Scand. **9**, 197—206 (1961).

Cavallucci, A.
1. Sulle proprietà differenziali delle soluzioni delle equazioni quasi-ellittiche. Ann. Mat. pura e appl. **4**, 67, 143—168 (1965).
2. Sulla regolarità delle soluzioni delle equazioni quasi-ellittiche in un semi-spazio. Atti Sem. Mat. Fis. Modena **17**, 1—18 (1967).

Chazarain, J.
1. Problèmes de Cauchy au sens des distributions vectorielles et applications. C. R. A. S. Paris **266**, 10—13 (1968).
2. Problèmes de Cauchy dans des espaces d'ultra distributions. C. R. A. S. Paris **266**, 564—566 (1968).
3. Un résultat de perturbation pour les générateurs de semi groupes et applications. C. R. A. S. Paris **267**, 13—15 (1968).
4. Problèmes de Cauchy abstraits et applications à quelques problèmes mixtes. J. Funct. Analysis **7**, 386—446 (1971).

Chou, C. C.
1. Problème de régularité universelle. C. R. Acad. Sc., Paris **260**, 4397—4399 (1965).

Cimmino, G.
1. Su alcuni esempi notevoli di dualità fra spazi lineari topologici. Rend. Sem. Mat. Fis. Milano **33**, 102—113 (1963).
2. Sulla totalità delle funzioni che in un dato campo verificano una equazione differenziale omogenea. Atti Acc. Scienze di Bologna **11**, 7 (1960).

Cioranescu, I.
1. La caractérisation spectrale d'operateurs générateurs de semi-groupes distributions d'ordre fini de croissance. J. Math. Anal. and Appl. **34**, 34—41 (1971).

Courant, R., Hilbert, D.
1. Methods of Mathematical Physics, Vol. 1 and 2, New York: Interscience Publishers 1953, 1962.

Da Prato, G.
1. Semi-gruppi regolarizzabili. Ricerche di Mat. **15**, 223—248 (1966).
2. R-semi-gruppi analitici ed equazioni di evoluzione in L^p. Ricerche di Mat. **16**, 233—249 (1967).

Da Prato, G., Mosco, U.
1. Semi-gruppi distribuzioni analitici. Ann. Sc. Norm. Sup. Pisa **18**, 367—396 and 563—576 (1965).

Delsarte, J.
1. Sur certaines transformations fonctionnelles relatives aux équations linéaires aux dérivées partielles du second ordre. C. R. Acad. Sc. Paris **206**, 1780—1782 (1938).

De Rham, G.
1. Variétés différentiables. Paris: Hermann 1955.

Derridj, M.
1. Un problème aux limites pour une classe d'opérateurs hypoelliptiques de second ordre. Ann. Inst. Fourier, Grenoble **21**, 4, 99—148 (1971).

Dieudonné, J., Schwartz, L.
1. La dualité dans les espaces $(\mathscr{F})$ et $(\mathscr{L}\mathscr{F})$. Ann. Inst. Fourier **1**, 61—101 (1949).

Douglas, J.
1. Approximate continuation of harmonic and parabolic functions. Proc. Symp. Numer. Solut. Part. Diff. Equat., New York: Academic Press 1966, 353—360.

Dzanasija, G. A.
1. Sur la composition des fonctions de Gevrey. Bull. Acad. Sc. Georgie, S. S. R. **33**, 2, 257—262 (1964).

2. Sur le problème de Carleman dans les espaces de Gevrey. Dokl. Akad. Nauk **145**, 259–262 (1962).

Ehrenpreis, L.

1. Analytic functions and the Fourier transform of distributions I. Annals. of Maths. **63**, 129–159 (1956).
2. Analytic uniform spaces and some applications. Trans. Amer. Mat. Soc. **101**, 52–74 (1961).
3. A fundamental principle for systems of linear differential equations with constant coefficients and some of its applications. Proc. Intern. Symp. on Linear Spaces. Jerusalem, 1961, 161–174.

Eidelman, S. D.

1. Majoration des solutions des systèmes paraboliques et applications. Mat. Sbornik **33**, 359–382 (1953).
2. Sur l'analyticité des solutions des systèmes paraboliques. Dokl. Akad. Nauk **103**, 359–382 (1955).
3. Sur la solution élémentaire des systèmes paraboliques. Math. Sbornik **38**, 51–92 (1956) (Amer. Math. Soc. Transl. (2), **41**, 1–48 (1964)).
4. Systèmes paraboliques, Moscow, 1964 (English translation; Amsterdam: North-Holland, 1969).

Fage, M. K.

1. Fonctions d'une variable analytiques par rapport à un opérateur. Tr. Moskov Mat. Obv. **7**, 227–268 (1958).

Fantappie, L.

1. I funzionali analitici. Mem. Acc. Naz. Lincei **6**, 3, 453–683 (1930).
2. Nuovi fondamentali della teoria dei funzionali analitici. Atti Acc. d'Italia, Mem. Cl. Sc. Fis. Mat. **12**, 617–706 (1942).

Fattorini, H. O.

1. A representation theorem for distribution semigroups. J. Funct. Anal. **6**, 1–25 (1970).

Foias, C.

1. Remarques sur les semi-groupes distributions d'opérateurs normaux. Portug. Math. **19**, 227–242 (1960).

Friberg, J.

1. Estimates for partially hypoelliptic differential operators. Medd. Lund's Univ. Math. Sem. **17**, 1–97 (1963).

Friedman, A.

1. Generalized functions and partial differential equations. Englewood-Cliffs, N.J.: Prentice-Hall 1963.
2. On the regularity of the solutions of non-linear elliptic and parabolic systems of partial differential equations. J. Math. Mech. **7**, 43–59 (1958).
3. On classes of solutions of elliptic linear differential equations. Proc. Amer, Math. Soc. **8**, 418–427 (1957).
4. Classes of solutions of linear systems of partial differential equations of parabolic type. Duke Math. J. **24**, 433–442 (1957).
5. Partial differential equations of parabolic type. New York: Prentice-Hall 1964.
6. Interior estimates for parabolic systems of partial differential equations. J. Math. Mech. **7**, 393–418 (1958).
7. Differentiability of solutions of ordinary differential equations in Hilbert space. Pacific J. of Math. **16**, 267–271 (1966).

Friedrichs, K. O., Lax, P.

1. Boundary value problems for first order operators. Comm. Pure Appl. Math. **18**, 355–388 (1965).

Fnjiwara, D.

1. A Characterization of exponential distribution semi-groups. J. Math. Soc. Japan, **18**, 3, 267—274 (1966).

Garnir, H. G., de Wilde, M., Schmets, J.

1. Analyse fonctionnelle, Théorie constructive. Basel: Birkhäuser 1968.

Gelfand, I. M., Shilov, G. E.

1. Fonctions généralisées, Vol. I, II, III. Moscow, 1958 (French translation: Les distributions, Dunod, Paris; English translation: New York: Academic Press).

Gelfand, I. M., Vilenkin, N. Ya.

1. Fonctions généralisées, Vol. IV. Moscow, 1961 (French translation: Dunod, Paris; English translation: New York: Academic Press).

Geymonat, G.

1. Su acluni spazi di funzioni infinitamente derivabile a valori vettoriali. Studia Ghisleriana, série spec. IV centenario, Pavia, 1967.
2. Proprietà di alcuni spazi di funzioni indefinitamente derivabili a valori vettoriali. Ann. Mat. Pura Appl. **4**, 76, 203—232 (1967).

Geymonat, G. Baouendi, M. S.

Cf. Baouendi-Geymonat.

Gevrey, M.

1. Sur la nature analytique des solutions des équations aux dérivées partielles. Ann. Ec. Norm. Sup. Paris **35**, 129—190 (1918).

Goodman, R.

1. Analytic domination by fractional powers of a positive operator. J. Funct. Anal. **3**, 246—264 (1969).

Goulaouic, C.

1. Prolongements de facteurs d'interpolation et applications. Ann. Inst. Fourier **18-1**, 1—98 (1968).
2. Interpolation entre espaces localement convexes construits à l'aide d'opérateurs non bornés; cas des espaces de Gevrey. Ann. Inst. Fourier **19-2**, 269 to 278 (1970).

Grisvard, P.

1. Équations différentielles abstraites. Ann. E. N. S., Paris (4) **2**, 311—395 (1969).

Grothendieck, A.

1. Produits tensoriels topologiques. Mémoires Amer. Math. Soc. **16** (1955).
2. Sur les espaces ($\mathscr{F}$) et ($\mathscr{DF}$). Summa Bras. Math. **3**, 57—123 (1954).
3. Sur certains espaces de fonctions holomorphes, I and II. J. reine angew. Math. **192**, 35—64 and 77—95 (1953).

Goulaouic, C., Baouendi, M. S.

Cf. Baouendi-Goulaouic.

Harvey, F. R.

1. Hyperfunctions and linear partial differential equations. Proc. Nat. Acad. Sc. U. S. A. **55**, 1042—1046 (1966).

Harvey, F. R., Komatsu, H.

1. Hyperfunctions and linear partial differential equations. Proc. Int. Congress of Math., Moscow, 1966.

Hasumi, M.

1. Note on the n-dimensional tempered ultra-distributions. Tôhoku Math. J. **2**, 13, 94—104 (1960).

Hille, E., Phillips, R. S.

1. Functional Analysis and Semi-groups. A. M. S. Coll. Pub., XXXI, 1957.

Hȯlmgren, E.

1. Sur l'équation de la propagation de la chaleur. Arkiv. for Math. Ast. Fys. **4**, 18, 1—28 (1908).

Hȯrmander, L.

1. Linear partial differential operators. Grundlehren B. 116, Springer, Berlin 1963.
2. On the theory of general partial differential operators. Acta Math. **94**, 161 to 248 (1955).
3. Pseudo-differential operators and hyperelliptic equations, Singular Integrals. Proceed. of Symp. on pure Math., Vol. 10; Amer. Math. Soc. 1968, 138—183.

Horvath, J.

1. Topological vector spaces and distributions, Vol. I. London: Addison-Wesley 1966.

Huet, D.

1. Perturbations singulières. C. R. Acad. Sc. Paris **260**, 6800—6801 (1965).

Ikawa, M.

1. On the mixed problem for the Wave equation with an oblique derivative boundary condition. Proc. Japan Acad. **44**, 1033—1037 (1968).

John, F.

1. Plane waves and spherical means applied to partial differential equations New York: Interscience Publishers 1955.

John, O.

1. Sulla regolarità delle soluzioni delle equazioni lineari ellittiche negli spazi di Beurling. Boll. U. M. I. **4**, 2, 183—195 (1969).

Johnson, G., jr.

1. Harmonic functions on the unit disc. I, II. Illinois J. Math. **12**, 366—385 (1968); ibid. **12**, 386—396 (1968).

Kantor, J. M.

1. Hyperfonctions cohérentes. C. R. Acad. Sc. Paris, 1. A. **269**, 18—20 (1969).

Kato, T.

1. Abstract evolution equations of parabolic type in Banach and Hilbert spaces. Nagoga Math. J. **19**, 93—125 (1961).

Kato, T., Tanabe, H.

1. On the analyticity of solutions of evolution equations. Osaka J. Math. **4**, 1—4 (1967).

Komatsu, H.

1. A proof of Kotake and Narasimhan's theorem. Proc. Japan Acad. **38**, 615 (1962).
2. Resolution by hyperfunctions of sheaves of solutions of differential equations with constant coefficients. Math. Ann. **176**, 77—86 (1968).
3. A characterization of real analytic functions. Proc. Jap. Acad. **36**, 90—93 (1960).
4. Boundary value for solutions of elliptic equations. Colloque sur l'Analyse fonctionnelle, Tokyo 1969.
5. Projective and injective limits of weakly compact sequences of locally convex spaces. J. Math. Soc. Japan. **19**, 366—383 (1967).
6. Abstract analyticity in time and unique continuation property of solutions of a parabolic equation. J. Sci. Univ. Tokyo, S. I, 9, p. I 1—11 (1961).

Komatsu, H., Harvey, R.

Cf. Harvey-Komatsu.

Kotake, T.
1. Analyticité du noyau élémentaire de l'opérateur parabolique. Colloques Intern. du C. N. R. S., N° 117, sur Les équations aux dérivées partielles, Paris, 1962, 53–60.
Kotake, T., Narasinham, M. S.
1. Fractional powers of a linear elliptic operator. Bull. Soc. Math. France **90**, 449–471 (1962).
Köthe, G.
1. Dualität in der Funktionentheorie. J. reine angew. Math. **191**, 30–49 (1953).
2. Die Randverteilungen analytischer Funktionen. Math. Zeitschr. **57**, 13–33 (1952).
3. Topologische lineare Räume. Grundlehren B. 107, Berlin/Göttingen/Heidelberg: Springer 1960.
Krée, P.
1. Problèmes aux limites en théorie des distributions. Ann. Mat. Pura Appl. **83**, 113–132 (1969).
2. A class of singular integrals. Pseudodifferential operators on non-quasi-analytic functions spaces. Proc. Symp. on Singular Integrals, A. M. S., Vol. 10, 1968, 208–212.
Krein, S.
1. Équations différentielles linéaires dans les espaces de Banach. Moscow, 1967.
Kriptun, V. G.
1. Sur les fonctions d'une variable quasi analytiques par rapport à un opérateur différentiel linéaire. Sibirsk. Mat. J. **6**, 1395–1406 (1965).
Larsson, E.
1. Generalized hyperbolicity. Ark. Mat. **7**, 1 (1967).
2. Generalized distribution semi-groups of bounded linear operators. Ann. Sc. Norm. Pisa **21**, 137–259 (1967).
Lax, P., Friedrichs, K. O.
Cf. Friedrichs-Lax.
Lax, P., Phillips, R. S.
1. Local boundary conditions for dissipative symmetric linear differential operators. Comm. Pure Appl. Math. **13**, 427–455 (1960).
Leray, J.
1. La solution unitaire d'un opérateur différential linéaire. Bull. Soc. Math. France **86**, 75–96 (1958).
2. Le calcul différentiel et intégral sur une variété analytique complexe. Bull. Soc. Math. France **87**, 81–180 (1959).
Leray, J.-Ohya, Y.
1. Systèmes linéaires hyperboliques non stricts. Séminaire Leray, Collège de France, 1964 and II. Coll. Anal. Fonct., Liège, C. B. R. M., Gauthier-Villars: Paris: 1964, 105–144.
Leray, J.,-Waelbroeck, L.
1. Norme formelle d'une fonction composée. II. Coll. Anal. Fonct., Liège, C. B. R. M., Paris: Gauthier-Villars 1964, 145–152.
Levi, E. E.
1. Sull'equazione del calore. Ann. Mat. Pura Appl. **3**, 14 (1908).
Lichnerowicz, A.
1. Théorèmes d'existence et d'unicité pour un fluide thermodynamique relativiste. C. R. A. S. Paris **260**, 3291–3295 (1965).
2. Étude mathématique des équations de la magnétohydrodynamique relativiste. C. R. A. S. Paris **260**, 4449–4453 (1965).

Lions, J. L.

1. Problèmes aux limites en théorie des distributions. Acta Math. **94**, 13—153 (1955).
2. Sur les problèmes mixtes pour certains systèmes paraboliques dans des ouverts non cylindriques. Ann. Inst. Fourier **7**, 143—182 (1957).
3. Une remarque sur les applications du théorème de Hille-Yosida. J. Math. Soc. Japan **9**, 62—70 (1957).
4. Les semi-groupes distributions. Port. Math. **19**, 141—164 (1960).
5. Équations différentielles opérationnelles et problèmes aux limites. Grundlehren Bd. 111, Berlin/Göttingen/Heidelberg: Springer 1961 (second edition in preparation).
6. Sur le contrôle optimal de systèmes décrits par des équations aux dérivées partielles linéaires, (I) Remarques générales, (II) Équations elliptiques, (III) Équations d'évolution. C. R. Acad. Sc. Paris **263**, 661—663; 713—715; 776—779 (1966).
7. Contrôle optimal de systèmes gouvernés par des équations aux dérivées partielles. Paris: Dunod, Gauthier-Villars 1968.
8. Opérateurs de Delsarte et problèmes mixtes. Bull. Soc. Math. France **84**, 9—95 (1956).
9. Quelques méthodes de résolution des problèmes aux limites non linéaires. Paris: Dunod, Gauthier-Villars 1969.
10. Problèmes aux limites non homogènes à données irrégulières; une méthode d'approximation. C. I. M. E. (1967), Roma: Cremonese 1968, 285—292.
11. Vectors of Gevrey classes; applications. Differential Equations and their applications. J. Acta Fac. rerum. natur. Univ. Comen, 1967, 93—100.

Lions, J. L., Magenes, E.

1. Problèmes aux limites non homogènes (VII). Ann. Math. Pura Appl. **4**, 63 (1963). 201—224
2. Sur certains aspects des problèmes aux limites non homogènes pour des opérateurs paraboliques. Ann. Sc. Norm. Sup., Pisa **28**, 303—344 (1964).
3. Espaces de fonctions et de distributions du type de Gevrey et problèmes aux limites paraboliques. Ann. Mat. Pura Appl. **4**, 68, 341—418 (1965).
4. Espaces du type de Gevrey et problèmes aux limites pour diverses classes d'équations d'évolution. Ann. Mat. Pura Appl. **4**, 72, 343—394 (1966).
5. Quelques remarques sur les problèmes aux limites linéaires elliptiques et paraboliques dans des classes d'ultradistributions, I and II. Rend. Acc. Lincei **8**, 43, 293—299 and 469—478 (1967).
6. Contrôle optimal et espaces du type de Gevrey, I and II. Rend. Acc. Lincei **8**, 44, 34—39 and 151—157 (1968).

Lions, J. L., Malgrange, B.

1. Sur l'unicité rétrograde. Math. Scand. **8**, 277—286 (1960).

Magenes, E.

1. Problèmes de traces et problèmes aux limites pour équations linéaires elliptiques et paraboliques. Colloque sur l'Analyse functionelle, Liège 1964, C. B. R. M., Paris: Gauthier-Villars 83—95.
2. Espaces de fonctions et de distributions vectorielles du type de Gevrey et équations différentielles. Séminaire Leray, Collège de France, Paris, 1965 to 1966.
3. Problèmes aux limites dans des espaces de fonctions et d'ultradistributions du type de Gevrey. Proc. Int. Congress of Math., Moscow, 1966.
4. Alcuni aspetti della teoria delle ultradistribuzioni e delle equazioni a derivate parziali. Symposia Math., Vol. II, Ist. Naz. Alta Mat., Roma, 1968, 235—254.

5. Semigruppi in spazi de tipo di Gevrey e applicazioni alle equazioni lineari di evoluzione. Atti VIII Congresso U. M. I., Triest, 2—7 October **1967**.

Magenes, E., Lions, J. L.
Cf. Lions-Magenes.

Magenes, E., Stampacchia, G.
1. I problemi al contorno per le equazioni differenziali del tipo ellittico. Ann. Soc. Norm. Sup. Pisa **12**, 247—357 (1958).

Malgrange, B.
1. Le théorème de préparation en géométrie différentielle, in Séminaire H. Cartan sur la Topologie différentielle, 1962—1963, Paris, exposé n° 12.
2. Ideals of Differentiable Functions. Tata Inst. of Fund. Res., Bombay, London: Oxford University Press 1966.
3. Existence et approximation des solutions des équations aux dérivées partielles et de convolution. Ann. Inst. Fourier **6**, 271—354 (1955).

Malgrange, B., Lions, J. L.
Cf. Lions-Malgrange.

Manaresi, F.
1. Sulla totalità delle funzioni armoniche entro un semispazio in relazione alle loro tracce di frontiera. Boll. U. M. I., **3**, 19, 311—326 (1964).

Mandelbrojt, S.
1. Séries adhérentes, régularisation des suites. Applications. Paris: Gauthier-Villars 1952.

Mantovani, F., Spagnolo, S.
1. Funzionali analitici reali e funzioni armoniche. Ann. Sc. Norm. Sup. Pisa **18**, 475—513 (1964).

Martineau, A.
1. Les hyperfonctions de M. Satô. Séminaire Bourbaki, 13e année, 1960—1961, n° 214, February 1961.
2. Distributions et valeurs au bord des fonctions holomorphes, extract from Theory of Distributions. Proc. Intern. Summer Inst., Lisbon 1964, 195—326.
3. Sur les fonctionnelles analytiques et la transformée de Fourier-Borel. Journal Analyse Math. **11**, 1—164 (1963).
4. Sur le théorème du graphe fermé. C. R. Acad. Sc. Paris **263**, 870—871 (1966).

Matagne, R.
1. Les espaces de Silva. Bull. Soc. Sc. Liège **12**, 754—768 (1964).

Matsuzawa, T.
1. On quasi elliptic boundary problems. Trans. Amer. Math. Soc. **133**, 241—265 (1968).
2. Sur les équations quasi-elliptiques et les classes de Gevrey. Bull. Soc. Math. France **96**, 243—263 (1968).
3. Regularity at the boundary for solutions of hypo-elliptic equations. Osaka Math. J. **3**, 313—334 (1965).

Masuda, K.
1. On the analyticity and the unique continuation theorem for solutions of the Navier-Stokes equation. Proc. Japan Acad. **43**, 827—832 (1967).

Mityagin, B. S.
1. Nucléarité et autres propriétés des espaces de type S. Trudy. Mosk. Math. **9**, 317—328 (1960).
2. Une fonction indéfiniment différentiable avec dérivées données en un point. Dokl. Akad. Nauk **138** (1961).

Mizohata, S.
1. Hypo-ellipticité des équations paraboliques. Bull. Soc. Math. France **85**, 15—50 (1957).
2. Quelques problèmes au bord, du type mixte, pour des équations hyperboliques, Séminaire Leray, Collège de France, 1966—1967, 23—60.

Miyatake, S.
1. On some mixed problems for fourth order hyperbolic equations. J. Math. of Kyoto University **8**, 285—311 (1968).

Morrey, C. B.
1. Multiple integrals in the calculus of variations. Grundlehren Bd. 130, Berlin/Heidelberg/New York: Springer 1966.

Morrey, C. B., Nirenberg, L.
1. On the analyticity of the solutions of linear elliptic systems of partial differential equations. Comm. Pure Appl. Math. **10**, 271—290 (1957).

Mosco, M.
Cf. Da Prato-Mosco.

Murthy, M. K. V.
1. A remark on the regularity at the boundary for solutions of elliptic equations. Ann. Sc. Norm. Sup. Pisa **15**, 355—370 (1961).

Narasimhan, M. S., Kotake, T.
Cf. Kotake-Narasimhan.

Nelson, E.
1. Analytic vectors. Ann. of Math. **70**, 572—615 (1959).

Nirenberg, L., Agmon, S.
Cf. Agmon-Nirenberg.

Nirenberg, L., Morrey, C. B.
Cf. Morrey-Nirenberg.

Nirenberg, L., Treves, F.
1. Solvability of a first order linear partial differential equation. Comm. Pure Appl. Math. **16**, 331—352 (1963).

Ohya, Y.
Cf. Leray-Ohya.

Oleinik, O. A.
1. Sur les équations de type elliptique qui dégénèrent à la frontière. Dokl. Akad. Nauk **87**, 885—888 (1952).

Ovciannikov, L. V.
1. Opérateurs singuliers dans des échelles d'espaces de Banach. Dokl. Akad. Nauk **163**, 819—822 (1965).

Peetre, J.
1. Sur la théorie des semi-groupes distributions. Séminaire Leray, Collège de France, 1963—1964, 79—98.
2. Elliptic partial differential equations of higher order, Lec. Series N° 40. Inst. For Fluid Dyn. and App. Math., University of Maryland 1962.

Pellegrino, F.
1. La théorie des fonctionnelles analytiques et ses applications, in P. Levy: Problèmes concrets d'analyse fonctionnelle. Paris: Gauthier-Villars, 1951, 357—477.

Petrowski, I. G.
1. Sur l'analyticité des solutions des systèmes d'équations différentielles. Mat. Sbornik **5**, 47, 3—70 (1939).

2. Über das Cauchysche Problem für ein System linearer partieller Differentialgleichungen im Gebiet der nichtanalytischen Funktionen. Bull. Univ. Moscow, **1**A, 7, 1—74 (1938).

Phillips, R. S., Hille, E.
Cf. Hille-Phillips.

Phillips, R. S., Lax, P.
Cf. Lax-Phillips.

Pini, B.
1. Proprietà locali delle soluzioni di una classe di equazioni ipoellittiche. Rend. Sem. Mat. Padova **32**, 222—238 (1962).
2. Una osservazione sulla natura delle soluzioni delle equazioni quasi-ellittiche in un semispazio. Rend. Acc. Lincei **8**, 43, 307—311 (1967).

Pozzi, G.
1. Problemi di Cauchy e problemi ai limiti per equazioni di evoluzione del tipo di Schroedinger lineari e non lineari. Annali Mat. Pura e Appl. **78**, 197—258 (1968) and **81**, 205—248 (1969).

Prouse, G., Amerio, L.
Cf. Amerio-Prouse.

Raikow, D. A.
1. Vollstetige Spektren von lokalkonvexen Räumen. Trud. Mosk. Mat. Ob-wa **7**, 413—438 (1959).

Roumieu, C.
1. Sur quelques extensions de la notion de distributions. Ann. Sc. École. Norm. Sup. **77**, 47—121 (1960).
2. Ultradistributions définies sur $\mathbf{R}^n$ et sur certaines classes de variétés différentiables. Journ. Analyse Math. **10**, 153—192 (1962—1963).

Rudin, W.
1. Division in algebra of infinitely differentiable functions. J. Math. Mech. **11**, 797—810 (1962).

Satô, M.
1. Theory of hyperfunctions, I and II. J. Fac. Sc. Tokyo **8**, 139—193 and 287 to 437 (1959—1960).

Schaefer, H. H.
1. Topological vector spaces. New York: Macmillan 1966.

Schaeffer, D. G.
1. The Dirichlet problem with generalized functions as data. Ann. Mat. Pura Appl. **83**, 153—174 (1969).

Schapira, P.
1. Sur les ultra-distributions. Ann. Sc. Ecole Norm. Sup. Paris **4**, 395—415 (1968).
2. Une équation aux dérivées partielles sans solutions dans l'espace des hyperfonctions. C. R. Acad. Sc. Paris **265** (1967) series A, 665—667.
3. Équations aux dérivées partielles dans l'espace des hyperfonctions. Séminaire Lelong, 8ᵉ année, 1967—1968. Berlin/Heidelberg/New York: Springer, Lectures Notes, n° 71.
4. Problème de Dirichlet et solutions hyperfonctions des équations elliptiques. Boll. U. M. I. **4**, 2, 367—372 (1969).
5. Théorie des hyperfonctions. Lectures Notes in Math. **126**. Berlin/Heidelberg/New York: Springer 1970.

Schmets, J., Garnir, H. G., de Wilde, M.
Cf. Garnir-de Wilde-Schmets.

Schwartz, L.
1. Théorie des distributions, Vol. I and II. Paris: Hermann 1950—1951 (second edition 1957).
2. Espaces de fonctions différentiables à valeurs vectorielles. Jour. Analyse Math. **4**, 88—148 (1954—1955).
3. Théorie des distributions à valeurs vectorielles, Vol. I and II. Ann. Inst. Fourier **7**, 1—141 (1957); **8**, 1—209 (1958).
4. Sur le théorème du graphe fermé. C. R. A. S. Paris **263**, 602—605 (1966).
5. Transformation de Laplace des distributions. Sem. Math. Univ. Lund, Volume dedicated to M. Riesz, 1952, 196—206.

Schwartz, L., Dieudonné, J.
Cf. Dieudonné-Schwartz.

Seeley, R. T.
1. Singular integrals and Boundary Value Problems. Amer. J. Math. **88**, 781 to 809 (1967).
2. Topics in pseudo differential operators. C. I. M. E., Session on Pseudodifferential operators (Stresa, Sept. 1968), Roma: Cremonese 1969, 167—305.

Shilov, G. E.
1. Propriétés locales des solutions des équations aux dérivées partielles à coefficients constants. Uspehi Math. Nauk **14**, 3—44 (1959).

Shilov, G. E., Gelfand, I. M.
Cf. Gelfand-Shilov.

Shimakura, N.
1. Sur une certaine classe d'opérateurs différentiels ordinaires, elliptiques dégénérés. Proc. Japan. Acad. **44**, 944—948 (1968).
2 Problèmes aux limites généraux du type elliptique dégénéré. J. of Math. Kyoto Univ. 9.2. (1969).
3. Problèmes aux limites variationnels du type elliptique. Ann. E. N. S. **2**, 255—310 (1969).

Shirota, T.
1. A remark on the abstract analyticity in time for solutions of a parabolic equation. Proc. Japan Acad. **35**, 367—369 (1959).

Shirota, T., Asano, K.
1. On mixed problems for hyperbolic equations of higher order (to be published).

Silva, J. S.
1. Su certe classi di spazi localmente convessi importanti per le applicazione. Rend. Mat. Univ. Roma **14**, 388—410 (1955).
2. Les fonctions analytiques comme ultra-distributions dans le calcul opérationnel. Math. Ann. **136**, 58—96 (1958).
3. O lemma de Weyl no quadro das ultradistribuçôes. Bol. Acad. Ciencias Lisboa **37**, 70—79 (1965).
4. Les séries de multipôles des physiciens et la theorie des ultra-distributions. Math. Ann. **174**, 109—142 (1967).
5. Sur la définition et la structure des distributions vectorielles. Portugaliae Math. **19**, 1—80 (1960).
6. As funçoes analyticas eaanalise functional. Portugaliae Math. **9**, 1—130 (1950).

da Silva Dias, C. L.
1. Espaços vectorials topologicos e sua applicaçao nos espaços funcionais analiticos. Bol. Soc. Mat. Saô Paulo **5**, 1—58 (1950).

Simon, L.
1. Generalizations of the Dirichlet problem for the equation $\Delta u = f$. Revue Roumaine de Math. pures et appl. **12**, 1087—1094 (1967).
Sobolevski, P. E.
1. Équations paraboliques dans un espace de Banach. Trudy Moskov Mat. Obs. **10**, 297—350 (1961). (Amer. Math. Soc. Transl. (2), **49**, 1—62 (1966)).
Sousa e Menoeses, A.
1. Sobre una construçâo axiomatica da teoria das ultra distribuçôes na recta e alguns dos sens possiveis modelos (to be published).
Stampacchia, G., Magenes, E.
Cf. Magenes-Stampacchia.
Steinberg, S., Treves, F.
1. Pseudo-Fokker-Planck equations and hyperdifferential operators. J. of Diff. Equations **8**, 333—366 (1970).
Suryanarayana, P.
1. The higher order differentiability of solutions of abstract evolution equations. Pacific J. of Math. **22**, 543—561 (1967).
Talenti, G.
1. Intorno alle classi funzionali di Gevrey. Ann. Mat. Pura Appl., LXIII **4**, 151—173 (1963).
2. Un problema di Cauchy. Ann. Sc. Norm. Sup. Pisa **18**, 165—186 (1964).
3. Osservazioni sulla nota: Un problema di Cauchy. Ann. Sc. Norm. Sup. Pisa **19**, 179—184 (1965).
4. Sul problema de Cauchy per le equazioni a derivate parziali. Ann. Mat. Pura Appl. **4**, 67, 365—394 (1965).
Tanabe, H.
1. On regularity of solutions of abstract differential equations of parabolic type in Banach space. Journal of the Math. Soc. of Japan **19**, 521—542 (1967).
2. On differentiability and analyticity of solutions of weighted elliptic boundary value problems. Osaka. J. of Math. **2**, 163—190 (1965).
3. On regularity of solutions of abstract differential equations. Colloque sur l'Analyse Fonctionnelle, Tokyo 1969.
Tanabe, H., Kato, T.
Cf. Kato-Tanabe.
Tilmann, H. G.
1. Randverteilungen analytischer Funktionen und Distributionen, I and II. Math. Zeitschr. **59**, 61—83 (1953); **76**, 5—21 (1961).
2. Darstellung der Schwartzschen Distributionen durch analytische Funktionen. Math. Zeitschr. **77**, 106—124 (1961).
3. Dualität in der Potentialtheorie. Portugaliae Math. **13**, 55—86 (1954).
Treves, F.
1. Topological vector spaces, distributions and Kernels. Academic Press, New York, 1967.
2. On the theory of linear partial differential operators with analytic coefficients. Trans. Amer. Math. Soc. 1968.
3. Locally convex Spaces and linear partial differential equations. Springer, Grundlehren Bd. 146, 1968.
4. The Ovcyannikov theorem and applications (to be published).
5. Ovcyannikov theorem and hyperdifferential operators. Notas de Matematica, IMPA, Rio de Janeiro 1968.
Treves, F., Steinberg, S.
Cf. Steinberg-Treves.

Triebel, H.
1. Erzeugung nuklearer lokalkonvexer Räume durch singuläre Differentialoperatoren zweiter Ordnung. Math. Annalen **174**, 163—176 (1967).
Vilenkin, N. Ya., Gelfand, I. M.
Cf. Gelfand-Vilenkin.
Vishik, I. M.
1. Problèmes aux limites pour les équations elliptiques dégénérant à la frontière. Mat. Sbornik **35**, 513—568 (1954) (Amer. Math. Soc. Transl. (2), **35**, 15—78 (1954).)
Volevich, L. R.
1. Propriétés locales des solutions des systèmes quasi elliptiques. Mat. Sbornik **59**, **101**, 3—52 (1962).
Whitney, H.
1. Analytic extension of differentiable functions defined in closed sets. Trans. Amer. Math. Soc. **36**, 63—89 (1934).
de Wilde, M.
1. Espaces de fonctions à valeurs dans un espace linéaire à semi-normes. Mém. Soc. Roy. Sc. Liège **13**, fasc. 2 (1966).
2. Réseaux dans les espaces linéaires à semi-normes. *Thesis*, Université de Liège, 1969.
de Wilde, M., Garnir, H. G., Schmets, J.
Cf. Garnir-de Wilde-Schmets.
Waelbroeck, L.,-Leray, J.
Cf. Leray-Waelbroeck.
Wloka, J.
1. Über die Gurewic-Hörmanderschen Distributionsräume. Math. Ann. **160**, 321—362 (1965).
Wohlers, M. R., Beltrami, E. J.
Cf. Beltrami-Wohlers.
Yoshinaga, K.
1. On a locally convex space introduced by J. S. Silva. J. Sc. Hiroshima Univ., ser. A, **21**, 89—98 (1957).
2. On spaces of distributions of exponential growth. Bull. Kyushu. Inst. Tech. **6**, 1—16 (1960).
3. Ultra distributions and semi-group distributions. Bull. Kyushu Inst. of Technology **10**, 1—24 (1963).
4. Values of vector valued distributions and smoothness of semi-group distributions. Bull. Kyushu Inst. Techn. **12**, 1—27 (1965).
Yosida, K.
1. An abstract analyticity in time for solutions of a diffusion equation. Proc. Japan. Acad. **35**, 109—113 (1959).
2. Functional analysis, Grundlehren Bd. 123, Berlin/Heidelberg/New York: Springer 1965.
Zerner, M.
1. Les fonctions holomorphes à valeurs vectorielles et leurs valeurs au bord. Paris Centre de Physique théorique, École Polytechnique, 1962.
Zuily, C.
1. Étude de la régularité d'une classe d'opérateurs elliptiques dégénérés d'ordre 2. C. R. A. S. Paris **268**, 532—534 (1969) and Rend. di Mat. November 1971.

Additional Bibliography

Chapter 7

For Gevrey and Beurling classes of functions and ultra-distributions, additional references are:

Berenstein, C. A., Dostal, M. A.

1. Fourier transforms of the Beurling classes $\mathscr{D}_\omega$, $\mathscr{E}'_\omega$. Bull. Amer. Math. Soc. **77**, 963–967 (1971).
2. Topological Properties of analytically uniform spaces. Trans. Amer. Math. Soc. **154**, 493–513 (1971).

Bernardi, M. L.

1. Ultradistribuzioni del tipo di Gevrey su varietà compatte; applicazioni al problema di Cauchy per equazioni lineari paraboliche. Rend. Ist. Lombardo Sc. Lettere (A) **104**, 534–579 (1970).

Björck, G.

1. Beurling distributions and linear partial differential equations. Ist Naz. Alta Math., Rome, Symposia Math. Vol. VII, 367–379 (1971).

Cerofolini, L.

1. Sulla nozione di ultradistribuzione su varietà compatte (to appear).

Dostal, M. A.

1. An analogue of a theorem of V. Bernstein and its applications to singular supports of distributions. Proc. London Math. Soc. XIX, 553–576 (1969).

Ehrenpreis, L.

1. Fourier transform in several complex variables. New York: Interscience Publishers 1970.

Komatsu, H.

1. Ultradistributions I. Structure theorems and a characterization. J. Fac. Sci. Univ. Tokyo (to appear).

Martineau, A.

1. Fonctionnelles analytiques. Actes Congrès Intern. Math. Paris: Gauthier-Villars, Vol. 2, 1970, 635–642.

Sato, M.

1. Regularity of hyperfunction solutions of partial differential equations. Actes Congrès Intern. Math. Paris: Gauthier-Villars, Vol. 2, 1970, 785–794.

A new class of "ultra-distributions" is introduced in:

Aronszajn, N.

1. Traces of analytic solutions of the heat equation; relation to distributions and hyperfunctions. Symp. C.N.R.S Paris, Sept. 1972.

Chapter 8

For elliptic boundary value problems in unbounded domains and in spaces of hyperfunctions (see also Problems 6.6 and 6.7) cf.

Schapira, P.
1. Hyperfonctions et problèmes aux limites elliptiques. Bull. Soc. Math. France, **99**, 113—141 (1971).

For questions related to trace problems and to Problem 6.4 we refer to

Baouendi, M. S., Geymonat, G.
1. Transposition des problèmes aux limites elliptiques. Ist. Naz. Alta Math., Rome, Symposia Math., Vol. VII, 421—441 (1971).

Boutet de Monvel, L., Geymonat, G.
1. Solutions irregulières d'un problème aux limites elliptiques. Ist. Naz. Alta Math., Rome, Symposia Math., Vol. VII, 381—402 (1971).

For the regularity of hyperfunction solutions for elliptic and other operators, cf. M. Sato, already quoted above.

For a theorem similar to Theorem 1.2 for degenerate elliptic operators, and applications to interpolation theory and to approximation theory, cf.

Baouendi, M. S., Goulaouic, C.
1. Régularité et itérés d'opérateurs elliptiques dégénérés, applications. J. Funct. Analysis.
2. Étude de l'analyticité et de la régularité Gevrey pour une classe d'opérateurs elliptiques dégénérés. Ann. Sc. Ec. Norm. Sup. **4**, 31—46 (1971).
3. Approximation polynomiale de fonctions C^∞ et analytiques. Annales Inst. Fourier **21**, 149—174 (1971).
4. Non-analytic-hypoellipticity for some degenerate elliptic operators. Bull. Amer. Math. Soc. **78**, 3 (May 1972).

Baouendi, M. S., Goulaouic, C., Hanouzet, B.
1. Caracterisation de classes de fonctions C^∞ et analytiques sur une variété irregulière à l'aide d'un opérateur différentiel. J. Math. Pures et Appliquées.

Derridj, M., Zuily, C.
1. Régularité analytique et Gevrey d'opérateurs elliptiques dégénérés. J. Math. Pures et appliquées, 1973.
2. Sur la regularité Gevrey des operateurs de Hörmander. J. Math. Pures et Appliquées. 1973.

For other results on degenerate operators, we refer to the book

Oleinik, O. A., Radkevitch.
1. Equations du 2ème ordre dont la forme caractéristique est non negative. In Russian, Moscow 1971.

Chapter 9

For general semigroups and applications to Cauchy problems, cf.

Beals, R.
1. Semi groups and abstract Gevrey spaces. To appear. J. Funct. Analysis.
2. On the abstract Cauchy problem (to appear).

Ouchi, S.
1. Hyperfunction solution of the abstract Cauchy problem. Proc. Jap. Acad. **47**, 541—544 (1971).

Simon, B.
1. The theory of semi-analytic vectors. A new proof of a theorem of Masson and McClary. To appear.

Ushijima, T.
1. On the abstract Cauchy problems and semi-groups of linear operators in locally convex spaces. Scient. Papers of Coll. Gen. Educ. Univ. Tokyo Vol. 21 (1971).

2. On the generation and smoothness of semi-groups of linear operators (to appear in J. Fac. Sci. Univ. Tokyo).

For Problem 14.2, cf.

Bernardi, M. L.

1. Equazioni differenziali lineari astratte del secondo ordine e di Schroedinger in spazi di Beurling. Ist. Lombardo (Rend. Sc.) **106** (1972).

Chapter 10

For parabolic problems on compact varieties in Gevrey classes, cf.

Bernardi, M. L.

already quoted in the additional bibliography for Chapter 7 and

Cerofolini, L.

1. Sulla totalità delle soluzioni di una equazione di tipo parabolico su di una varietà. Boll. U. M. I. IV (1969) 65—70.
2. Sulle tracce sulla varieta M delle soluzioni in $M_x(0, +\infty)$ di una equazione omogenea di tipo parabolico. Rend. Accad. Lincei, XLVII (1969) 236—239.

Chapter 11

For Problem 6.10, cf.

Pozzi, G. A.

1. Problemi di limiti per l'equazione della corda vibrante nell'ambito delle distribuzioni. Rend. Ist. Lombardo Sc. Lett. (A) **104**, 534—579 (1970).
2. Ulteriori osservazioni sui problemi ... Id. **105**, 306—316 (1971).

For mixed problems for general hyperbolic operators, cf.

Balaban, T.

1. On the mixed problem for a hyperbolic equation. Mem. Amer. Math. Soc. 112 (1971).

Beals, R.

1. Hyperbolic equations and systems with multiple characteristics (to appear).

Chazarain, J., Piriou, A.

1. Caractérisation des problémes mixtes hyperboliques bien posés (to appear in Ann. Inst. Fourier).

Ikawa, M.

1. Remarques sur les problèmes mixtes pour l'équation des ondes. Symp. C.N.R.S., Sept. 1972, Paris.

Kasahara, K.

1. On weak well posedness of mixed problems for hyperbolic systems. Publ. R. I. M. S. Kyoto Univ. Vol. 6, 3, 1970.

Kato, J.

1. Mixed problems of hyperbolic equations in a general domain. Proc. Jap. Acad. **47**, 67—70 (1971).

Kreiss, H. O.

1. Initial boundary value problems for hyperbolic systems. C. P. A. M. **13** (1970).

Rauch, J.

1. L_2 is a continuable initial condition for Kreiss'mixed problems. C. P. A. M. (to appear).

Sakamoto, R.

1. Mixed problems for hyperbolic equations (I) and (II). J. Math. Kyoto Univ. Vol. 10, 2 and 3.

Shirota, T., Agemi, R.

1. On necessary and sufficient conditions for L_2 well-posedness of mixed problems for hyperbolic equations. J. Fac. Sci. Hokkaido Univ. Vol. 21, 2, 1970.

For the application of hyperfunctions, ultra-distributions and trace functions to various problems for general partial differential operators ("trace" problems, Cauchy problems etc.), cf.

Baouendi, M. S.
1. Solution of P.D.E. in trace functions. Symp. C.N.R.S., Sept. 1972, Paris.

Bony, J. M., Schapira, P.
1. Problème de Cauchy, existence et prolongement pour les hyperfonctions solutions des équations hyperboliques non strictes. C. R. A. S. Paris **274**, 188–191 (1972).
2. Solutions hyperfonctions du problème de Cauchy (to appear).

Kawai, T.
1. Construction of elementary solutions of I-hyperbolic operators and solutions with small singularities. Proc. Japan Acad. **46**, 912–915 (1970).

Komatsu, H., Kawai, T.
1. Boundary values of hyperfunction solutions of linear partial differential equations. Publ. R. I. M. S. Kyoto Univ. **7**, 95–104 (1971/1972).

For the global existence of real analytic solutions of general partial differential equations and the properties of analyticity of the solutions, cf.

Anderson, K. G.
1. Propagation of analyticity of solutions of partial differential equations with constant coefficients. Ark. Mat. Vol. 8, 277–302 (1971).

de Giorgi, E., Cattabriga, L.
1 Una dimostrazione diretta dell'esistenza di soluzioni analitiche nel piano reale di equazioni a derivate parziali a coefficienti costanti. Boll. dell'U. M. I. **4**, 1015–1027 (1971).

Hörmander, L.
3. Uniqueness theorems and wave front sets for solutions of linear differential equations with analytic coefficients. Comm. Pure Appl. Math. **24**, 671–703 (1971).

Kawai, T.
1. On the global existence of real analytic solutions of linear differential equations (I), (II). Proc. Japan Acad. **47**, 537–540 (1971).

For finite element method approaches in numerical analysis using variational methods of partial differential equations, we already gave some basic references in the Additional Bibliography of Vol. I, p. 354. To these references, we wish to add here, for *elliptic problems*:

Babuska, I.
1. The finite element method with Lagrangian multipliers. Technical Note BN-724. Inst. Fluid Dyn. and Appl. Maths. Univ. of Maryland, January 1972.

Babuska, I., Kellog, R. B.
1. Numerical solution of the neutron diffusion equation in the presence of corners and interfaces. Technical Note BN-720. Inst. Fluid Dyn. and Appl. Maths. Univ. of Maryland, December 1971.

Bramble, J. H.
1. Variational methods for the numerical solution of elliptic problems. Lecture Notes Chalmers Inst. Technology 1970.

Bramble, J. H., Zlamal, M.
1. Triangular elements in the finite element method. Math. Comp. **24**, 112 809–820 (1970).

Brezzi, F.
1. Sull 'analisi numerica del problema di Dirichlet per le equazioni lineari ellittiche. Publicazione 18, Laboratorio di Analisi Numerica, Pavia, 1971.

Ciarlet, P. G., Raviart, P. A.

1. General Lagrange and Hermite interpolation in $\mathbf{R}^n$ with applications to finite element methods. Arch. Rat. Mech. Anal. (to appear).
2. Interpolation theory over curved elements with applications to finite element methods. Comp. Math. Appl. Mech. Eng. (to appear).

Strang, G.

1. Approximation in the finite element method. Num. Math. **19**, 81–98 (1972).

For *evolution* problems, we mention

Douglas, J., Dupont, T.

1. Galerkin methods for parabolic equations. SIAM J. Num. Anal. **7**, 4, 575–626 (1970).

Dupont, T.

1. L^2-estimates for Galerkin methods for second order hyperbolic equations (to appear).

Raviart, P. A.

1. The use of numerical integration in finite element methods for solving parabolie equations, Conference on Num. Anal., Royal Irish Academy, Dublin, 1972.

721/8/73

Die Grundlehren der mathematischen Wissenschaften in Einzeldarstellungen mit besonderer Berücksichtigung der Anwendungsgebiete

Eine Auswahl

---

67. Byrd/Friedman: Handbook of Elliptic Integrals for Engineers and Scientists. 2nd edition
68. Aumann: Reelle Funktionen. 2. Aufl.
74. Boerner: Darstellungen von Gruppen. 2. Aufl.
76. Tricomi: Vorlesungen über Orthogonalreihen. 2. Aufl.
77. Behnke/Sommer: Theorie der analytischen Funktionen einer komplexen Veränderlichen. Nachdruck der 3. Aufl.
78. Lorenzen: Einführung in die operative Logik und Mathematik. 2. Aufl.
86. Richter: Wahrscheinlichkeitstheorie. 2. Aufl.
87. van der Waerden: Mathematische Statistik. 3. Aufl.
94. Funk: Variationsrechnung und ihre Anwendung in Physik und Technik. 2. Aufl.
97. Greub: Linear Algebra. 3rd edition
99. Cassels: An Introduction to the Geometry of Numbers. 2nd printing
104. Chung: Markov Chains with Stationary Transition Probabilities. 2nd edition
107. Köthe: Topologische lineare Räume I. 2. Aufl.
114. MacLane: Homology. Reprint of 1st edition
116. Hörmander: Linear Partial Differential Operators. 3rd printing
117. O'Meara: Introduction to Quadratic Forms. 2nd printing
120. Collatz: Funktionalanalysis und numerische Mathematik. Nachdruck der 1. Aufl.

121./122. Dynkin: Markov Processes

123. Yosida: Functional Analysis. 3rd edition
124. Morgenstern: Einführung in die Wahrscheinlichkeitsrechnung und mathematische Statistik. 2. Aufl.
125. Itô/McKean jr.: Diffusion Processes and Their sample Paths
126. Lehto/Virtanen: Quasiconformal Mappings in the Plane.
127. Hermes: Enumerability, Decidability, Computability. 2nd edition
128. Braun/Koechler: Jordan-Algebren
129. Nikodým: The Mathematical Apparatus for Quantum Theories
130. Morrey jr.: Multiple Integrals in the calculus of Variations
131. Hirzebruch: Topological Methods in Algebraic Geometry. 3rd edition
132. Kato: Perturbation Theory for Linear Operators
133. Haupt/Künneth: Geometrische Ordnungen
134. Huppert: Endliche Gruppen I
135. Handbook for Automatic Computation. Vol. 1/Part a: Rutishauser: Description of ALGOL 60
136. Greub: Multilinear Algebra
137. Handbook for Automatic Computation. Vol. 1/Parth b: Grau/Hill/Langmaack: Translation of ALGOL 60
138. Hahn: Stability of Motion
139. Doetsch/Schäfke/Tietz: Mathematische Hilfsmittel des Ingenieurs. 1. Teil
140. Collatz/Nicolovius/Törnig: Mathematische Hilfsmittel des Ingenieurs. 2. Teil
141. Mathematische Hilfsmittel des Ingenieurs. 3. Teil
142. Mathematische Hilfsmittel des Ingenieurs. 4. Teil
143. Schur: Vorlesungen über Invariantentheorie
144. Weil: Basic Number Theory

152. Hewitt/Ross: Abstract Harmonic Analysis. Vol. 2: Structure and Analysis for Compact Groups, Analysis on Locally Compact Abelian Groups
154. Singer: Bases in Banach Spaces I
155. Müller: Foundations of the Methematical Theory of Electromagnetic Waves
156. van der Waerden: Mathematical Statistics
157. Prohorov/Rozanov: Probability Theory
158. Constantinescu/Cornea: Potential Theory on Harmonic Spaces
159. Köthe: Topological Vector Spaces I
160. Agrest/Maksimov: Theory of Incomplete Cylindrical Functions and Their Applications
161. Bhatia/Szegö: Stability Theory of Dynamical Systems
162. Nevanlinna: Analytic Functions
163. Stoer/Witzgall: Convexity and Optimization in Finite Dimensions I
164. Sario/Nakai: Classification Theory of Riemann Surfaces
165. Mitrinović: Analytic Inequalities
166. Grothendieck/Dieudonné: Eléments de Géométrie Algebriqué I
167. Chandrasekharan: Arithmetical Functions
168. Palamodov: Linear Differential Operators with Constant Coefficients
169. Rademacher: Topics in Analytic Number Theory
170. Lions: Optimal Control of Systems Governed by Partial Differential Equations
171. Singer: Best Approximation on Normed Linear Spaces by Elements of Linear Subspaces
172. Bühlmann: Mathematical Methods in Risk Theory
173. F. Maeda/S. Maeda: Theory of Symmetric Lattices
174. Stiefel/Scheifele: Linear and Regular Celestial Mechanics. Perturbed two-body Motion—Numerical Methods—Canonical Theory
175. Larsen: An Introduction of the Theory of Multipliers
176. Grauert/Remmert: Analytische Stellenalgebren
177. Flügge: Practical Quantum Mechanics I
178. Flügge: Practical Quantum Mechanics II
179. Giraud: Cohomologie non abélienne
180. Landkof: Foundations of Modern Potential Theory
181. Lions/Magenes: Non-Homogeneous Boundary Value Problems and Applications I
182. Lions/Magenes: Non-Homogeneous Boundary Value Problems and Applications II
183. Lions/Magenes: Non-Homogeneous Boundary Value Problems and Applications III
184. Rosenblatt: Markov Processes. Structure and Asymptotic Begavior
185. Rubinowicz: Sommerfeldsche Polynommethode
186. Wilkinson/Reinsch: Handbook for Automatic Computation II, Linear Algebra
187. Siegel/Moser: Lectures in Celestial Mechanics
188. Warner: Harmonic Analysis on Semi-Simple Lie Groups I
189. Warner: Harmonic Analysis on Semi-Simple Lie Groups II
190. Faith: Algebra: Rings, Modules, and Categories I
191. Faith: Algebra: Rings, Modules, and Categories II
192. Mal'cev: Algebraic-Systems
193. Pólya/Szegö: Problems and Theorems in Analysis. Vol. 1
194. Igusa: Theta Functions
195. Berberian: Baer*-Rings
196. Athreya: Branching Processes
197. Benz: Vorlesungen über Geometrie der Algebren
200. Dold: Lectures on Algebraic Topology